# COMPUTER-AIDED MULTIVARIATE ANALYSIS

**Second Edition**

# COMPUTER-AIDED MULTIVARIATE ANALYSIS

## Second Edition

A.A. Afifi
Virginia Clark

Professors of Biostatistics
and Biomathematics
University of California, Los Angeles

 Van Nostrand Reinhold Company
New York

Library of Congress Catalog Card Number 90-34524
ISBN 0-442-23944-0

Printed in the United States of America

Van Nostrand Reinhold
115 Fifth Avenue
New York, New York 10003

Van Nostrand Reinhold International Company Limited
11 New Fetter Lane
London EC4P 4EE, England

Van Nostrand Reinhold
102 Dodds Street
South Melbourne, Victoria 3205, Australia

Nelson Canada
1120 Birchmount Road
Scarborough, Ontario M1K 5G4, Canada

16   15   14   13   12   11   10   9   8   7   6   5   4   3   2   1

**Library of Congress Cataloging-in-Publication Data**

Afifi, A. A. (Abdelmonem A.), 1939–
    Computer-aided multivariate analysis / A. A. Afifi, Virginia Clark.
    — 2nd ed.
    p.   cm.
    Includes bibliographical references.
    ISBN 0-442-23944-0
    1. Multivariate analysis—Data processing.   I. Clark, Virginia,
    1928–   .  II. Title.
    QA278.A33   1990
    519.5'35'0285—dc20                                          90-34524
                                                                    CIP

# CONTENTS

# PREFACE TO THE SECOND EDITION

The emphasis in the second edition of *Computer-Aided Multivariate Analysis* continues to be on performing and understanding multivariate analysis, not on the necessary mathematical derivations. We added new features while keeping the descriptions brief and to the point in order to maintain the book at a reasonable length.

In this new edition, additional emphasis is placed on using personal computers, particularly in data entry and editing. While continuing to use the BMDP, SAS, and SPSS packages for working out the statistical examples, we also describe their data editing and data management features. New material has also been included to enable the reader to make the choice of transformations more straightforward.

Beginning with Chapter 6, a new section, entitled "What to Watch Out For," is included in each chapter. These sections attempt to warn the reader about common problems related to the data set that may lead to biased statistical results. We relied on our own experience in consulting rather than simply listing the assumptions made in deriving the techniques. Statisticians sometimes get nervous about nonstatisticians performing multivariate analyses without necessarily being well versed in their mathematical derivations. We hope that the new sections will provide some warning of when the results should not be blindly trusted and when it may be necessary to consult an expert statistical methodologist.

Since the first edition, many items on residual analysis have been added to the output of most multivariate and especially regression analysis programs. Accordingly, we expanded and updated our discussion of the use of residuals to detect outliers, check for lack of independence, and assess normality. We also explained the use of principal components in regression analysis in the presence of multicollinearity. In discussing logistic regression, we presented information on evaluating how well the equation predicts outcomes and how to use the receiver operating characteristic (ROC) curves in this regard.

We added a chapter in this edition on regression analysis using survival data. Here the dependent variable is the length of time until a defined event occurs. Both the loglinear, or accelerated failure time model, and the Cox proportional hazards model are presented along with examples of their use. We compare the interpretation of the coefficients resulting from the two models. For survival data, the outcome can be classified into one of two possibilities, success or failure, and logistic regression can be used for analyzing the data. We present a comparison of this type of analysis with the other two regression methods for survival data. To compensate for the addition of this chapter, we deleted from the first edition the chapter on nonlinear regression. We also placed the chapter on canonical correlation immediately after the regression chapters, since this technique may be viewed as a generalization of multiple correlation analysis.

Descriptions of computer output in all chapters have been updated to include new features appearing since the first edition. New problems have been added to the problem sets, and references were expanded to incorporate recent literature.

We thank Stella Grosser for her help in writing new problems, updating the references, deciphering new computer output, and checking the manuscript. We also thank Mary Hunter, Evalon Witt, and Jackie Champion for their typing and general assistance with the manuscript for the second edition.

*A.A. Afifi*
*Virginia Clark*

# PREFACE TO THE FIRST EDITION

This book has been written for investigators, specifically behavioral scientists, biomedical scientists, econometricians, and industrial users who wish to perform *multivariate statistical analyses* on their data and understand the results. In addition, we believe that the book will be helpful to many statisticians who have been trained in conventional mathematical statistics (where applications of the techniques were not discussed) and who are now working as statistical consultants. Statistical consultants overall should find the book useful in assisting them in giving explanations to clients who lack sufficient background in mathematics.

We do not present mathematical derivation of the techniques in this book but, rather, rely on geometric and graphical arguments and on examples to illustrate them. The mathematical level has been kept deliberately low, with no mathematics beyond high-school level required. Ample references are included for those who wish to see the derivations of the results.

We have assumed that you have taken a basic course in statistics and are familiar with statistics such as the mean and the standard deviation. Also, tests of hypotheses are presented in several chapters, and we assume you are familiar with the basic concept of testing a null hypothesis. Many of the computer programs utilize analysis of variance, and that part of the programs can only be understood if you are familiar with one-way analysis of variance.

## APPROACH OF THE BOOK

The content and organizational features of this book are discussed in detail in Chapter 1, Section 1.4. Because no university-level mathematics is assumed, some topics often found in books on multivariate analysis have not been included. For example, there is no theoretical discussion of sampling from a multivariate normal distribution or the Wishart distribution, and the usual chapter on matrices has not been included. Also, we point out that the book is not intended to be a comprehensive text on multivariate analysis.

The choice of topics included reflects our preferences and experience as consulting statisticians. For example, we deliberately excluded multivariate analysis of variance (see, e.g., Anderson 1958; Morrison 1976) because we felt that it is not as commonly used as the other topics we describe. On the other hand, we would have liked to include the log-linear model for analyzing multivariate categorical data (see, e.g., Bishop, Fienberg, and Holland 1975; Upton 1978), but we decided against this for fear that doing so would have taken us far afield and would have added greatly to the length of the book.

The multivariate analyses have been discussed more as separate techniques than as special cases arising from some general framework. The advantage of the approach used here is that we can concentrate on explaining how to analyze a certain type of data by using output from readily available computer programs in order to answer realistic questions. The disadvantage is that the theoretical interrelationships among some of the techniques are not highlighted.

## USES OF THE BOOK

This book can be used as a text in an applied statistics course or in a continuing education course. We have used preliminary versions of this book in teaching behavioral scientists, epidemiologists, and applied statisticians. It is possible to start with Chapter 6 or 7 and cover the remaining chapters easily in one semester if the students have had a solid background in basic statistics. Two data sets are included that can be used for homework, and a set of problems is included at the end of each chapter except the first.

## COMPUTER ORIENTATION

The original derivations for most of the current multivariate techniques were done over forty years ago. We feel that the application of these techniques to real life problems is now the "fun" part of this field. The presence of computer packages and the availability of computers have removed many of the tedious aspects of this discipline so that we can concentrate on thinking about the nature of the scientific problem itself and on what we can learn from our data by using multivariate analysis.

Because the multivariate techniques require a lot of computations, we have assumed that packaged programs will be used. Here we bring together discussions of data entry, data screening, data reduction, and data analysis aimed at helping you to perform these functions, understand the basic methodology, and determine what insights into your data you can gain from multivariate analyses. Examples of control statements are given for various computer runs used in the text. Also included are discussions of the options available in the different statistical packages and how they can be used to achieve the desired output.

## ACKNOWLEDGMENTS

Our most obvious acknowledgment is due to those who have created the computer-program packages for statistical analysis. Many of the applications of multivariate analysis that are now prevalent in the literature are directly due to the development and availability of these programs. The efforts of the programmers and statisticians who managed and wrote the BMDP, SAS, SPSS, and other statistical packages have made it easier for statisticians and researchers to actually use the methods developed by Hotelling, Wilks, Fisher, and others in the 1930s or even earlier. While we have enjoyed longtime associations with members of the BMDP group, we also regularly use SAS and SPSS-X programs in our work, and we have endeavored to reflect this interest in our book. We recognize that other statistical packages are in use, and we hope that our book will be of help to their users as well.

We are indebted to Dr. Ralph Frerichs and Dr. Carol Aneshensel for the use of a sample data set from the Los Angeles Depression Study, which appears in many of the data examples. We wish to thank Dr. Roger Detels for the use of lung function data taken from households from the UCLA population studies of chronic obstructive respiratory disease in Los Angeles. We also thank *Forbes Magazine* for allowing us to use financial data from their publications.

We particularly thank Welden Clark and Nanni Afifi for reviewing drafts and adding to the discussion of the financial data. Helpful reviews and suggestions were also obtained from Dr. Mary Ann Hill, Dr. Roberta Madison, Mr. Alexander Kugushev, and several anonymous reviewers. Welden Clark has, in addition, helped with programming and computer support in the draft revisions and checking, and in preparation of the bibliographies and data sets. We would further like to thank Dr. Tim Morgan, Mr. Steven Lewis, and Dr. Jack Lee for their help in preparing the data sets. The BMDP Statcat (trademark of BMDP Statistical Software, Inc.) desktop computer with the UNIX (trademark of Bell Telephone Laboratories, Inc.) program system has served for text processing as well as further statistical analyses. The help provided by Jerry Toporek of BMDP and Howard Gordon of Network Research Corporation is appreciated.

In addition we would like to thank our copyeditor, Carol Beal, for her efforts in making the manuscript more readable both for the readers and for the typesetter. The major portion of the typing was performed by Mrs. Anne Eiseman; we thank her also for keeping track of all the drafts and teaching material derived in its production. Additional typing was carried out by Mrs. Judie Milton, Mrs. Indira Moghaddan, and Mrs. Esther Najera.

*A.A. Afifi*
*Virginia Clark*

# COMPUTER-
# AIDED
# MULTIVARIATE
# ANALYSIS

## Second Edition

# Part One
# PREPARATION FOR ANALYSIS

# Chapter One

# *WHAT IS MULTIVARIATE ANALYSIS?*

## *1.1 HOW IS MULTIVARIATE ANALYSIS DEFINED?*

The expression *multivariate analysis* is used to describe analyses of data that are multivariate in the sense that numerous observations or variables are obtained for each individual or unit studied. In a typical survey 30 to 100 questions are asked of each respondent. In describing the financial status of a company, an investor may wish to examine five to ten measures of the company's performance. Commonly, the answers to some of these measures are interrelated. The challenge of disentangling complicated interrelationships among various measures on the same individual or unit and of interpreting these results is what makes multivariate analysis a rewarding activity for the investigator. Often results are obtained that could not be attained without multivariate analysis.

In the next section of this chapter several studies are described in which the use of multivariate analyses is essential to understanding the underlying problem. Section 1.3 gives a listing and a very brief description of the multivariate analysis techniques discussed in this book. Section 1.4 then outlines the organization of the book.

# 1.2 EXAMPLES OF STUDIES IN WHICH MULTIVARIATE ANALYSIS IS USEFUL

The studies described in the following subsections illustrate various multivariate analysis techniques. Some are used later in the book as examples.

## Depression Study Example

The data for the depression study have been obtained from a complex, random, multiethnic sample of 1000 adult residents of Los Angeles County. The study was a *panel* or *longitudinal* design where the same respondents were interviewed four times between May 1979 and July 1980. About three-fourths of the respondents were reinterviewed for all four interviews. The field work for the survey was conducted by professional interviewers from the Institute for Social Science Research at UCLA.

This research is an epidemiological study of depression and help-seeking behavior among free-living (noninstitutionalized) adults. The major objectives are to provide estimates of the prevalence and incidence of depression and to identify causal factors and outcomes associated with this condition. The factors examined include demographic variables, life events stressors, physical health status, health care utilization, medication use, life-style, and social support networks. The major instrument used for classifying depression is the Depression Index (CESD) of the National Institute of Mental Health, Center for Epidemiologic Studies. A discussion of this index and the resulting prevalence of depression in this sample is given in Frerichs, Aneshensel, and Clark (1981).

The longitudinal design of the study offers advantages for assessing causal priorities since the time sequence allows us to rule out certain potential causal links. Nonexperimental data of this type cannot directly be used to establish causal relationships, but models based on an explicit theoretical framework can be tested to determine if they are consistent with the data. An example of such model testing is given in Aneshensel and Frerichs (1982).

Data from the first time period of the depression study are presented in Chapter 3. Only a subset of the factors measured on a sample of the respondents is included in order to keep the data set easily comprehensible.

These data are used several times in subsequent chapters to illustrate some of the multivariate techniques presented in this book.

## Bank Loan Study

The managers of a bank need some way to improve their prediction of which borrowers will successfully pay back a type of bank loan. They have data from the past of the characteristics of persons to whom the bank has lent money and the subsequent record of how well the person has repaid the loan. Loan payers can be classified into several types: those who met all of the terms of the loan, those who eventually repaid the loan but often did not meet deadlines, and those who simply defaulted. They also have information on age, sex, income, other indebtedness, length of residence, type of residence, family size, occupation, and the reason for the loan. The question is, can a simple rating system be devised that will help the bank personnel improve their prediction rate and lessen the time it takes to approve loans? The methods described in Chapters 11 and 12 can be used to answer this question.

## Chronic Respiratory Disease Study

The purpose of the ongoing respiratory disease study is to determine the effects of various types of smog on lung function of children and adults in the Los Angeles area. Because they could not randomly assign people to live in areas that had different levels of pollutants, the investigators were very concerned about the interaction that might exist between the locations where persons chose to live and their values on various lung function tests. The investigators picked four areas of quite different types of air pollution and are measuring various demographic and other responses on all persons over seven years old who live there. These areas were chosen so that they are close to an air-monitoring station.

The researchers are taking measurements at two points in time and are using the change in lung function over time as well as the levels at the two periods as outcome measures to assess the effects of air pollution. The investigators have had to do the lung function tests by using a mobile unit in the field, and much effort has gone into problems of validating the accuracy of the field observations. A discussion of the particular lung function measurements used for one of the four areas can be found in Detels et al. (1975). In the analysis of the data, adjustments must be made for sex, age, height, and smoking status of each person.

Over 15,000 respondents have been examined and interviewed in this study. The original data analyses were restricted to the first collection period, but now analyses include both time periods. This data set is being used to answer numerous questions concerning effects of air pollution, smoking, occupation, etc. on different lung function measurements. For example, since the investigators obtained measurements on all family members 7 years old and older, it is possible to assess the effects of having parents who smoke on the lung function of their children (see Tashkin et al. 1984). Studies of this type require multivariate analyses so that investigators can arrive at plausible scientific conclusions that could explain the resulting lung function levels.

A subset of this data set is included in Appendix B. Lung function and associated data are given for nonsmoking families for the father, mother, and up to three children ages 7–17.

## Assessor Office Example

Local civil laws often require that the amount of property tax a home-owner pays be a percentage of the current value of the property. Local assessor's offices are charged with the function of estimating current value. Current value can be estimated by finding comparable homes that have been recently sold and using some sort of an average selling price as an estimate of the price of those properties not sold.

Alternatively, the sample of sold homes can indicate certain relationships between selling price and several other characteristics such as the size of the lot, the size of the livable area, the number of bathrooms, the location, etc. These relationships can then be incorporated into a mathematical equation used to estimate the current selling price from those other characteristics. Multiple regression analysis methods discussed in Chapters 7–9 can be used by many assessor's offices for this purpose (see Tchira 1973).

## 1.3 MULTIVARIATE ANALYSES DISCUSSED IN THIS BOOK

In this section a brief description of the major multivariate techniques covered in this book is presented. To keep the statistical vocabulary to a minimum, we illustrate the descriptions by examples.

## Simple Linear Regression

A nutritionist wishes to study the effects of early calcium intake on the bone density of postmenopausal women. She can measure the bone density of the arm (radial bone), in grams per square centimeter, by using a noninvasive device. Women who are at risk of hip fractures because of too low a bone density will show low arm bone density also. The nutritionist intends to sample a group of elderly churchgoing women. For women over 65 years of age, she will plot calcium intake as a teenager (obtained by asking the women about their consumption of high-calcium foods during their teens) on the horizontal axis and arm bone density (measured) on the vertical axis. She expects the radial bone density to be lower in women who had a lower calcium intake. The nutritionist plans to fit a simple linear regression equation and test whether the slope of the regression line is zero. In this example a single outcome factor is being predicted by a single predictor factor.

Simple linear regression as used in this case would not be considered multivariate by some statisticians, but it is included in this book to introduce the topic of multiple regression.

## Multiple Linear Regression

A manager is interested in determining which factors predict the dollar value of sales of the firm's personal computers. Aggregate data on population size, income, educational level, proportion of population living in metropolitan areas, etc. have been collected for 30 areas. As a first step, a multiple linear regression equation is computed, where dollar sales is the outcome factor and the other factors are considered as candidates for predictor factors. A linear combination of the predictor factors is used to predict the outcome or response factor.

## Canonical Correlation

A psychiatrist wishes to correlate levels of both depression and physical well-being from data on age, sex, income, number of contacts per month with family and friends, and marital status. This problem is different from the one posed in the multiple linear regression example because more than one outcome factor is being predicted. The investigator wishes to determine the linear function of age, sex, income, contacts per month, and marital status that is most highly correlated with a linear function of depression and physical well-being. After these two linear functions, called canonical variables, are determined, the investigator will test to see whether there is a statistically significant (canonical) correlation between scores from the two linear functions and whether a reasonable interpretation can be made of the two sets of coefficients from the functions.

## Discriminant Function Analysis

A large sample of initially disease-free men over 50 years of age from a community has been followed to see who subsequently has a diagnosed heart attack. At the initial visit blood was drawn from each man, and numerous determinations were made from it, including serum cholesterol, phospholipids, and blood glucose. The investigator would like to determine a linear function of these and possibly other measurements that would be useful in predicting who would and who would not get a heart attack within ten years. That is, the investigator wishes to derive a classification (discriminant) function that would help determine whether or not a middle-aged man is likely to have a heart attack.

## Logistic Regression

A television station staff has classified movies according to whether they have a high or low proportion of the viewing audience when shown. The staff has also measured factors such as the length and the type of story and the characteristics of the actors. Many of the characteristics are discrete yes-no or categorical types of data. The investigator may use logistic regression because some of the data do not meet the assumptions for statistical inference used in discriminant function analysis, but they do meet the assumptions for logistic regression. In logistic regression we derive an equation to estimate the probability of capturing a high proportion of the audience.

## Survival Analysis

An administrator of a large health maintenance organization (HMO) has collected data since 1970 on length of employment in years for their physicians who are either family practitioners or internists. Some of the physicians are still employed, but many have left. For those still employed, the administrator can only know that their ultimate length of employment will be greater than their current length of employment. The administrator wishes to describe the distribution of length of employment for each type of physician, determine the possible effects of factors such as gender and location of work, and test whether or not the length of employment is the same for the two specialties. Survival analysis, or event history analysis (as it is often called by behavioral scientists), can be used to analyze the distribution of time to an event such as quitting work, having a relapse of a disease, or dying of cancer.

## Principal Components Analysis

An investigator has made a number of measurements of lung function on a sample of adult males who do not smoke. In these tests each man is told to inhale deeply and then blow out as fast and as much as possible into a spirometer, which makes a trace of the volume of air expired over time. The maximum or forced vital capacity (FVC) is measured as the difference between maximum inspiration and maximum expiration. Also, the amount of air expired in the first second (FEV1), the forced mid-expiratory flow rate (FEF 25–75), the maximal expiratory flow rate at 50% of forced vital capacity (V50), and other measures of lung function are calculated from this trace. Since all these measures are made from the same flow-volume curve for each man, they are highly interrelated. From past experience it is known that some of these measures are more interrelated than others and that they measure airway resistance in different sections of the airway.

The investigator performs a principal components analysis to determine whether a new set of measurements called principal components can be obtained. These principal components will be linear functions of the original lung function measurements and will be uncorrelated with each other. It is hoped that the first two or three principal components will explain most of the variation in the original lung function measurements among the men. Also, it is anticipated that some operational meaning can be attached to these linear functions that will aid in their interpretation. The investigator may decide to do future analyses on these uncorrelated principal components rather than on the original data. One advantage of this method is that often fewer principal components are needed than original variables. Also, since the principal components are uncorrelated, future computations and explanations can be simplified.

## Factor Analysis

An investigator has asked each respondent in a survey whether he or she strongly agrees, agrees, is undecided, disagrees, or strongly disagrees with 15 statements concerning attitudes toward inflation. As a first step, the investigator will do a factor analysis on the resulting data to determine which statements belong together in sets that are uncorrelated with other sets. The particular statements that form a single set will be examined to obtain a better understanding of attitudes toward inflation. Scores derived from each set or factor will be used in subsequent analysis to predict consumer spending.

## Cluster Analysis

Investigators have made numerous measurements on a sample of patients who have been classified as being depressed. They wish to determine, on the basis of their measurements, whether these patients can be classified by type of depression. That is, is it possible to determine distinct types of depressed patients by performing a cluster analysis on patient scores on various tests?

Unlike the investigator of men who do or do not get heart attacks, these investigators do not possess a set of individuals whose type of depression can be known before the analysis is performed (see Andreasen and Grove 1982 for an example). Nevertheless, the investigators want to separate the patients into separate groups and to examine the resulting groups to see whether distinct types do exist and, if so, what their characteristics are.

## 1.4 ORGANIZATION AND CONTENT OF THE BOOK

This book is organized into three major parts. Part 1 (Chapters 1–5) deals with data preparation, entry, screening, transformations, and decisions about likely choices for analysis. Part 2 (Chapters 6–9) deals with regression analysis. Part 3 (Chapters 10–16) deals with a number of multivariate analyses. Statisticians disagree on whether or not regression is properly considered as part of multivariate analysis. We have tried to avoid this argument by including regression in the book, but as a separate part. Statisticians certainly agree that regression is an important technique for dealing with problems having multiple variables. In Part 2 on regression analysis we have included various topics, such as dummy variables, that are used in Part 3.

Chapters 2 through 5 are concerned with data preparation and the choice of what analysis to use. First, *variables* and how they are classified are discussed in Chapter 2. The next two chapters concentrate on the practical problems of getting data into the computer, getting rid of erroneous values, checking assumptions of normality and independence, creating new variables, and preparing a useful code book. The use of personal computers for data entry and analysis is discussed as well as the use of mainframe computers. The choice of appropriate statistical analyses is discussed in Chapter 5.

Readers who are familiar with handling data sets on computers could skip these initial chapters and go directly to Chapter 6. However, formal coursework in statistics often leaves an investigator unprepared for the complications and difficulties involved in real data sets. The material in Chapters 2–5 was deliberately included to fill this gap in preparing investigators for real world data problems.

For a course limited to multivariate analysis, Chapters 2–5 can be omitted if a carefully prepared data set is used for analysis. The depression data set, presented in Section 3.6, has been modified to make it directly usable for multivariate data analysis. Also, the lung function data presented in Appendix B and the lung cancer data presented in Appendix C can be used directly, although the former has values that some investigators may question.

In Chapters 6–16 we follow a standard format. The topics discussed in each chapter are listed, followed by a discussion of when the technique is used. Then the basic concepts and formulas used are explained. Further interpretations and data examples follow, with topics chosen that relate directly to the technique. Finally, a summary of the available computer output that may be obtained from three widely used statistical package programs is presented, and examples of output from at least one of the packages are presented. We conclude each chapter with a discussion of pitfalls to avoid when performing the analyses described.

We recommend reading the material on regression in Chapters 6–9 before proceeding with the remaining chapters.

As much as possible, we tried to make each chapter self-contained. However, Chapters 11 and 12, on discriminant analysis and logistic regression, are somewhat interrelated, as are Chapters 14 and 15, covering principal components and factor analyses.

References for further information on each topic are given at the end of each chapter. Most of the references at the ends of the chapters do require more mathematics than this book, but special emphasis has been placed on references that include examples. References requiring a strong mathematical background are preceded by an asterisk. If you wish primarily to learn the concepts involved in the multivariate techniques and are not as interested in performing the analysis, then a conceptual introduction to multivariate analysis can be found in Kachigan (1986).

We believe that the best way to learn multivariate analysis is to do it on data that the investigator is familiar with. No book can illustrate all of the features found in computer output for a real life data set. Learning multivariate analysis is similar to learning to swim: You can go to lectures, but the real learning occurs when you get into the water.

## BIBLIOGRAPHY

Andreasen, N. C., and Grove, W. M. 1982. The classification of depression: Traditional versus mathematical approaches. *American Journal of Psychiatry* 139:45–52.

Aneshensel, C. S., and Frerichs, R. R. 1982. Stress, support, and depression: A longitudinal causal model. *Journal of Community Psychology* 10:363–376.

Detels, R.; Coulson, A.; Tashkin, D.; and Rokaw, S. 1975. Reliability of plethysmography, the single breath test, and spirometry in population studies. *Bulletin de Physiopathologie Respiratoire* 11:9–30.

Frerichs, R. R., Aneshensel, C. S., and Clark, V. A. 1981. Prevalence of depression in Los Angeles County. *American Journal of Epidemiology* 113:691–699.

Kachigan, S. K. 1986. *Statistical analysis, an interdisciplinary introduction to univariate and multivariate methods.* New York: Radius Press.

Tashkin, D. P., Clark, V. A., Simmons, M., Reems, C., Coulson, A. H., Bourque, L. B., Sayre, J. W., Detels, R., and Rokaw, S. 1984. The UCLA population studies of chronic obstructive respiratory disease. VII. Relationship between parents smoking and children's lung function. *American Review of Respiratory Disease* 129:891–897.

Tchira, A. A. 1973. Stepwise regression applied to a residential income valuation system. *Assessors Journal* 8:23–35.

# Chapter Two

# CHARACTERIZING DATA FOR FUTURE ANALYSES

## 2.1 WHAT WILL YOU LEARN FROM THIS CHAPTER?

From this chapter you will learn:

- The definition of the word *variable* (2.2).
- About the Stevens system for classification of variables (2.3).
- How variables are used in statistical analyses (2.4).

Chapter 2 is an introductory chapter in which concepts and vocabulary used later in the book are introduced.

## 2.2 DEFINING STATISTICAL VARIABLES

The word *variable* is used in statistically oriented literature to indicate a characteristic or a property that it is possible to measure. When we measure something, we make a numerical model of the thing being measured. We follow some rule for assigning a number to each level of the particular characteristic being measured. For example, height of a person is a variable. We assign a numerical value to correspond to each person's height. Two people who are equally tall are assigned the same numerical value. On the other hand, two people of different heights are assigned two different values. Measurements of a variable gain their meaning from the fact that there exists a unique correspondence between the assigned numbers and the levels of the property being measured. Thus two people with different assigned heights are not equally tall. Conversely, if a variable has the same assigned value for all individuals in a group, then this variable does not convey useful information about individuals in the group.

Physical measurements, such as height and weight, can be measured directly by using physical instruments. On the other hand, properties such as reasoning ability or the state of depression of a person must be measured indirectly. We might choose a particular intelligence test and define the variable "intelligence" to be the score achieved on this test. Similarly, we may define the variable "depression" as the number of positive responses to a series of questions. Although what we wish to measure is the degree of depression, we end up with a count of yes answers to some questions. These examples point out a fundamental difference between direct physical measurements and abstract variables.

Often the question of how to measure a certain property can be perplexing. For example, if the property we wish to measure is the cost of keeping the air clean in a particular area, we may be able to come up with a reasonable estimate, although different analysts may produce different dollar estimates. The problem becomes much more difficult if we wish to estimate the benefits of clean air.

On any given individual or thing we may measure several different characteristics. We would then be dealing with several variables, such as age, height, annual income, race, sex, and level of depression of a certain individual. Similarly, we can measure characteristics of a corporation, such as various financial measures. In this book we are concerned with analyzing data sets consisting of measurements on several variables for each individual in a given sample. We use the symbol $P$ to denote the number of *variables* and the symbol $N$ to denote the number of *individuals, observations, cases,* or *sampling units*.

## 2.3 HOW VARIABLES ARE CLASSIFIED: STEVENS'S CLASSIFICATION SYSTEM

In the determination of the appropriate statistical analysis for a given set of data, it is useful to classify variables by type. One method for classifying variables is by the degree of sophistication evident in the way they are measured. For example, we can measure height of people according to whether the top of their head exceeds a mark on the wall: If yes, they are tall; and if no, they are short. On the other hand, we can also measure height in centimeters or inches. The latter technique is a more sophisticated way of measuring height. As a scientific discipline advances, measurements of the variables with which it deals become more sophisticated.

Various attempts have been made to formalize variable classification. A commonly accepted system is that proposed by Stevens (1951). In this system measurements are classified as *nominal, ordinal, interval,* or *ratio.* In deriving his classification, Stevens characterized each of the four types by a transformation that would not change a measurement's classification. In the subsections that follow, rather than discuss the mathematical details of these transformations, we present the practical implications for data analysis.

### Nominal Variables

With *nominal variables* each observation belongs to one of several distinct categories. The categories are not necessarily numerical, although numbers may be used to represent them. For example, "sex" is a nominal variable. An individual's sex is either male or female. We may use any two symbols, such as $M$ and $F$, to represent the two categories. In computerized data analysis, numbers are used as the symbols since many computer programs are designed to handle only numerical symbols. Since the categories may be arranged in any desired order, any set of numbers can be used to represent them. For example, we may use 0 and 1 to represent males and females, respectively. We may also use 1 and 2 to avoid the use of zero since some programs do not distinguish zeros from blanks. Any two other numbers can be used as long as they are used consistently.

**TABLE 2.1.** Stevens's Measurement System

| Type of Measurement | Basic Empirical Operation | Examples |
|---|---|---|
| Nominal | Determination of equality of categories | Company names<br>Race<br>Religion<br>Basketball players' numbers |
| Ordinal | Determination of greater than or less than (ranking) | Hardness of minerals<br>Socioeconomic status<br>Rankings of wines |
| Interval | Determination of equality of differences between levels | Temperature, in degrees Fahrenheit<br>Calendar dates |
| Ratio | Determination of equality of ratios of levels | Height<br>Weight<br>Density<br>Difference in time |

An investigator may rename the categories, thus performing a numerical operation. In so doing, the investigator must preserve the uniqueness of each category. Stevens expressed this last idea as a "basic empirical operation" that preserves the category to which the observation belongs. For example, two males must have the same value on the variable "sex," regardless of the two numbers chosen for the categories. Table 2.1 summarizes these ideas and presents further examples. Nominal variables with more than two categories, such as race or religion, may present special challenges to the multivariate data analyst. Some ways of dealing with these variables are presented in Chapter 9.

## Ordinal Variables

Categories are used for *ordinal variables* as well, but there also exists a known order among them. For example, in the Mohs Hardness Scale minerals and rocks are classified according to ten levels of hardness. The hardest mineral is diamond and the softest is talc (see Pough 1976). Any ten numbers may be used to represent the categories, as long as they are ordered in magnitude. For instance, the integers 1 to 10 would be natural to use. On the other hand, any sequence of increasing numbers may also be used. Thus the basic empirical operation defining ordinal variables is whether one observation is greater than another. For example, we must be able to determine whether one mineral is harder than another. Hardness can be tested easily by noting which mineral can scratch the other. Note that for most ordinal variables there is an underlying continuum being approximated by artificial categories. For example, in the above hardness scale fluorite is defined as having a hardness of 4, and calcite, 3. However, there is a range of hardness between these two numbers not accounted for by the scale.

## Interval Variables

An *interval variable* is a special ordinal variable in which the differences between successive values are always the same. For example, the variable "temperature," in degrees Fahrenheit, is measured on the interval scale since the difference between 12° and 13° is the same as the difference between 13° and 14° or the difference between any two successive temperatures. In contrast, the Mohs Hardness Scale does not satisfy this condition since the intervals between successive categories are not necessarily the same. The scale must satisfy the basic empirical operation of preserving the equality of intervals.

## Ratio Variables

*Ratio variables* are interval variables with a natural point representing the origin of measurement, i.e., a natural zero point. For instance, height is a ratio variable since zero height is a naturally defined point on the scale. We may change the unit of measurement (e.g., centimeters to inches), but we would still preserve the zero point and also the ratio of any two values of height. Temperature in degrees Fahrenheit is not a ratio variable since we may choose the zero point arbitrarily, thus not preserving ratios.

There is an interesting relationship between interval and ratio variables. For example, although time of day is measured on the interval scale, the length of a time period is a ratio variable since it has a natural zero point.

## Other Classifications

Other methods of classifying variables have also been proposed (Coombs 1964). Many authors use the term *categorical* to refer to nominal and ordinal variables where categories are used. We mention, in addition, that variables may be classified as discrete or continuous.

A variable is called *continuous* if it can take on any value in a specified range. Thus the height of an individual may be 70 in. or 70.4539 in. Any numerical value in a certain range is a conceivable height.

A variable that is not continuous is called *discrete*. A discrete variable may take on only certain specified values. For example, counts are discrete variables since only zero or positive integers are allowed. In fact, all nominal and ordinal variables are discrete. Interval and ratio variables can be continuous or discrete. This latter classification carries over to the possible distributions assumed in the analysis. For instance, the normal distribution is often used to describe the distribution of continuous variables.

Statistical analyses have been developed for various types of variables. In Chapter 5 a guide to selecting the appropriate descriptive measures and multivariate analyses will be presented. The choice depends on how the variables are used in the analysis, a topic that is discussed next.

## 2.4 HOW VARIABLES ARE USED IN DATA ANALYSIS

The type of data analysis required in a specific situation is also related to the way in which each variable in the data set is used. Variables may be used to measure outcomes or to explain why a particular outcome resulted. For example, in the treatment of a given disease a specific drug may be used. The outcome may be a discrete variable classified as "cured" or "not cured." Also, the outcome may depend on several characteristics of the patient such as age, genetic background, and severity of the disease. These characteristics are sometimes called *explanatory variables*. Equivalently, we may call the outcome the *dependent variable* and the characteristics the *independent variables*. The latter terminology is very common in statistical literature. This choice of terminology is unfortunate in that the "independent" variables do not have to be statistically independent of each other. Indeed, these independent variables are usually interrelated in a complex way. Another disadvantage of this terminology is that the common connotation of the words implies a causal model, an assumption not needed for the multivariate analyses described in this book. In spite of these drawbacks, the widespread use of these terms forces us to adopt them.

In other situations the dependent variable may be treated as a continuous variable. For example, in household survey data we may wish to relate monthly expenditure on cosmetics per household to several explanatory or independent variables such as the number of individuals in the household, their sex, and household income.

In some situations the roles that the various variables play are not obvious and may also change, depending on the question being addressed. Thus a data set for a certain group of people may contain observations on their sex, age, diet, weight, height, and blood pressure. In one analysis we may use weight as a dependent variable with height, sex, age, and diet as the independent variables. In another analysis blood pressure might be the dependent variable with weight and the other variables considered as independent variables.

In certain exploratory analyses all the variables may be used as one set with no regard to whether they are dependent or independent. For example, in the social sciences a large number of variables may be defined initially, followed by attempts to combine them into a smaller number of summary variables. In this analysis the original variables are not classified as dependent or independent. The summary variables may later be used to possibly explain certain outcomes or dependent variables. In Chapter 5 multivariate analyses described in this book will be characterized by situations in which they apply according to the types of variables analyzed and the roles they play in the analysis.

## 2.5 EXAMPLES OF CLASSIFYING VARIABLES

In the depression data example several variables are measured on the nominal scale: sex, marital status, employment, and religion. The general health scale is an example of an ordinal variable. Income and age are both ratio variables. No interval variable is included in the data set. A listing and code book for this data set are given in Chapter 3.

One of the questions that may be addressed in analyzing this data set is, What are the factors related to the degree of psychological depression of a person? The variable "cases" may be used as the dependent or outcome variable since an individual is considered a case if his or her score on the depression scale exceeds a certain level. "Cases" is an ordinal variable, although it can be considered nominal because it has only two categories. The independent variable could be any or all of the other variables (except ID and measures of depression). Examples of analyses without regard to variable roles are given in Chapters 14 and 15, using the variables $C_1$ to $C_{20}$ in an attempt to summarize them into a small number of components or factors.

Sometimes, the Stevens classification system is difficult to apply, and two investigators could disagree on a given variable. For example, there may be disagreement about the ordering of the categories of a socioeconomic status variable. Thus the status of blue-collar occupations with respect to the status of certain white-collar occupations might change over time or from culture to culture. So such a variable might be difficult to justify as an ordinal variable, but we would be throwing away valuable information if we used it as a nominal variable. Despite these difficulties, the Stevens system is useful in making decisions on appropriate statistical analysis, as will be discussed in Chapter 5.

## SUMMARY

In this chapter statistical variables were defined. Their types and the roles they play in data analysis were discussed.

These concepts can affect the choice of analyses to be performed, as will be discussed in Chapter 5.

## BIBLIOGRAPHY

Churchman, C. W., and Ratoosh, P., eds. 1959. *Measurement: Definition and theories.* New York: Wiley.

*Coombs, C. H. 1964. *A theory of data.* New York: Wiley.

Ellis, B. 1966. *Basic concepts of measurement.* London: Cambridge University Press.

Pough, F. H. 1976. *Field guide to rocks and minerals.* 4th ed. Boston: Houghton Mifflin.

*References preceded by an asterisk require strong mathematical background.

Stevens, S. S. 1951. Mathematics, measurement, and psychophysics. In Stevens, S. S., ed. *Handbook of experimental psychology.* New York: Wiley.

Torgerson, W. S. 1958. *Theory and methods of scaling.* New York: Wiley.

## PROBLEMS

2.1 Classify the following types of data by using Stevens's measurement system: decibels of noise level, father's occupation, parts per million of an impurity in water, density of a piece of bone, rating of a wine by one judge, net profit of a firm, and score on an attitude test.

2.2 In a survey of users of a walk-in mental health clinic, data have been obtained on sex, age, household roster, race, educational level (number of years of school), family income, reason for coming to the clinic, symptoms, and scores on screening examination. The investigator wishes to determine what variables affect whether or not coercion by the family, friends, or a governmental agency was used to get the patient to the clinic. Classify the data according to Stevens's measurement system. What would you consider to be possible independent variables? Dependent variables? Do you expect the independent variables to be independent of each other?

2.3 For the chronic respiratory study data presented in Appendix B, classify each variable according to the Stevens scale and according to whether it is discrete or continuous. Pose two possible research questions and decide on the appropriate dependent and independent variables.

2.4 From a field of statistical application (perhaps your own field of specialty), describe a data set and repeat the procedures described in Problem 2.3.

2.5 If the RELIG variable described in Table 3.2 of this text was recoded 1 = Catholic, 2 = Protestant, 3 = Jewish, 4 = none, and 5 = other, would this meet the basic empirical operation as defined by Stevens?

2.6 Give an example of nominal, ordinal, interval, and ratio variables from a field of application you are familiar with.

2.7 The variables C1, C2, ..., C20 as described in Table 3.2 are ordinal variables. How could you redefine them so that they measure the same characteristics but on an interval scale? Why would you not want to do this?

## Chapter Three

# PREPARING
# FOR DATA
# ANALYSIS

### 3.1 WHAT WILL YOU LEARN FROM THIS CHAPTER?

From this chapter you will learn what type of computer to use to perform statistical analyses, how to choose a statistical package, how to enter data into computers, how to prepare data and perform initial data screening, how to describe your data set, and what is available in the depression data set. In particular, you will learn:

- How to choose between using a mainframe computer and a personal computer (3.2).
- What factors affect the choice of a computer package and which packages are described in this book (3.3).
- What are some of the other packages available to you (3.3).
- How to enter your data into the computer for use in a statistical package (3.4).
- How to use data management features of statistical packages to combine and select different sets of data (3.5).
- How to perform initial data screening (3.4, 3.5).

- What a code book is and what information should be included in it (3.6).
- What variables are available in the depression data set (3.6).

## 3.2 CHOICE OF COMPUTER FOR STATISTICAL ANALYSIS

In this book we are mainly concerned with analyzing data on several variables simultaneously using computers. We describe how the computer is used for this purpose and the steps taken prior to performing multivariate statistical analyses. As an introduction, we first discuss how to decide what type of computer to use for the three main operations: data entry, data management, and statistical analysis.

At opposite extremes, one can use either a personal computer on a desk or a mainframe computer from a terminal for all three operations. By mainframe computer, we mean a large computer typically in a computer center serving an institution or corporation and staffed by systems programmers, analysts, and consultants. A mainframe computer generally has large memory and many data storage devices, and provides services to many users at one time. By personal computer (or PC) we mean a machine of desk or laptop size designed to be used by one person at a time and generally for one task at a time. (Personal computers are sometimes referred to as microcomputers.) Between these extremes there are several alternatives. The desktop computer may be linked to other desktop computers in a local area network to share data, programs, and peripherals such as printers and file servers. The personal computers or terminals might be high-capability machines commonly known as technical workstations, which provide faster computation, more memory, and high resolution graphics. Any of these personal computers, terminals, or workstations might be linked to larger minicomputers, often known as departmental computers, which might in themselves be linked to computer-center mainframes.

Many investigators find it convenient to perform data entry and data management on personal computers even if they perform the statistical analysis on the mainframe. Using a personal computer has the advantage of making it possible to place the machine close to the location of the data and the people. In laboratories, personal computers are often used to capture data directly from laboratory equipment. In telephone surveys, interviewers can enter the data directly into the PC as they receive the answers from respondents. In offices, staff members can enter data into a PC without leaving the office. If mainframe or departmental computer terminals are conveniently located, many of the same advantages apply. Further discussion on data entry is given in Section 3.4.

For data management and preliminary data screening, some of the same arguments apply. These activities are often done together with or soon after data entry, so often the same equipment is used.

For multivariate statistical analysis, either mainframe or personal computers can be used. However, performing elaborate analyses of large data sets on personal computers requires either well-equipped personal computers, programs that are written to use small memories, or a lot of patience. For the analyses discussed in this book, it is recommended that the PC user have an IBM-compatible machine. It should include at least 640 kilobytes (KB) of memory if DOS is used, a math coprocessor chip, and a hard disk with at least 40 megabytes (MB) of storage.

The advantages of using a personal computer include the ability to do things your own way without being dependent on the structure of a central computer center, the multiplicity of programs available to perform statistical analyses and supporting graphics on personal computers, and the ease of going from data entry to analysis to report writing all on the same personal computer with the printer close at hand. Many programs for the personal computer are menu-driven; that is, after each step a choice of subsequent actions appears on the screen. This saves the user the trouble of remembering and typing the necessary commands. Also, while a complex analysis on a large data set can take noticeably more time on a PC, it may be less frustrating to wait in your own office for the program to finish than to wait in a shared facility.

One advantage of using a mainframe facility is that much of the work is done for you. Somebody else enters the statistical packages into the computer, makes sure the system is running, provides extra back-up copies of your data, and provides consultation. Also, the actual running time for each analysis you submit will tend to be much shorter. Storage space is available for very large data sets. Shared printer facilities often mean more expensive printers with superior features. The central facility often does the purchasing of the statistical software and arranges for obtaining updates. If you are willing to follow the procedures without learning why you are doing them, less time and money will probably be spent in computer-related activities if you use a mainframe facility.

Another factor to consider in deciding between the PC and the mainframe is cost. The PC expenses include purchase price for the hardware and software, as well as the costs of maintenance and upgrading. This must be compared with the mainframe cost which is usually based on use.

## 3.3 CHOICE OF A STATISTICAL PACKAGE

Whether the investigator decides to use a PC or a mainframe, there is a wide choice of statistical packages available. Unlike the situation for word processing where a handful of packages have captured a large share of the market, there are numerous statistical packages available, some written for a particular area of application (such as survey analysis) and others quite general. One feature that distinguishes among the statistical packages is whether they were originally written for mainframe computers or for the PC. Packages written for mainframe computers tend to be more general and comprehensive. They also take more computer memory to store and are often more expensive. Originally, the programs written for the mainframe computers were just adapted for the PC, but recent versions include more interactive features and menu-driven options. The cost of a PC package is less of a factor if a site license is purchased by the school, business, or governmental unit where one works; in which case the cost can be shared by numerous users.

### Packages Used in this Book

In this book, we make specific reference to three general purpose statistical packages that are available on both the mainframe and personal computer. They are BMDP, SAS, and SPSS. We use the generic names of the packages without distinguishing between the mainframe and PC versions. Thus, for example, we write SPSS to refer to the package. When necessary, we write SPSS-X for the mainframe version and SPSS/PC+ for the PC version to distinguish between the two. Each of the three packages offers a comprehensive set of programs or procedures that allows the user to enter data, edit them, manipulate them, screen them for erroneous values, perform statistical analyses, and display the results. In addition, the computing algorithms in these packages have been widely tested and, although errors do occasionally appear, there are relatively few. The packages offer the features needed to perform multivariate analysis.

Both BMDP and SPSS adopt the philosophy of offering a number of *comprehensive* programs, each with its own options and variants for performing portions of the analyses. The SAS package, on the other hand, offers a large number of *limited-purpose* procedures, some with a number of options. The SAS philosophy is that the user should string together a sequence of the procedures to perform a major analysis.

When purchasing these packages or entering them on a personal computer, the differences referred to in the previous paragraph become obvious. With SAS you need to buy entire sets of procedures. At a minimum you need the basic package and the statistics package to perform the statistical analyses discussed in this book. BMDP programs can be purchased either in sets or individually. SPSS is similar to SAS. There is a Base Package set of programs and an Advanced Statistics set which contains many of the procedures discussed in this book. To run SAS you need nearly 10 MB disk storage for basic SAS and statistics, and less than 1 MB for data entry. For BMDP about one third of one megabyte is needed for storing BMDP system files; each program takes about 0.75 to 1 MB. SPSS requires a minimum of 4.5 MB of storage space on the hard disk for the programs in the Base Package, with an additional 1.5 MB for the Advanced Statistics module.

### User's Manuals

Each of the three computer packages has multiple manuals. In the following we list the manuals used in this book. (The editions listed also refer to the versions of the programs used.)

### BMDP

*BMDP PC Supplement: Installation and Special Features,* 1988 (installing BMDP programs and running them on a PC)
*BMDP Data Entry,* 1990 (data entry procedures)
*BMDP Data Management Manual,* 1988 (data management procedures)
*BMDP Statistical Software Manual, Volumes 1 and 2,* 1988 (general statistical programs)

## SAS

*SAS/FSP Guide Version 6*, 1987 (data entry)

*SAS Introductory Guide for Personal Computers Release 6.03*, 1988 (useful for new users)

*SAS Procedures Guide Release 6.03*, 1988 (provides descriptions of elementary statistics, reporting, scoring, and utility procedures)

*SAS Language Guide for Personal Computers Release 6.03*, 1988 (describes DATA and PROC steps, syntax and use of SAS statements, and other options)

*SAS/STAT Guide for Personal Computers Version 6*, 1987 (describes multivariate statistical procedures)

*SUGI Supplemental Library User's Guide Version 5*, 1986 (describes additional SAS statistical procedures mostly available on mainframe computers)

## SPSS

*SPSS Data Entry II*, 1988 (data entry)

*SPSS-X User's Guide*, 3rd edition, 1988 (complete guide, from syntax to statistical procedures, for the mainframe version)

*SPSS/PC + V2.0, Base Manual*, 1988 (for the PC—descriptions of syntax, data management and elementary statistical procedures)

*SPSS/PC + Advanced Statistics V2.0*, 1988 (for the PC—descriptions of multivariate statistical procedures)

*SPSS/PC + V3.0 Update Manual*, 1988 (for the PC—describes changes and new features added in Version 3, compared to Version 2 of SPSS/PC + )

Further information and manuals can be obtained by writing to either:

BMDP Statistical Software, Inc.
1440 Sepulveda Blvd.
Los Angeles, CA 90025

SAS Institute Inc.
SAS Circle Box 8000
Cary, NC 27512-8000

SPSS Inc.
Suite 3000
444 North Michigan Ave
Chicago, IL 60611

When you are learning to use a package for the first time, there is no substitute for reading the manuals. For any of the packages, the examples shown in the manuals are often very helpful. However, at times the sheer number of options presented in these general purpose programs may seem confusing and advice from an experienced user may save you time. Many of the programs offer what are called default options, and it often helps to use these when you run a program for the first time. In this book, we frequently recommend which options to use.

It is not necessary to purchase all of the manuals to perform the analyses in this book. Most investigators decide to use only one of the packages mentioned above or choose a less extensive package. It is often possible to borrow manuals from a computer center or from libraries. However, you will find it convenient to have your own copy.

A listing of other personal computer packages is given in Table 3.1. It should be noted that this list is not comprehensive and some fine packages may have been omitted because we were unaware of them. There are numerous new programs appearing continually, as well as major improvements being made in existing ones. It is a very fluid situation.

Several packages listed in Table 3.1 tend to have shorter manuals; to take considerably less storage space on the computer; to handle fewer variables and observations; to cost less; to have a less rich set of options for data management, missing values, and transformations; and to make more use of menus and window displays. The important consideration here is to check that they have the options you need before you purchase them. Reviews of these packages can help (see personal computing journals, e.g., *PC Magazine*, 1989 and the statistical computing software reviews commonly printed in *The American Statistician*). Calls to the vendors with specific questions may assist you in your choice. Some packages have been written with a specific type of user in mind and, if you are of that type, then they make an excellent choice.

**TABLE 3.1.** Additional Statistical Packages for IBM-PC and Compatible Machines

| Package (Latest Version) | Source | Version reviewed in The American Statistician Volume: Number (Year) |
|---|---|---|
| BASS (88.10) | BASS Institute Inc.<br>P.O. Box 349<br>Chapel Hill, NC 27514<br>(919) 489-0729 | |
| CRISP | Crunch Software Corporation<br>2547 22nd Avenue<br>San Francisco, CA 94116<br>(415) 564-7337 | v. 3<br>41:2 (1987) p. 139 |
| CSS (2.1) | StatSoft Inc.<br>2325 E. 13th Street<br>Tulsa, OK 74104<br>(918) 583-4149 | |
| Minitab (7.0) | Minitab Inc.<br>3081 Enterprise Dr.<br>State College, PA 16801<br>(814) 238-3280 | v. 5.1.1<br>42:3 (1988) p. 220 |
| NCSS (5.01) | NCSS Inc.<br>865 E. 400 North<br>Kaysville, UT 84037<br>(801) 546-0445 | v. 4.1<br>39:4 (1985) p. 315 |
| Prodas | Conceptual Software Inc.<br>P.O. Box 56627<br>Houston, TX 77256-6627<br>(713) 667-4222 | |
| P-Stat (2.10) | P-Stat Inc.<br>P.O. Box AH<br>Princeton, NJ 08542<br>(609) 924-9100 | |
| Sigstat | Significant Statistics<br>3336 N. Canyon Road<br>Provo, UT 84604<br>(801) 377-4860 | |

**TABLE 3.1.**  *(Continued)*

| Package (Latest Version) | Source | Version reviewed in The American Statistician Volume: Number (Year) |
|---|---|---|
| SOLO | BMDP Statistical Software 1440 Sepulveda Blvd. #316 Los Angeles, CA 90025 (213) 479-7799 | |
| Stata (2.0) | Computing Resource Center 10801 National Blvd. Los Angeles, CA 90064 (800) STATAPC (213) 470-4341 | v. 1.0 41:1 (1987) p. 68 |
| Statgraphics (3.0) | STSC Inc. 2115 E. Jefferson Street Rockville, MD 20852 (301) 984-5000 (800) 592-0050 | v. 2.0 41:1 (1987) p. 64 |
| Statistix (2.0) | NH Analytical Software 1958 Eldridge Avenue Roseville, MN 55113 (612) 631-2852 | v. 1.1 41:3 (1987) p. 229 |
| StatPac Gold (3.0) | Walonick Associates Inc. 6500 Nicollet Ave. South Minneapolis, MN 55423 (612) 866-9022 | v. 1.1 mentioned in Dallal, 1988 |
| Systat (4.0) | Systat Inc. 1800 Sherman Ave. Evanston, IL 60201 (312) 864-5670 | v. 3.1 41:4 (1987) p. 318 v. 3.0 mentioned in Dallal, 1988 |
| —condensed version MYSTAT | | described in 41:4 (1987) p. 334 |
| Turbo Spring-Stat (2.9) | Spring Systems P.O. Box 10073 Chicago, IL 60610 (312) 275-5273 | |

Dallal, G. E. 1988. Statistical microcomputing—like it is. *The American Statistician* 42:3, pp. 212–216.

## 3.4 TECHNIQUES FOR ENTERING DATA

Appropriate techniques for entering data for analysis by BMDP, SAS, or SPSS depend mainly on the size of the data set and the form in which the data are stored. Each package has its own way of representing data at the machine level. The advantage of such files, sometimes called system files, is that data can be accessed easily and quickly in subsequent analyses that use the same statistical package. Variable location, formats, labels, and other identifying characteristics do not need to be respecified. This last feature becomes especially convenient when data are shared among users and institutions. Data stored in system files, however, cannot be directly accessed by any other statistical package, listed on a terminal screen, or printed in a legible form directly from the operating system.

ASCII is the name of another and more common way that data is stored. (ASCII files are sometimes called DOS files.) An ASCII data set must be converted to a package-specific format with some sort of input statement in your program before it can be used in an analysis.

Several methods of data entry are possible. We will discuss three of them:

1. entering the data along with the program or procedure statements;
2. entering the data from an outside file which is constructed without the use of the above statistical packages;
3. using the data entry features of the statistical package you intend to use.

The first method is only recommended for very small data sets and its use is illustrated in the respective manuals. Using SAS for example, suppose you have a small data set consisting of the first three variables (id, sex, and age) for the first four persons in the depression data set discussed in Section 3.6 (Table 3.3). Here the data are listed in what we will call the spreadsheet format, that is the three variables are listed in three columns and each row comprises the data for one case or individual. For the variable sex, 1 = male and 2 = female, and age is given in years.

| ID | SEX | AGE |
|----|-----|-----|
| 1  | 2   | 68  |
| 2  | 1   | 58  |
| 3  | 2   | 45  |
| 4  | 2   | 50  |

A SAS data set called depress could be made on a personal computer by stating,

```
data depress;
    input id sex age;
    cards;
1    2    68
2    1    58
3    2    45
4    2    50
;
run;
```

Similar types of statements can be used for the other programs utilizing the spreadsheet type of format.

The disadvantage of this type of data entry is that there are only limited editing features available to the person entering the data. No checks are made as to whether or not the data are within reasonable ranges for this data set. For example, all respondents were supposed to be 18 years old or older but there is no automatic check to verify that the age of the third person, who was 45 years old, was not erroneously entered as 15 years. Another disadvantage is that the data set disappears after the program is run unless additional statements are made (the actual statements depend on the type of computer used, mainframe or PC). In small data sets, the ability to save the data set, edit typing, and have range checks performed is not as important as in larger data sets.

The second possibility is to create an ASCII file using a word processor such as WordPerfect, a spreadsheet program such as LOTUS 123, a data base program that has data entry features such as dBaseIII, or data entry programs such as Survey Mate. The advantage of this strategy is that you can use familiar programs to obtain a data file. This ASCII file can then be converted to a file in a form directly accessible by the statistical package you are planning to use (e.g., a SAS file for use in SAS programs). Instructions for creating an ASCII file are given in the manual for the program used to enter the data. The instructions for reading an outside ASCII file are given in the manuals for the respective statistical packages. If the process for doing this appears difficult, there is a special-purpose program called DBMS/COPY that will copy data files created by a wide range of programs and put them into the right form for access by any of a wide range of statistical packages.

An example of copying a LOTUS 123 file into a BMDP program is given in *BMDP Software Communications 1985*. First the data are entered into the LOTUS program. The user enters /PRINT and chooses the File option to specify the variables with a Range command. To void page numbers, headers, or footers, the sub commands OptionsOtherUnformatted are used. This is followed by a Go and finally a Quit. An ASCII file will be made with the LOTUS file name followed by .PRN. This file could be read by BMDP using the input paragraph in any BMDP program. Suppose there were three variables named id, sex, and age, and the LOTUS file was called DEPRESS. The BMDP instructions would be

```
/input   variables = 3. format = free.
         file is 'depress.prn'.
```

To name the three variables, we would type

```
/variable names are id,sex,age.
```

In reading in an external file each program requires you to tell it the name of the file (assuming you are in the directory of the file), the number of variables, and the format they are in. If you are not in the directory of the file, then a pathway to that directory must be included with the file name. The programs tend to describe formats in slightly different ways, so one must consult the relevant manual.

The third strategy is to use the data entry packages belonging to the statistical program you wish to use. In the proposed BMDP data entry system to be released in 1990, there are two options for entering data. One is the spreadsheet (or table mode), which was already discussed. The second is the form mode. In the BMDP form mode, data for one case or person is displayed vertically on the screen. For example, for the first case in the depression data set, we would have

```
case: 1
id    1
sex   2
age   68
```

This standard form includes variable labels, the data entry area, the name of the data file, a status line for error messages, and a set of options listed in a menu.

For each variable used you describe its attributes, including the type of variable (numerical, categorical, label, or date); various properties of that variable, for example, the range of values that you expect if it is a numerical variable; column width, whether missing values are permitted, and how to identify them; long variable name; and comments concerning that variable. The use of minimum and maximum possible values to define ranges is a valuable option, as it prevents data that are clearly out of range from ever entering the data set. Thus, the data entry program acts as the first line of defense in data screening. The data can be edited to remove errors. Errors can also be found by writing IF statements that allow for checking several variables simultaneously. For example, you would not expect males to be pregnant. Subgroups can be selected that meet stated conditions and saved in a separate file. The program leads you by the use of highlighting and menus.

SAS uses the form system of data entry in their FSEDIT procedure and also operates in an interactive fashion. Here there are two major options: A standard form can be used or the user can design a special form. The second option has the advantage that a form can be devised which mimics the placement on the page of the data being entered. Different colors can be used to make the display very readable. It has a wide range of options that assist in data screening, including special locating options for finding specific cases. Maximum and minimum values can be set.

SPSS has a data entry module called SPSS Data Entry II, which offers data screening as well. One available option that is especially useful in survey work is "skip and fill," which allows the user to specify values in particular data fields that will automatically determine data values in other, dependent fields. Data Entry II allows data to be entered into either a spreadsheet form or a user-defined form. It stores the entered data as an SPSS file but also offers the option of storing the data directly in any of a choice of formats, including ASCII, dBase, and LOTUS, or translating between such formats.

Finally, if you have a very large data set to enter, it often is sensible to use the services of a professional data entering service. They tend to be very fast and can offer different levels of data checking and advice on which data entry form to use. But whether or not a professional service is used, the following suggestions may be helpful for data entry:

1. Whenever possible, code information in numbers, not letters.
2. Code information in the most detailed form you will ever need. You can use the computer to aggregate the data into coarser groupings later. For example, it is better to record age as exact age at the last birthday rather than to record the ten-year age interval into which it falls.
3. If the data are stored on a personal computer, make a backup copy on a floppy disk or tape. Backups should be updated regularly as changes are made in the data set.
4. For each variable, use a code to indicate missing values. The various programs each have their own way to indicate missing values. The manuals should be consulted so that you can match what they require with what you do.

To summarize, there are three important considerations in data entry: accuracy, cost, and ease of use of the data file. Whatever system is used, the investigator should ensure that the data file is free of typing errors, that time and money are not wasted, and that the data file is easy to use for data management or analysis. The use of maximum and minimum data value limits and logical checks can detect some extreme errors but cannot prevent all errors, e.g., those that are within the acceptable range.

## 3.5 DATA MANAGEMENT FOR STATISTICS

Prior to statistical analysis, it is often necessary to make some changes in the data set. This manipulation of the data set is called *data management.* The term data management is also used in other fields to describe a somewhat different set of activities that are infrequently used in statistical data management. For example, large companies often have very large sets of data that are stored in a complex system of files and can be accessed in a differential fashion by different types of employees. This type of data management is not a common problem in statistical data management. Here we will limit the discussion to the more commonly used options in statistical data management programs. The manuals describing the programs should be read for further details.

## Combining Data Sets

Combining separate data sets is an operation that is widely used. For example, in biomedical studies it is common to have data taken from medical history forms, a questionaire, and laboratory results all concerning the same patient. These need to be combined together into a single record. In longitudinal studies of voting intentions, the records for each respondent must be combined together in order to analyze change in voting intentions of an individual over time.

There are essentially two steps in this operation. The first is sorting on some variable (called a key variable in BMDP and SPSS) which is often an id variable in data sets containing information from individuals. The second step is combining the separate data sets side by side and matching the correct records with the correct person. Sometimes part of the data is missing. Perhaps the interviewer could not locate the respondent for one of the time periods. In such a case, a symbol or symbols indicating missing values can be inserted into the spaces for the missing data items.

Data sets can be combined in the manner described above in SAS by using the MERGE statement followed by a BY statement and the variable(s) to be used to match the records. (The data must be sorted by the values of the matching variables.) In BMDP this is done with a JOIN paragraph after the data are sorted. In SPSS, the MATCH FILES command will combine two data sets side by side. If the structures of the two files are not parallel, by which SPSS means that there are cases that have duplicate records in one data set or are missing from the other, then the data sets must be sorted by, and matched on, a key variable.

Another common need is to combine separate sets of records where one set is put at the end of the other, or interleaved together based on some key variable. For example, an investigator may have data sets that are collected at different places and then combined together. In an education study, student records could be combined from two high schools, with one set simply placed at the bottom of the other set. This is done using the APPEND procedure (or proc) in SAS. It would be followed by a SORT proc if you want the data interleaved. The MERGE paragraph in BMDP can be used to either combine end to end or interleaved. In SPSS, the ADD FILES command is used to concatenate files or, with the specification of a key variable, to interleave them.

It is also possible to update the data files with later information. Thus a single file can be obtained that contains the latest information, if this is desired for analysis. This option can also be used to replace data that were originally entered incorrectly.

## Missing Values

As will be discussed in later chapters, most multivariate analyses require complete data on all the variables used in the analysis. To determine which observations are complete, the user is advised to examine patterns of missing data in order to select variables and cases to be included. Most statisticians agree on the following guidelines:

**1.** If a variable is missing in a high proportion of the cases, then that variable should be deleted.

**2.** If a case is missing variables that you wish to analyse, then that case should be deleted.

Following these guidelines will give you a data set with a minimum of missing values. Further discussion on other options available when there are missing data is given in Section 9.2.

Missing values can occur in two ways. First, the data may be missing at the start. In this case, the investigator enters a code for missing values at the time the data set is entered into the computer. Commonly, a numerical value is used that is outside the range of possible values. For example, for the variable sex (with 1 = male and 2 = female) a missing code could be 9. A string of 9s is often used; thus, for the weight of a person 999 could be used as a missing code. Then that value is declared to be missing. For example, in SAS, one could state

if sex = 9 then sex = .;

In SAS, using a period for missing values is recommended. Missing values are indicated in print with a period. In BMDP, we state MISS = (sex)9 to instruct the computer that 9 is the missing value code for sex. SPSS recognizes both user-defined and system missing values. System missing values appear as a period. They are assigned automatically when a blank or a non-numeric character is encountered as a value for a numeric variable, or when transformations of such characters are attempted. User-defined missing values arise when the user specifies the value that represents missing information for a given variable. For example, MISSING VALUE SEX(9) specifies that the value 9 indicates that information on sex is missing.

The second way in which values can be considered missing is if data values are beyond the range of the stated maximum or minimum values. In BMDP, when the data set is printed, the variables are either assigned their true numerical value or called MISSING, GT MAX (greater than the maximum), or LT MIN (less than the minimum). Cases that have missing or out of range values can be printed out for examination. The pattern of missing values can be determined from the BMDPAM program.

## *Detection of Outliers*

Outliers are observations that appear inconsistent with the remainder of the data set (Barnett and Lewis 1984). One method for determining outliers has already been discussed, namely setting minimum and maximum values. By applying these limits, extreme or unreasonable outliers are prevented from entering the data set.

Often, observations are obtained that seem quite high or low but are not impossible. These values are the most difficult ones to cope with. Should they be removed or not? Statisticians differ in their opinions, from "if in doubt, throw it out" to the point of view that it is unethical to remove an outlier for fear of biasing the results. The investigator may wish to eliminate these outliers from the analyses but report them along with the statistical analysis. Another possibility is to run the analyses twice, both with the outliers and without them, to see if they make an appreciable difference in the results. Most investigators would hesitate, for example, to report rejecting a null hypothesis if the removal of an outlier would result in the hypothesis not being rejected.

A review of formal tests for detection of outliers is given in Dunn and Clark (1987) and in Barnett and Lewis (1984). To make the formal tests you usually must assume normality of the data. Some of the formal tests are known to be quite sensitive to nonnormality and should only be used when you are convinced that this assumption is reasonable. Often an alpha level of 0.10 or 0.15 is used for testing if it is suspected that outliers are not extremely unusual. Smaller values of alpha can be used if outliers are thought to be rare.

The data can be examined one variable at a time by using histograms if the variable is measured on the interval or ratio scale. A questionable value would be one that is separated from the remaining observations. For nominal or ordinal data, the frequency of each outcome can be noted. If a recorded outcome is impossible, it can be declared missing. If a particular outcome occurs only once or twice, the investigator may wish to consolidate that outcome with a similar one. We will return to the subject of outliers in connection with the statistical analyses starting in Chapter 6, but mainly the discussion in this book is not based on formal tests.

## Transformations of the Data

Transformations are commonly made either to create new variables with a form more suitable for analysis or to achieve an approximate normal distribution. Here we discuss the first possibility. Transformations to achieve approximate normality are discussed in Chapter 4.

Transformations to create new variables can either be performed as a step in data management or can be included later when the analyses are being performed. It is recommended that they be done as a part of data management. The advantage of this is that the new variables are created once and for all, and sets of instructions for running data analysis from then on do not have to include the data transformation statements. This results in shorter sets of instructions with less repetition and chance for errors when the data are being analyzed. This is almost essential if several investigators are analyzing the same data set.

One common use of transformations occurs in the analysis of questionnaire data. Often the results from several questions are combined to form a new variable. For example, in studying the effects of smoking on lung function it is common to ask a first question such as,

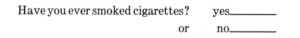

Have you ever smoked cigarettes?     yes_____

                or      no_____

If the subjects answer no, they skip a set of questions and go on to another topic. If they answer yes, they are questioned further about the amount in terms of packs per day and the length of time they smoked (in years). From this information, a new pack-year variable is created that is the number of years times the average number of packs. For the person who never smoked, the answer is zero. Transformation statements are used to create the new variable.

Each package offers a slightly different set of transformation statements, but some general options exist. The programs allow you to select cases that meet certain specifications using IF statements. Here for instance, if the response is no to whether the person ever smoked, the new variable should be set to zero. If the response is yes, then pack-years is computed by multiplying the average amount smoked by the length of time smoked. This sort of arithmetic operation is provided for and the new variable is added to the end of the data set.

Additional options include taking means of a set of variables or the maximum value of a set of variables. Another common arithmetic transformation involves simply changing the numerical values coded for a nominal or ordinal variable. For example, for the depression data set, sex was coded male = 1 and female = 2. In some of the analyses used in this book, we recoded that to male = 0 and female = 1 by simply subtracting one from the given value.

## Saving the Results

After the data have been screened for outliers, recoding is done for some of the variables, and new variables are created; the results are saved in a file that can be used for analysis. It is also possible to keep only a subset of the variables. For example, from the smoking question perhaps only the pack-years, or the pack-years and whether a person smokes, may be kept in the file. It is always advisable to save the original file in case a different recoding becomes necessary.

When considerable manipulation is done, it is recommended that the control language be stored along with the data sets. Then, should the need arise, the manipulation can be re-done by simply editing the control language instructions rather than completely recreating them.

## 3.6 DATA EXAMPLE: LOS ANGELES DEPRESSION STUDY

In this section we discuss a data set that will be used in several of the succeeding chapters to illustrate multivariate analyses. The depression study itself is described in Chapter 1.

The data given in this chapter are from a subset of 294 respondents randomly chosen from the original 1000. This subset of the observations is large enough to provide a good illustration of the statistical techniques but small enough to save computer costs. Only data from the first time period are included. Variables were chosen so that they would be easily understood and would be sensible to use in the multivariate statistical analyses described in Chapters 6 through 16.

The code book, the variables used, and the data set are described below.

## Code Book

An important step in making a data set understandable to its users is to create a data code book. The code book should contain a description and information on the location of each variable in the data set after data management is finished. It serves as a guide and record for all users of the data set, and for future documentation of the results.

Table 3.2 contains a code book for the depression data set. In the first column the variable number is listed, since that is often the simplest way to refer to the variables in the computer. The location of each variable is listed next. A variable name is given next, and this name is used in later data analyses. These names were chosen to be eight characters or less in length so that they could be used by all three packaged programs. It is helpful to choose variable names that are easy to remember but are short enough to reduce entry time when running programs; then use them consistently.

Finally, a description of the outcome of each variable is given in the last column of Table 3.2. For nominal or ordinal data the numbers used to code each answer are listed. For ratio data such as age the units used are included. Note that income is given in thousands of dollars per year for the household; thus an income of 15 would be $15,000 per year.

## Depression Variables

The 20 items used in the depression scale are variables 9–28 and are named C1, C2, . . . , C20. (The wording of each item is given later in the text, in Table 14.2.) Each item was written on a card and the respondent was asked to tell the interviewer the number that best describes how often he or

**TABLE 3.2.** Code Book for Depression Data

| Variable Number | Variable Location (Columns) | Variable Name | Description |
|---|---|---|---|
| 1 | 6–8 | ID | Identification number from 1 to 294 |
| 2 | 9 | SEX | 1 = male; 2 = female |
| 3 | 10–11 | AGE | Age in years at last birthday |
| 4 | 12 | MARITAL | 1 = never married; 2 = married; 3 = divorced; 4 = separated; 5 = widowed |
| 5 | 13 | EDUCAT | 1 = less than high school; 2 = some high school; 3 = finished high school; 4 = some college; 5 = finished bachelor's degree; 6 = finished master's degree; 7 = finished doctorate |
| 6 | 14 | EMPLOY | 1 = full time; 2 = part time; 3 = unemployed; 4 = retired; 5 = houseperson; 6 = in school; 7 = other |
| 7 | 15–16 | INCOME | Thousands of dollars per year |
| 8 | 17 | RELIG | 1 = Protestant; 2 = Catholic; 3 = Jewish; 4 = none; 5 = other |
| 9–28 | 18–37 | C1–C20 | "Please look at this card and tell me the number that best describes how often you felt or behaved this way during the past week." 20 items from depression scale (already reflected; see text) 0 = rarely or none of the time (less than 1 day); 1 = some or a little of the time (1–2 days); 2 = occasionally or a moderate amount of the time (3–4 days); 3 = most or all of the time (5–7 days) |
| 29 | 38–39 | CESD | Sum of C1–C20; 0 = lowest level possible; 60 = highest level possible |
| 30 | 40 | CASES | 0 = normal; 1 = depressed, where depressed is CESD ≥ 16 |
| 31 | 41 | DRINK | Regular drinker? 1 = yes; 2 = no |
| 32 | 42 | HEALTH | General health? 1 = excellent; 2 = good; 3 = fair; 4 = poor |
| 33 | 43 | REGDOC | Have a regular physician? 1 = yes; 2 = no |
| 34 | 44 | TREAT | Has a doctor prescribed or recommended that you take medicine, medical treatments, or change your way of living in such areas as smoking, special diet, exercise, or drinking? 1 = yes; 2 = no |
| 35 | 45 | BEDDAYS | Spent entire day(s) in bed in last two months? 0 = no; 1 = yes |
| 36 | 46 | ACUTEILL | Any acute illness in last two months? 0 = no; 1 = yes |
| 37 | 47 | CHRONILL | Any chronic illnesses in last year? 0 = no; 1 = yes |

she felt or behaved this way during the past week. Thus respondents who answered item C2, "I felt depressed," could respond 0 through 3, depending on whether this particular item applied to them rarely or none of the time (less than 1 day:0), some or a little of the time (1–2 days:1), occasionally or a moderate amount of the time (3–4 days:2), or most or all of the time (5–7 days:3).

Most of the items are worded in a negative fashion, but items C8 through C11 are positively worded. For example, C8 is "I felt that I was as good as other people." For positively worded items the scores are *reflected*; that is, a score of 3 is changed to be 0, 2 is changed to 1, 1 is changed to 2, and 0 is changed to 3. In this way, when the total score of all 20 items is obtained by summation of variables C1 through C20, a large score indicates a person who is depressed. This sum is the 29th variable, named CESD.

Persons whose CESD score is greater than or equal to 16 are classified as depressed since this value is the common cutoff point used in the literature (see Frerichs, Aneshensel, and Clark 1981). These persons are given a score of 1 in variable 30, the CASES variable. The particular depression scale employed here was developed for use in community surveys of noninstitutionalized respondents (see Comstock and Helsing 1976; Radloff 1977).

## Data Set

As can be seen by examining the code book given in Table 3.2, demographic data (variables 2–8), depression data (variables 9–30), and general health data (variables 32–37) are included in this data set. Variable 31, drinking habits, was included so that it would be possible to determine if an association exists between drinking and depression. Frerichs et al. (1981) have already noted a lack of association between smoking and scores on the depression scale.

The actual data for the 294 respondents are listed in Table 3.3. The headings at the top of Table 3.3 correspond to the variable names given in Table 3.2. For example, the first respondent is a female, aged 68, who is widowed, has had some high school education, is retired with a current income of $4000 per year, and is a Protestant. She has a CESD score of 0, so she would not be considered depressed. She is not a regular drinker and considers herself in good health. She has a regular physician who prescribed or recommended a treatment to her. She has had no bed days or acute illnesses in the last two months, but she has had at least one chronic illness. Similar statements could be made about each of the 294 respondents by reading the data in Table 3.3.

## SUMMARY

In this chapter we discussed the steps necessary before statistical analysis can begin. The first of these is the decision of what computer and software packages to use. Once this decision is made, data entry and data management can be started. Figure 3.1 summarizes the steps taken in data entry and data management.

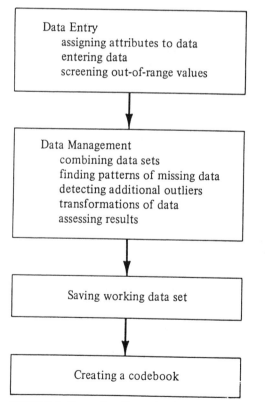

**FIGURE 3.1.** Preparing Data for Statistical Analysis

**TABLE 3.3.** Depression Data

| OBS | ID | SEX | AGE | MARITAL | EDUC | EMPLOY | INCOME | RELIG | C1 | C2 | C3 | C4 | C5 | C6 | C7 | C8 | C9 | C10 | C11 | C12 | C13 | C14 | C15 | C16 | C17 | C18 | C19 | C20 | CESD | CASES | DRINK | HEALTH | REGDOC | TREAT | BEDDAYS | ACUTEILL | CHRONILL |
|---|---|---|---|---|---|---|---|---|---|---|---|---|---|---|---|---|---|---|---|---|---|---|---|---|---|---|---|---|---|---|---|---|---|---|---|---|---|
| 1 | 1 | 2 | 68 | 5 | 2 | 4 | 4 | 1 | 0 | 0 | 0 | 0 | 0 | 0 | 0 | 0 | 0 | 0 | 0 | 0 | 0 | 0 | 0 | 0 | 0 | 0 | 0 | 0 | 0 | 0 | 2 | 2 | 1 | 1 | 0 | 0 | 1 |
| 2 | 2 | 1 | 58 | 3 | 3 | 1 | 15 | 1 | 0 | 0 | 0 | 0 | 0 | 0 | 0 | 0 | 0 | 0 | 0 | 0 | 0 | 0 | 0 | 0 | 0 | 0 | 0 | 0 | 4 | 0 | 1 | 1 | 1 | 1 | 0 | 0 | 1 |
| 3 | 3 | 2 | 45 | 2 | 3 | 1 | 28 | 1 | 0 | 0 | 0 | 0 | 1 | 0 | 0 | 0 | 0 | 0 | 0 | 0 | 0 | 0 | 1 | 0 | 0 | 0 | 0 | 0 | 4 | 0 | 1 | 2 | 1 | 2 | 0 | 0 | 0 |
| 4 | 4 | 2 | 50 | 3 | 3 | 3 | 9 | 1 | 0 | 0 | 0 | 0 | 0 | 0 | 0 | 0 | 0 | 0 | 0 | 0 | 0 | 0 | 0 | 0 | 0 | 0 | 0 | 0 | 5 | 0 | 2 | 1 | 1 | 1 | 0 | 0 | 1 |
| 5 | 5 | 1 | 33 | 2 | 3 | 1 | 35 | 1 | 0 | 0 | 0 | 0 | 0 | 1 | 0 | 0 | 0 | 0 | 0 | 0 | 0 | 0 | 0 | 0 | 0 | 0 | 0 | 0 | 6 | 0 | 1 | 1 | 1 | 1 | 0 | 0 | 0 |
| 6 | 6 | 2 | 24 | 1 | 2 | 5 | 11 | 1 | 0 | 0 | 0 | 0 | 0 | 0 | 0 | 0 | 0 | 0 | 0 | 0 | 0 | 0 | 0 | 0 | 0 | 0 | 0 | 0 | 7 | 0 | 2 | 1 | 1 | 2 | 0 | 0 | 1 |
| 7 | 7 | 1 | 58 | 2 | 3 | 1 | 19 | 2 | 2 | 2 | 1 | 1 | 1 | 0 | 0 | 0 | 3 | 0 | 0 | 0 | 0 | 1 | 1 | 0 | 0 | 0 | 0 | 0 | 15 | 1 | 1 | 2 | 1 | 1 | 0 | 0 | 1 |
| 8 | 8 | 2 | 22 | 2 | 3 | 4 | 23 | 4 | 0 | 0 | 0 | 0 | 0 | 0 | 0 | 0 | 0 | 0 | 0 | 0 | 0 | 0 | 0 | 0 | 0 | 0 | 0 | 0 | 0 | 0 | 1 | 1 | 1 | 1 | 0 | 0 | 0 |
| 9 | 9 | 1 | 47 | 2 | 2 | 3 | 35 | 1 | 0 | 0 | 0 | 0 | 1 | 0 | 0 | 0 | 0 | 0 | 0 | 0 | 0 | 0 | 0 | 0 | 0 | 0 | 0 | 0 | 6 | 0 | 2 | 1 | 1 | 1 | 0 | 0 | 0 |
| 10 | 10 | 2 | 30 | 5 | 2 | 2 | 24 | 1 | 0 | 0 | 0 | 0 | 0 | 0 | 0 | 0 | 0 | 0 | 0 | 0 | 0 | 0 | 0 | 0 | 0 | 0 | 0 | 0 | 0 | 0 | 2 | 1 | 1 | 2 | 0 | 0 | 0 |
| 11 | 11 | 2 | 20 | 2 | 3 | 1 | 28 | 1 | 0 | 0 | 0 | 0 | 0 | 1 | 0 | 0 | 0 | 0 | 0 | 0 | 0 | 0 | 0 | 0 | 0 | 0 | 0 | 0 | 8 | 0 | 1 | 2 | 1 | 2 | 0 | 0 | 0 |
| 12 | 12 | 2 | 57 | 1 | 2 | 4 | 13 | 1 | 0 | 0 | 0 | 0 | 0 | 0 | 0 | 0 | 0 | 0 | 0 | 0 | 0 | 0 | 0 | 0 | 0 | 0 | 0 | 0 | 4 | 0 | 1 | 3 | 1 | 2 | 0 | 0 | 0 |
| 13 | 13 | 1 | 39 | 1 | 3 | 1 | 15 | 1 | 0 | 0 | 0 | 0 | 0 | 0 | 0 | 0 | 0 | 0 | 0 | 0 | 0 | 0 | 0 | 0 | 0 | 0 | 0 | 0 | 8 | 0 | 1 | 1 | 1 | 1 | 0 | 0 | 1 |
| 14 | 14 | 2 | 61 | 4 | 2 | 1 | 6 | 1 | 0 | 0 | 0 | 0 | 0 | 0 | 0 | 0 | 0 | 0 | 0 | 0 | 0 | 0 | 0 | 0 | 0 | 0 | 0 | 0 | 4 | 0 | 1 | 2 | 1 | 2 | 0 | 0 | 1 |
| 15 | 15 | 2 | 23 | 5 | 6 | 1 | 19 | 1 | 1 | 0 | 1 | 0 | 0 | 0 | 0 | 0 | 0 | 0 | 0 | 0 | 0 | 0 | 0 | 0 | 0 | 0 | 0 | 0 | 8 | 0 | 1 | 1 | 1 | 1 | 0 | 0 | 0 |
| 16 | 16 | 2 | 21 | 1 | 3 | 3 | 15 | 1 | 0 | 0 | 0 | 0 | 0 | 0 | 0 | 0 | 0 | 0 | 0 | 0 | 0 | 0 | 0 | 0 | 0 | 0 | 0 | 0 | 5 | 0 | 1 | 2 | 1 | 2 | 0 | 0 | 0 |
| 17 | 17 | 2 | 23 | 2 | 3 | 1 | 9 | 1 | 0 | 0 | 0 | 3 | 0 | 0 | 0 | 0 | 0 | 0 | 0 | 0 | 0 | 0 | 0 | 0 | 0 | 0 | 0 | 0 | 6 | 0 | 1 | 1 | 1 | 2 | 0 | 0 | 0 |
| 18 | 18 | 1 | 55 | 5 | 3 | 3 | 37 | 1 | 0 | 0 | 0 | 0 | 0 | 0 | 0 | 0 | 0 | 0 | 0 | 0 | 0 | 0 | 0 | 0 | 0 | 0 | 0 | 0 | 0 | 0 | 1 | 1 | 1 | 1 | 0 | 0 | 1 |
| 19 | 19 | 2 | 26 | 2 | 3 | 1 | 9 | 4 | 0 | 0 | 0 | 0 | 0 | 0 | 0 | 0 | 0 | 2 | 0 | 0 | 0 | 0 | 0 | 0 | 0 | 0 | 0 | 0 | 9 | 0 | 1 | 1 | 1 | 2 | 0 | 0 | 0 |
| 20 | 20 | 1 | 64 | 3 | 2 | 1 | 19 | 2 | 0 | 0 | 2 | 0 | 2 | 0 | 0 | 0 | 2 | 2 | 0 | 0 | 0 | 0 | 0 | 0 | 0 | 0 | 0 | 0 | 6 | 0 | 1 | 3 | 1 | 2 | 0 | 0 | 1 |
| 21 | 21 | 2 | 72 | 2 | 3 | 5 | 13 | 1 | 0 | 0 | 0 | 0 | 0 | 0 | 0 | 0 | 0 | 0 | 0 | 0 | 0 | 0 | 0 | 0 | 0 | 0 | 0 | 0 | 3 | 0 | 2 | 2 | 1 | 2 | 0 | 0 | 1 |
| 22 | 22 | 2 | 61 | 2 | 2 | 5 | 15 | 2 | 0 | 0 | 0 | 0 | 0 | 0 | 0 | 0 | 0 | 0 | 0 | 0 | 0 | 0 | 0 | 0 | 0 | 0 | 0 | 0 | 0 | 0 | 1 | 3 | 1 | 2 | 0 | 0 | 1 |
| 23 | 23 | 2 | 43 | 2 | 3 | 4 | 19 | 1 | 0 | 0 | 0 | 0 | 0 | 0 | 0 | 0 | 0 | 0 | 0 | 0 | 0 | 0 | 0 | 0 | 0 | 0 | 0 | 0 | 11 | 0 | 2 | 2 | 1 | 1 | 0 | 0 | 1 |
| 24 | 24 | 2 | 52 | 2 | 2 | 2 | 20 | 2 | 0 | 0 | 0 | 0 | 0 | 0 | 0 | 0 | 0 | 0 | 0 | 0 | 0 | 0 | 0 | 0 | 0 | 0 | 0 | 0 | 2 | 0 | 2 | 2 | 1 | 1 | 0 | 0 | 1 |
| 25 | 25 | 1 | 23 | 5 | 5 | 1 | 45 | 1 | 0 | 0 | 0 | 0 | 0 | 0 | 0 | 0 | 0 | 0 | 0 | 0 | 0 | 0 | 0 | 0 | 0 | 0 | 0 | 0 | 60 | 1 | 1 | 2 | 1 | 1 | 0 | 0 | 0 |
| 26 | 26 | 2 | 73 | 2 | 3 | 1 | 23 | 1 | 0 | 0 | 2 | 0 | 2 | 0 | 2 | 3 | 3 | 0 | 0 | 0 | 0 | 2 | 2 | 0 | 0 | 0 | 0 | 0 | 3 | 0 | 2 | 2 | 1 | 2 | 0 | 0 | 1 |
| 27 | 27 | 1 | 34 | 2 | 4 | 1 | 35 | 2 | 0 | 0 | 0 | 0 | 0 | 0 | 0 | 0 | 0 | 0 | 0 | 0 | 0 | 0 | 0 | 0 | 0 | 0 | 0 | 0 | 3 | 0 | 2 | 2 | 1 | 1 | 0 | 0 | 1 |
| 28 | 28 | 2 | 34 | 2 | 3 | 1 | 15 | 2 | 0 | 0 | 0 | 0 | 0 | 0 | 0 | 0 | 0 | 0 | 0 | 0 | 0 | 0 | 0 | 0 | 0 | 0 | 0 | 0 | 0 | 0 | 1 | 2 | 1 | 1 | 0 | 0 | 1 |
| 29 | 29 | 1 | 47 | 2 | 3 | 2 | 23 | 1 | 0 | 0 | 0 | 0 | 0 | 0 | 0 | 0 | 0 | 0 | 0 | 0 | 0 | 0 | 0 | 0 | 0 | 0 | 0 | 0 | 3 | 0 | 2 | 3 | 1 | 2 | 0 | 0 | 1 |
| 30 | 30 | 2 | 31 | 2 | 2 | 1 | 11 | 4 | 0 | 0 | 0 | 0 | 0 | 0 | 0 | 0 | 1 | 0 | 0 | 0 | 0 | 0 | 0 | 0 | 0 | 0 | 0 | 0 | 7 | 0 | 2 | 3 | 1 | 2 | 0 | 0 | 0 |
| 31 | 31 | 1 | 60 | 2 | 3 | 1 | 23 | 4 | 0 | 0 | 0 | 0 | 0 | 0 | 0 | 0 | 0 | 0 | 0 | 0 | 0 | 0 | 0 | 0 | 0 | 0 | 0 | 0 | 0 | 0 | 1 | 2 | 1 | 1 | 0 | 0 | 1 |
| 32 | 32 | 2 | 35 | 2 | 1 | 1 | 11 | 1 | 0 | 0 | 0 | 0 | 0 | 0 | 0 | 0 | 0 | 0 | 0 | 0 | 0 | 0 | 0 | 0 | 0 | 0 | 0 | 0 | 4 | 0 | 1 | 2 | 1 | 1 | 0 | 0 | 1 |
| 33 | 33 | 2 | 56 | 5 | 3 | 1 | 55 | 1 | 1 | 0 | 0 | 0 | 0 | 0 | 0 | 3 | 0 | 0 | 0 | 0 | 0 | 0 | 0 | 0 | 0 | 0 | 0 | 0 | 8 | 0 | 1 | 2 | 1 | 1 | 0 | 0 | 0 |
| 34 | 34 | 2 | 40 | 1 | 7 | 1 | 28 | 1 | 0 | 0 | 0 | 0 | 0 | 0 | 0 | 0 | 0 | 0 | 0 | 0 | 0 | 0 | 0 | 0 | 0 | 0 | 0 | 0 | 1 | 0 | 1 | 3 | 1 | 2 | 0 | 0 | 0 |
| 35 | 35 | 2 | 33 | 6 | 6 | 2 | 23 | 1 | 0 | 0 | 0 | 0 | 0 | 0 | 0 | 0 | 0 | 0 | 0 | 0 | 0 | 0 | 0 | 0 | 0 | 0 | 0 | 0 | 7 | 0 | 1 | 2 | 1 | 2 | 0 | 0 | 0 |
| 36 | 36 | 2 | 39 | 5 | 3 | 4 | 9 | 1 | 0 | 0 | 0 | 0 | 0 | 0 | 0 | 0 | 0 | 0 | 0 | 0 | 0 | 0 | 0 | 0 | 0 | 0 | 0 | 0 | 0 | 0 | 1 | 3 | 1 | 2 | 0 | 0 | 0 |
| 37 | 37 | 1 | 42 | 3 | 2 | 2 | 35 | 1 | 0 | 0 | 0 | 0 | 0 | 0 | 0 | 0 | 0 | 0 | 0 | 0 | 0 | 0 | 0 | 0 | 0 | 0 | 0 | 0 | 1 | 0 | 2 | 2 | 1 | 2 | 0 | 0 | 1 |
| 38 | 38 | 1 | 19 | 2 | 3 | 1 | 55 | 1 | 0 | 0 | 0 | 0 | 0 | 0 | 0 | 0 | 0 | 0 | 0 | 0 | 0 | 0 | 0 | 0 | 0 | 0 | 0 | 0 | 0 | 0 | 1 | 3 | 1 | 2 | 0 | 0 | 0 |
| 39 | 39 | 1 | 32 | 2 | 5 | 1 | 51 | 1 | 0 | 0 | 0 | 0 | 0 | 0 | 0 | 0 | 0 | 0 | 0 | 0 | 0 | 0 | 0 | 0 | 0 | 0 | 0 | 0 | 2 | 0 | 2 | 2 | 1 | 2 | 0 | 0 | 1 |
| 40 | 40 | 1 | 47 | 2 | 6 | 1 | 66 | 1 | 0 | 0 | 0 | 0 | 0 | 0 | 0 | 0 | 0 | 0 | 0 | 0 | 0 | 0 | 0 | 0 | 0 | 0 | 0 | 0 | 3 | 0 | 2 | 2 | 1 | 2 | 0 | 0 | 0 |
| 41 | 41 | 2 | 19 | 2 | 3 | 1 | 23 | 1 | 0 | 0 | 0 | 0 | 0 | 0 | 0 | 0 | 0 | 0 | 0 | 0 | 0 | 0 | 0 | 0 | 0 | 0 | 0 | 0 | 5 | 0 | 1 | 5 | 1 | 2 | 0 | 0 | 0 |
| 42 | 42 | 1 | 51 | 2 | 2 | 1 | 28 | 1 | 0 | 0 | 0 | 0 | 0 | 0 | 0 | 0 | 0 | 0 | 0 | 0 | 0 | 0 | 0 | 0 | 0 | 0 | 0 | 0 | 9 | 0 | 2 | 2 | 1 | 1 | 0 | 0 | 0 |
| 43 | 43 | 2 | 66 | 5 | 3 | 1 | 9 | 4 | 0 | 0 | 0 | 0 | 0 | 0 | 0 | 3 | 0 | 0 | 0 | 0 | 0 | 0 | 0 | 0 | 0 | 0 | 0 | 0 | 24 | 1 | 2 | 3 | 1 | 2 | 0 | 0 | 0 |
| 44 | 44 | 2 | 53 | 1 | 7 | 1 | 35 | 4 | 0 | 0 | 0 | 0 | 0 | 0 | 0 | 0 | 0 | 0 | 0 | 0 | 0 | 0 | 0 | 2 | 2 | 0 | 0 | 0 | 13 | 0 | 1 | 2 | 1 | 2 | 0 | 0 | 0 |
| 45 | 45 | 2 | 32 | 6 | 6 | 2 | 55 | 1 | 0 | 0 | 0 | 0 | 0 | 0 | 0 | 0 | 0 | 0 | 0 | 0 | 0 | 0 | 0 | 2 | 1 | 1 | 1 | 2 | 12 | 0 | 1 | 2 | 1 | 1 | 0 | 0 | 0 |
| 46 | 46 | 2 | 56 | 5 | 3 | 4 | 9 | 1 | 0 | 0 | 0 | 0 | 0 | 0 | 0 | 0 | 0 | 0 | 0 | 0 | 0 | 0 | 0 | 0 | 0 | 0 | 0 | 0 | 0 | 0 | 2 | 2 | 1 | 2 | 0 | 0 | 0 |
| 47 | 47 | 1 | 28 | 3 | 2 | 2 | 35 | 1 | 0 | 0 | 0 | 0 | 0 | 0 | 0 | 0 | 0 | 0 | 0 | 0 | 0 | 0 | 0 | 0 | 0 | 0 | 0 | 0 | 0 | 0 | 1 | 2 | 1 | 1 | 0 | 0 | 0 |
| 48 | 48 | 1 | 43 | 2 | 3 | 1 | 15 | 1 | 0 | 0 | 0 | 0 | 0 | 0 | 0 | 0 | 0 | 0 | 0 | 0 | 0 | 0 | 0 | 0 | 0 | 0 | 0 | 0 | 0 | 0 | 1 | 2 | 1 | 1 | 0 | 0 | 0 |
| 49 | 49 | 2 | 56 | 2 | 6 | 1 | 55 | 1 | 0 | 0 | 0 | 0 | 0 | 1 | 0 | 0 | 2 | 2 | 1 | 1 | 0 | 0 | 0 | 0 | 0 | 1 | 0 | 2 | 12 | 0 | 1 | 2 | 1 | 1 | 0 | 0 | 1 |

**TABLE 3.3.** Depression Data (*Continued*)

| OBS | ID | SEX | AGE | MARITAL | EDUC | EMPLOY | INCOME | RELIG | C1 | C2 | C3 | C4 | C5 | C6 | C7 | C8 | C9 | C10 | C11 | C12 | C13 | C14 | C15 | C16 | C17 | C18 | C19 | C20 | CESD | CASES | DRINK | HEALTH | REGDOC | TREAT | BEDDAYS | ACUTEILL | CHRONILL |
|---|---|---|---|---|---|---|---|---|---|---|---|---|---|---|---|---|---|---|---|---|---|---|---|---|---|---|---|---|---|---|---|---|---|---|---|---|---|
| 50 | 50 | 2 | 30 | 3 | 3 | 7 | 15 | 4 | 0 | 0 | 1 | 2 | 0 | 2 | 0 | 0 | 0 | 1 | 1 | 0 | 2 | 0 | 0 | 0 | 0 | 0 | 0 | 0 | 9 | 0 | 2 | 3 | 2 | 1 | 1 | 0 | 1 |
| 51 | 51 | 1 | 77 | 3 | 4 | 1 | 28 | 4 | 1 | 0 | 0 | 0 | 0 | 0 | 1 | 0 | 2 | 0 | 0 | 0 | 0 | 0 | 0 | 0 | 0 | 1 | 1 | 0 | 5 | 0 | 1 | 2 | 1 | 2 | 2 | 0 | 1 |
| 52 | 52 | 2 | 24 | 3 | 2 | 2 | 5 | 2 | 0 | 1 | 0 | 0 | 2 | 0 | 0 | 3 | 0 | 2 | 0 | 0 | 0 | 0 | 0 | 0 | 0 | 0 | 0 | 8 | 0 | 1 | 2 | 1 | 2 | 2 | 0 | 0 |
| 53 | 53 | 2 | 37 | 2 | 4 | 1 | 23 | 1 | 0 | 0 | 0 | 0 | 0 | 0 | 0 | 1 | 0 | 0 | 0 | 0 | 0 | 0 | 0 | 0 | 0 | 0 | 0 | 0 | 0 | 1 | 1 | 1 | 2 | 1 | 1 | 0 |
| 54 | 54 | 1 | 81 | 2 | 4 | 4 | 20 | 1 | 0 | 0 | 0 | 0 | 0 | 0 | 0 | 0 | 0 | 0 | 0 | 0 | 0 | 0 | 0 | 0 | 0 | 0 | 0 | 0 | 0 | 1 | 3 | 1 | 1 | 1 | 0 | 1 |
| 55 | 55 | 2 | 66 | 2 | 2 | 4 | 15 | 1 | 0 | 0 | 0 | 1 | 0 | 0 | 0 | 0 | 0 | 0 | 0 | 0 | 0 | 0 | 0 | 0 | 0 | 0 | 0 | 0 | 0 | 1 | 3 | 1 | 1 | 2 | 0 | 1 |
| 56 | 56 | 2 | 26 | 1 | 5 | 2 | 26 | 4 | 1 | 0 | 1 | 1 | 2 | 2 | 1 | 3 | 3 | 1 | 3 | 3 | 0 | 2 | 2 | 2 | 3 | 0 | 0 | 40 | 1 | 1 | 2 | 2 | 1 | 1 | 1 | 0 |
| 57 | 57 | 2 | 48 | 2 | 2 | 1 | 35 | 1 | 0 | 0 | 0 | 1 | 1 | 1 | 0 | 0 | 1 | 0 | 0 | 0 | 0 | 0 | 0 | 0 | 0 | 1 | 0 | 18 | 0 | 2 | 1 | 2 | 1 | 1 | 0 | 0 |
| 58 | 58 | 1 | 55 | 2 | 3 | 5 | 8 | 1 | 0 | 0 | 0 | 3 | 2 | 2 | 0 | 0 | 0 | 0 | 0 | 0 | 0 | 0 | 0 | 0 | 2 | 0 | 0 | 26 | 1 | 1 | 3 | 2 | 1 | 1 | 0 | 0 |
| 59 | 59 | 2 | 20 | 2 | 3 | 5 | 28 | 4 | 0 | 0 | 0 | 0 | 3 | 0 | 0 | 0 | 0 | 0 | 0 | 0 | 0 | 0 | 0 | 0 | 0 | 0 | 0 | 28 | 1 | 1 | 2 | 1 | 2 | 1 | 0 | 0 |
| 60 | 60 | 2 | 68 | 1 | 4 | 1 | 11 | 1 | 0 | 0 | 0 | 0 | 0 | 2 | 0 | 0 | 0 | 0 | 0 | 0 | 0 | 0 | 0 | 0 | 0 | 0 | 0 | 0 | 0 | 1 | 2 | 1 | 1 | 0 | 0 | 0 |
| 61 | 61 | 2 | 23 | 3 | 3 | 4 | 45 | 4 | 2 | 0 | 0 | 0 | 2 | 2 | 0 | 0 | 0 | 0 | 3 | 0 | 0 | 0 | 0 | 0 | 0 | 0 | 0 | 21 | 0 | 1 | 2 | 1 | 2 | 1 | 1 | 0 |
| 62 | 62 | 1 | 63 | 2 | 1 | 3 | 14 | 1 | 0 | 0 | 0 | 0 | 0 | 1 | 0 | 0 | 0 | 0 | 0 | 0 | 0 | 0 | 0 | 0 | 0 | 0 | 0 | 16 | 0 | 1 | 2 | 1 | 1 | 1 | 0 | 0 |
| 63 | 63 | 2 | 34 | 2 | 2 | 1 | 17 | 1 | 1 | 0 | 0 | 0 | 0 | 0 | 0 | 0 | 0 | 0 | 0 | 0 | 0 | 0 | 0 | 0 | 0 | 0 | 0 | 42 | 1 | 2 | 2 | 2 | 1 | 1 | 0 | 1 |
| 64 | 64 | 2 | 58 | 2 | 3 | 1 | 12 | 3 | 0 | 0 | 0 | 0 | 2 | 3 | 0 | 0 | 0 | 0 | 0 | 0 | 0 | 0 | 0 | 0 | 0 | 0 | 0 | 18 | 0 | 1 | 1 | 2 | 1 | 1 | 1 | 1 |
| 65 | 65 | 2 | 60 | 2 | 7 | 1 | 12 | 2 | 0 | 0 | 0 | 0 | 0 | 0 | 0 | 0 | 0 | 0 | 0 | 0 | 0 | 0 | 0 | 0 | 0 | 0 | 0 | 1 | 0 | 1 | 3 | 1 | 2 | 1 | 0 | 1 |
| 66 | 66 | 1 | 34 | 4 | 4 | 1 | 65 | 2 | 0 | 1 | 0 | 0 | 0 | 0 | 0 | 0 | 1 | 0 | 0 | 0 | 0 | 0 | 0 | 0 | 0 | 0 | 3 | 7 | 1 | 1 | 4 | 1 | 1 | 1 | 0 | 0 |
| 67 | 67 | 1 | 68 | 3 | 4 | 1 | 11 | 2 | 0 | 0 | 3 | 0 | 0 | 3 | 0 | 0 | 0 | 3 | 0 | 0 | 0 | 0 | 0 | 0 | 0 | 0 | 0 | 5 | 0 | 1 | 1 | 1 | 2 | 1 | 0 | 0 |
| 68 | 68 | 2 | 70 | 5 | 5 | 4 | 5 | 1 | 0 | 0 | 0 | 0 | 0 | 0 | 0 | 0 | 0 | 0 | 0 | 0 | 0 | 0 | 0 | 0 | 0 | 0 | 0 | 1 | 0 | 1 | 2 | 1 | 1 | 1 | 0 | 0 |
| 69 | 69 | 2 | 46 | 5 | 5 | 3 | 47 | 4 | 0 | 0 | 0 | 2 | 2 | 2 | 0 | 0 | 0 | 2 | 0 | 0 | 0 | 0 | 0 | 0 | 0 | 0 | 0 | 7 | 0 | 1 | 2 | 2 | 2 | 1 | 1 | 1 |
| 70 | 70 | 2 | 40 | 1 | 3 | 2 | 13 | 4 | 0 | 0 | 3 | 0 | 0 | 0 | 0 | 0 | 0 | 0 | 0 | 0 | 0 | 0 | 0 | 0 | 0 | 0 | 0 | 5 | 0 | 1 | 4 | 2 | 1 | 1 | 1 | 1 |
| 71 | 71 | 2 | 34 | 2 | 3 | 1 | 35 | 3 | 0 | 0 | 0 | 1 | 0 | 1 | 1 | 2 | 0 | 0 | 0 | 0 | 0 | 0 | 0 | 0 | 0 | 0 | 0 | 50 | 0 | 2 | 3 | 2 | 2 | 1 | 0 | 0 |
| 72 | 72 | 2 | 24 | 1 | 3 | 4 | 13 | 3 | 0 | 2 | 2 | 1 | 2 | 2 | 0 | 0 | 0 | 0 | 0 | 0 | 0 | 0 | 0 | 0 | 0 | 0 | 0 | 0 | 0 | 1 | 1 | 1 | 2 | 1 | 0 | 0 |
| 73 | 73 | 2 | 32 | 3 | 2 | 3 | 28 | 4 | 0 | 0 | 0 | 1 | 0 | 0 | 0 | 0 | 0 | 0 | 0 | 0 | 0 | 0 | 0 | 0 | 0 | 0 | 0 | 26 | 0 | 1 | 1 | 1 | 2 | 1 | 1 | 1 |
| 74 | 74 | 2 | 26 | 1 | 2 | 1 | 19 | 1 | 0 | 0 | 0 | 0 | 0 | 2 | 0 | 0 | 0 | 0 | 0 | 0 | 0 | 0 | 0 | 0 | 0 | 0 | 0 | 0 | 0 | 1 | 1 | 1 | 2 | 1 | 1 | 0 |
| 75 | 75 | 2 | 65 | 5 | 2 | 2 | 8 | 1 | 0 | 1 | 0 | 0 | 0 | 0 | 0 | 0 | 0 | 0 | 0 | 0 | 0 | 0 | 0 | 0 | 0 | 0 | 0 | 7 | 0 | 1 | 3 | 1 | 2 | 1 | 0 | 0 |
| 76 | 76 | 2 | 71 | 5 | 3 | 5 | 19 | 1 | 0 | 0 | 0 | 0 | 0 | 0 | 0 | 0 | 0 | 0 | 3 | 0 | 0 | 0 | 0 | 0 | 0 | 0 | 0 | 48 | 1 | 2 | 1 | 2 | 2 | 1 | 1 | 0 |
| 77 | 77 | 1 | 20 | 3 | 6 | 2 | 65 | 3 | 2 | 0 | 0 | 0 | 0 | 0 | 0 | 3 | 0 | 0 | 0 | 0 | 0 | 0 | 0 | 0 | 0 | 0 | 0 | 0 | 0 | 1 | 2 | 1 | 2 | 1 | 0 | 0 |
| 78 | 78 | 2 | 70 | 5 | 5 | 1 | 7 | 3 | 0 | 0 | 0 | 0 | 0 | 0 | 0 | 0 | 0 | 0 | 0 | 0 | 0 | 0 | 0 | 0 | 0 | 0 | 0 | 27 | 1 | 2 | 1 | 1 | 1 | 0 | 1 |
| 79 | 79 | 2 | 83 | 3 | 3 | 5 | 8 | 4 | 0 | 0 | 0 | 0 | 2 | 0 | 0 | 0 | 0 | 0 | 0 | 0 | 0 | 0 | 0 | 0 | 0 | 0 | 0 | 4 | 0 | 1 | 2 | 2 | 2 | 1 | 1 | 0 |
| 80 | 80 | 1 | 75 | 2 | 7 | 4 | 4 | 1 | 0 | 0 | 0 | 0 | 0 | 0 | 0 | 0 | 0 | 0 | 0 | 0 | 0 | 0 | 0 | 0 | 0 | 0 | 0 | 3 | 0 | 1 | 1 | 1 | 2 | 1 | 1 | 1 |
| 81 | 81 | 2 | 24 | 5 | 3 | 3 | 13 | 1 | 0 | 0 | 0 | 0 | 0 | 0 | 0 | 0 | 0 | 0 | 0 | 0 | 0 | 0 | 0 | 0 | 0 | 0 | 0 | 4 | 0 | 1 | 2 | 1 | 2 | 1 | 0 | 0 |
| 82 | 82 | 2 | 71 | 3 | 3 | 4 | 35 | 1 | 1 | 0 | 0 | 0 | 0 | 0 | 0 | 3 | 0 | 0 | 3 | 0 | 0 | 0 | 0 | 0 | 2 | 0 | 0 | 13 | 0 | 1 | 3 | 1 | 2 | 1 | 0 | 1 |
| 83 | 83 | 2 | 67 | 2 | 2 | 2 | 13 | 1 | 0 | 0 | 0 | 0 | 0 | 0 | 0 | 0 | 0 | 0 | 0 | 0 | 2 | 0 | 0 | 0 | 0 | 0 | 0 | 6 | 0 | 1 | 1 | 1 | 2 | 1 | 0 | 1 |
| 84 | 84 | 1 | 59 | 2 | 4 | 1 | 28 | 3 | 0 | 0 | 0 | 0 | 0 | 0 | 0 | 0 | 0 | 0 | 0 | 0 | 2 | 0 | 0 | 0 | 0 | 0 | 0 | 8 | 0 | 1 | 1 | 2 | 1 | 1 | 0 | 0 |
| 85 | 85 | 1 | 48 | 2 | 2 | 1 | 19 | 1 | 0 | 0 | 0 | 0 | 0 | 0 | 0 | 0 | 0 | 0 | 0 | 0 | 0 | 0 | 0 | 0 | 0 | 0 | 0 | 9 | 0 | 1 | 1 | 2 | 2 | 1 | 0 | 0 |
| 86 | 86 | 2 | 45 | 2 | 5 | 3 | 12 | 1 | 0 | 0 | 0 | 0 | 0 | 0 | 0 | 0 | 0 | 0 | 0 | 0 | 0 | 0 | 0 | 0 | 0 | 0 | 0 | 3 | 0 | 1 | 2 | 1 | 2 | 1 | 0 | 0 |
| 87 | 87 | 2 | 43 | 3 | 2 | 1 | 19 | 2 | 0 | 0 | 0 | 0 | 0 | 0 | 0 | 0 | 0 | 0 | 0 | 0 | 0 | 0 | 0 | 0 | 0 | 0 | 0 | 9 | 0 | 1 | 3 | 1 | 2 | 1 | 0 | 0 |
| 88 | 88 | 2 | 32 | 1 | 4 | 1 | 65 | 1 | 0 | 1 | 0 | 0 | 0 | 0 | 0 | 0 | 0 | 0 | 0 | 0 | 0 | 0 | 0 | 0 | 0 | 0 | 0 | 2 | 0 | 1 | 1 | 1 | 2 | 1 | 0 | 0 |
| 89 | 89 | 1 | 38 | 2 | 3 | 1 | 7 | 1 | 0 | 0 | 0 | 0 | 0 | 0 | 0 | 0 | 0 | 0 | 0 | 0 | 0 | 0 | 0 | 0 | 0 | 0 | 0 | 12 | 0 | 2 | 1 | 1 | 1 | 0 | 0 |
| 90 | 90 | 2 | 22 | 1 | 2 | 2 | 8 | 4 | 0 | 0 | 0 | 0 | 0 | 0 | 0 | 0 | 0 | 0 | 0 | 0 | 0 | 0 | 0 | 0 | 0 | 0 | 0 | 5 | 0 | 1 | 1 | 2 | 2 | 1 | 0 | 0 |
| 91 | 91 | 2 | 23 | 1 | 5 | 5 | 19 | 1 | 0 | 0 | 0 | 0 | 0 | 0 | 0 | 0 | 0 | 0 | 0 | 0 | 0 | 0 | 0 | 0 | 0 | 0 | 0 | 0 | 0 | 2 | 2 | 1 | 2 | 1 | 1 | 1 |
| 92 | 92 | 2 | 59 | 2 | 4 | 4 | 65 | 3 | 0 | 0 | 0 | 0 | 0 | 0 | 0 | 0 | 0 | 0 | 0 | 0 | 0 | 0 | 0 | 0 | 0 | 0 | 0 | 0 | 0 | 1 | 2 | 1 | 2 | 1 | 0 | 0 |
| 93 | 93 | 2 | 65 | 2 | 3 | 2 | 7 | 1 | 0 | 0 | 0 | 0 | 0 | 0 | 0 | 0 | 0 | 0 | 0 | 0 | 0 | 0 | 0 | 0 | 0 | 0 | 0 | 0 | 0 | 1 | 2 | 2 | 2 | 1 | 0 | 1 |
| 94 | 94 | 2 | 21 | 5 | 3 | 1 | 8 | 1 | 0 | 0 | 0 | 0 | 0 | 0 | 0 | 0 | 0 | 0 | 0 | 0 | 0 | 0 | 0 | 0 | 0 | 0 | 0 | 0 | 0 | 1 | 3 | 1 | 1 | 1 | 0 | 0 |
| 95 | 95 | 2 | 82 | 4 | 2 | 2 | 15 | 4 | 0 | 0 | 0 | 0 | 0 | 0 | 0 | 0 | 0 | 0 | 0 | 0 | 0 | 0 | 0 | 0 | 0 | 0 | 0 | 0 | 0 | 1 | 2 | 2 | 1 | 1 | 1 | 1 |
| 96 | 96 | 2 | 54 | 2 | 2 | 5 | 65 | 1 | 0 | 0 | 0 | 0 | 0 | 0 | 0 | 0 | 0 | 0 | 0 | 0 | 0 | 0 | 0 | 0 | 0 | 0 | 0 | 0 | 0 | 1 | 2 | 1 | 2 | 1 | 0 | 0 |
| 97 | 97 | 2 | 64 | 5 | 4 | 4 | 7 | 1 | 0 | 0 | 0 | 0 | 0 | 0 | 0 | 0 | 0 | 0 | 0 | 0 | 0 | 0 | 0 | 0 | 0 | 0 | 0 | 0 | 0 | 2 | 3 | 1 | 2 | 1 | 0 | 1 |
| 98 | 98 | 1 | 64 | 2 | 3 | 2 | 4 | 2 | 0 | 0 | 0 | 1 | 1 | 0 | 0 | 0 | 0 | 0 | 0 | 0 | 0 | 0 | 0 | 0 | 0 | 0 | 3 | 5 | 0 | 1 | 2 | 1 | 1 | 0 | 0 |

**TABLE 3.3.** Depression Data (*Continued*)

| OBS | ID | SEX | AGE | MARITAL | EDUC | EMPLOY | INCOME | RELIG | C1 | C2 | C3 | C4 | C5 | C6 | C7 | C8 | C9 | C10 | C11 | C12 | C13 | C14 | C15 | C16 | C17 | C18 | C19 | C20 | CESD | CASES | DRINK | HEALTH | REGDOC | TREAT | BEDDAYS | ACUTEILL | CHRONILL |
|---|---|---|---|---|---|---|---|---|---|---|---|---|---|---|---|---|---|---|---|---|---|---|---|---|---|---|---|---|---|---|---|---|---|---|---|---|---|
| 99 | 99 | 1 | 72 | 5 | 2 | 4 | 15 | 1 | 0 | 0 | 2 | 0 | 2 | 2 | 2 | 0 | 3 | 1 | 0 | 0 | 3 | 3 | 0 | 0 | 3 | 0 | 0 | 0 | 25 | 1 | 2 | 2 | 1 | 1 | 0 | 0 | 0 |
| 100 | 100 | 1 | 52 | 2 | 5 | 1 | 45 | 3 | 0 | 0 | 0 | 0 | 0 | 1 | 0 | 0 | 0 | 0 | 0 | 0 | 0 | 0 | 0 | 0 | 0 | 0 | 0 | 5 | 0 | 3 | 2 | 1 | 1 | 0 | 0 | 1 |
| 101 | 101 | 2 | 22 | 1 | 5 | 1 | 19 | 1 | 1 | 0 | 0 | 0 | 0 | 0 | 0 | 0 | 0 | 0 | 0 | 0 | 0 | 0 | 0 | 0 | 0 | 0 | 0 | 6 | 0 | 2 | 1 | 1 | 2 | 0 | 0 | 1 |
| 102 | 102 | 1 | 21 | 3 | 3 | 1 | 8 | 1 | 1 | 0 | 1 | 0 | 1 | 0 | 0 | 0 | 0 | 0 | 0 | 0 | 1 | 0 | 0 | 0 | 1 | 0 | 1 | 17 | 0 | 2 | 1 | 1 | 1 | 0 | 0 | 1 |
| 103 | 103 | 2 | 37 | 2 | 6 | 1 | 65 | 3 | 1 | 0 | 0 | 0 | 1 | 0 | 0 | 0 | 0 | 0 | 0 | 0 | 0 | 0 | 0 | 0 | 0 | 0 | 0 | 16 | 0 | 1 | 1 | 1 | 1 | 0 | 0 | 1 |
| 104 | 104 | 1 | 60 | 3 | 3 | 2 | 35 | 3 | 1 | 0 | 0 | 0 | 0 | 0 | 0 | 0 | 0 | 0 | 0 | 0 | 0 | 0 | 0 | 0 | 0 | 0 | 0 | 13 | 0 | 3 | 3 | 1 | 2 | 1 | 0 | 1 |
| 105 | 105 | 1 | 32 | 1 | 3 | 1 | 13 | 1 | 2 | 0 | 0 | 0 | 0 | 0 | 0 | 0 | 0 | 0 | 0 | 0 | 0 | 0 | 0 | 0 | 0 | 0 | 0 | 8 | 0 | 2 | 1 | 1 | 2 | 1 | 0 | 0 |
| 106 | 106 | 2 | 27 | 5 | 5 | 1 | 7 | 2 | 1 | 0 | 0 | 0 | 0 | 0 | 0 | 0 | 0 | 0 | 0 | 0 | 0 | 0 | 0 | 0 | 0 | 0 | 0 | 19 | 1 | 2 | 2 | 1 | 1 | 0 | 0 | 0 |
| 107 | 107 | 2 | 28 | 2 | 4 | 1 | 18 | 4 | 2 | 0 | 0 | 0 | 0 | 0 | 0 | 0 | 0 | 0 | 0 | 0 | 0 | 0 | 0 | 0 | 0 | 0 | 1 | 22 | 1 | 2 | 2 | 1 | 1 | 1 | 1 | 0 |
| 108 | 108 | 1 | 20 | 2 | 2 | 2 | 23 | 3 | 2 | 0 | 0 | 0 | 0 | 3 | 0 | 0 | 0 | 3 | 0 | 0 | 0 | 0 | 0 | 0 | 3 | 0 | 0 | 37 | 1 | 2 | 2 | 1 | 2 | 0 | 1 | 1 |
| 109 | 109 | 1 | 18 | 3 | 3 | 2 | 17 | 4 | 0 | 0 | 0 | 0 | 0 | 0 | 0 | 0 | 0 | 0 | 0 | 0 | 0 | 0 | 0 | 0 | 0 | 0 | 0 | 2 | 0 | 3 | 1 | 1 | 2 | 1 | 1 | 0 |
| 110 | 110 | 2 | 24 | 5 | 5 | 6 | 9 | 4 | 1 | 0 | 1 | 3 | 1 | 0 | 0 | 0 | 0 | 0 | 0 | 0 | 3 | 0 | 0 | 0 | 0 | 0 | 1 | 23 | 1 | 2 | 3 | 1 | 2 | 0 | 0 | 1 |
| 111 | 111 | 2 | 79 | 3 | 6 | 4 | 15 | 1 | 0 | 0 | 0 | 0 | 0 | 0 | 0 | 0 | 0 | 0 | 0 | 0 | 0 | 0 | 0 | 0 | 0 | 0 | 1 | 6 | 0 | 3 | 2 | 1 | 2 | 0 | 0 | 1 |
| 112 | 112 | 2 | 50 | 6 | 5 | 1 | 23 | 3 | 0 | 0 | 0 | 0 | 0 | 0 | 0 | 0 | 0 | 0 | 0 | 0 | 0 | 0 | 0 | 0 | 0 | 0 | 0 | 6 | 0 | 3 | 3 | 1 | 2 | 0 | 0 | 1 |
| 113 | 113 | 2 | 51 | 5 | 4 | 5 | 23 | 4 | 1 | 1 | 1 | 0 | 1 | 0 | 0 | 0 | 0 | 0 | 0 | 0 | 0 | 0 | 0 | 0 | 0 | 0 | 1 | 12 | 0 | 2 | 3 | 1 | 2 | 1 | 0 | 1 |
| 114 | 114 | 1 | 52 | 4 | 2 | 1 | 19 | 2 | 0 | 0 | 0 | 0 | 0 | 0 | 0 | 0 | 0 | 0 | 0 | 0 | 0 | 0 | 0 | 0 | 0 | 0 | 0 | 0 | 0 | 2 | 1 | 1 | 1 | 0 | 0 | 0 |
| 115 | 115 | 2 | 28 | 2 | 3 | 1 | 15 | 1 | 3 | 3 | 3 | 3 | 3 | 3 | 3 | 3 | 3 | 3 | 3 | 1 | 0 | 0 | 3 | 0 | 0 | 0 | 0 | 39 | 1 | 2 | 3 | 1 | 1 | 1 | 0 | 1 |
| 116 | 116 | 1 | 41 | 2 | 3 | 1 | 8 | 1 | 1 | 0 | 2 | 0 | 3 | 2 | 2 | 0 | 2 | 2 | 0 | 0 | 0 | 0 | 0 | 0 | 0 | 0 | 0 | 29 | 1 | 2 | 3 | 1 | 1 | 0 | 0 | 0 |
| 117 | 117 | 2 | 70 | 3 | 2 | 2 | 9 | 2 | 0 | 0 | 0 | 0 | 0 | 0 | 0 | 0 | 0 | 0 | 0 | 0 | 0 | 0 | 0 | 0 | 0 | 0 | 1 | 13 | 0 | 2 | 2 | 1 | 2 | 1 | 0 | 1 |
| 118 | 118 | 2 | 62 | 2 | 5 | 4 | 65 | 4 | 1 | 0 | 1 | 0 | 1 | 0 | 0 | 0 | 0 | 0 | 0 | 0 | 0 | 0 | 0 | 0 | 0 | 0 | 0 | 6 | 0 | 3 | 2 | 1 | 2 | 0 | 1 | 0 |
| 119 | 119 | 2 | 81 | 5 | 5 | 5 | 19 | 3 | 1 | 1 | 0 | 0 | 0 | 2 | 0 | 0 | 0 | 0 | 0 | 0 | 0 | 0 | 0 | 0 | 0 | 0 | 0 | 21 | 1 | 2 | 2 | 1 | 1 | 0 | 0 | 1 |
| 120 | 120 | 2 | 83 | 5 | 5 | 1 | 24 | 2 | 2 | 0 | 0 | 0 | 0 | 0 | 0 | 0 | 0 | 0 | 0 | 0 | 0 | 0 | 0 | 0 | 0 | 0 | 0 | 29 | 1 | 1 | 2 | 1 | 2 | 1 | 0 | 0 |
| 121 | 121 | 1 | 46 | 5 | 5 | 3 | 9 | 3 | 0 | 0 | 0 | 0 | 0 | 0 | 0 | 0 | 0 | 0 | 0 | 0 | 0 | 0 | 0 | 0 | 0 | 0 | 0 | 0 | 0 | 3 | 3 | 1 | 1 | 0 | 1 | 1 |
| 122 | 122 | 1 | 18 | 2 | 5 | 4 | 7 | 1 | 0 | 0 | 0 | 0 | 0 | 0 | 0 | 0 | 0 | 0 | 0 | 0 | 0 | 0 | 0 | 0 | 0 | 0 | 0 | 11 | 0 | 2 | 2 | 1 | 1 | 0 | 1 | 0 |
| 123 | 123 | 1 | 25 | 1 | 2 | 5 | 15 | 4 | 3 | 0 | 3 | 0 | 3 | 3 | 1 | 3 | 2 | 3 | 0 | 3 | 0 | 3 | 0 | 0 | 0 | 0 | 3 | 43 | 1 | 2 | 4 | 1 | 1 | 0 | 0 | 1 |
| 124 | 124 | 2 | 78 | 5 | 3 | 3 | 19 | 4 | 0 | 0 | 1 | 0 | 0 | 0 | 0 | 0 | 0 | 0 | 0 | 0 | 0 | 0 | 0 | 0 | 0 | 0 | 0 | 3 | 0 | 3 | 2 | 1 | 2 | 0 | 1 | 1 |
| 125 | 125 | 2 | 71 | 3 | 6 | 3 | 11 | 1 | 1 | 0 | 1 | 0 | 1 | 0 | 0 | 0 | 0 | 0 | 0 | 0 | 0 | 0 | 0 | 0 | 0 | 0 | 1 | 21 | 1 | 2 | 3 | 1 | 1 | 0 | 0 | 1 |
| 126 | 126 | 1 | 73 | 2 | 5 | 1 | 7 | 3 | 0 | 0 | 0 | 0 | 0 | 0 | 0 | 0 | 0 | 0 | 0 | 0 | 0 | 0 | 0 | 0 | 0 | 0 | 0 | 0 | 0 | 2 | 2 | 1 | 1 | 0 | 0 | 1 |
| 127 | 127 | 2 | 60 | 1 | 3 | 4 | 8 | 4 | 0 | 0 | 0 | 0 | 2 | 2 | 0 | 3 | 3 | 3 | 0 | 0 | 0 | 0 | 0 | 0 | 2 | 0 | 0 | 38 | 1 | 2 | 4 | 1 | 2 | 0 | 1 | 1 |
| 128 | 128 | 1 | 42 | 5 | 1 | 2 | 13 | 1 | 1 | 0 | 0 | 0 | 0 | 0 | 0 | 0 | 0 | 0 | 0 | 0 | 0 | 0 | 0 | 0 | 0 | 0 | 0 | 13 | 0 | 2 | 3 | 1 | 1 | 1 | 0 | 1 |
| 129 | 129 | 2 | 51 | 5 | 4 | 1 | 35 | 2 | 0 | 2 | 0 | 0 | 0 | 0 | 0 | 0 | 0 | 0 | 0 | 0 | 0 | 0 | 0 | 0 | 0 | 0 | 0 | 19 | 1 | 2 | 3 | 1 | 1 | 0 | 0 | 1 |
| 130 | 130 | 1 | 34 | 2 | 5 | 7 | 15 | 4 | 0 | 0 | 0 | 0 | 0 | 0 | 0 | 0 | 0 | 0 | 0 | 0 | 0 | 0 | 0 | 0 | 0 | 0 | 0 | 0 | 0 | 3 | 2 | 1 | 2 | 0 | 0 | 1 |
| 131 | 131 | 1 | 33 | 2 | 3 | 1 | 45 | 4 | 0 | 0 | 0 | 0 | 0 | 0 | 0 | 0 | 0 | 0 | 0 | 0 | 0 | 0 | 0 | 0 | 0 | 0 | 0 | 0 | 0 | 3 | 2 | 1 | 2 | 0 | 0 | 1 |
| 132 | 132 | 2 | 48 | 3 | 6 | 2 | 9 | 3 | 0 | 1 | 0 | 0 | 0 | 3 | 0 | 0 | 0 | 0 | 0 | 0 | 0 | 0 | 0 | 0 | 0 | 0 | 0 | 27 | 1 | 3 | 3 | 1 | 2 | 0 | 0 | 1 |
| 133 | 133 | 2 | 34 | 1 | 3 | 1 | 7 | 1 | 1 | 1 | 0 | 0 | 0 | 0 | 0 | 0 | 0 | 0 | 0 | 0 | 0 | 0 | 0 | 0 | 0 | 0 | 1 | 20 | 1 | 2 | 3 | 1 | 2 | 0 | 0 | 1 |
| 134 | 134 | 1 | 32 | 1 | 5 | 1 | 13 | 3 | 0 | 0 | 0 | 0 | 0 | 0 | 0 | 0 | 0 | 0 | 0 | 0 | 0 | 0 | 0 | 0 | 0 | 0 | 0 | 0 | 0 | 2 | 2 | 1 | 2 | 0 | 0 | 1 |
| 135 | 135 | 2 | 29 | 1 | 5 | 1 | 45 | 4 | 1 | 0 | 0 | 0 | 0 | 0 | 0 | 0 | 0 | 0 | 0 | 0 | 0 | 0 | 0 | 0 | 0 | 0 | 0 | 3 | 0 | 2 | 1 | 1 | 1 | 0 | 0 | 0 |
| 136 | 136 | 1 | 24 | 5 | 5 | 1 | 20 | 3 | 1 | 0 | 0 | 0 | 0 | 0 | 0 | 0 | 0 | 0 | 0 | 0 | 0 | 0 | 0 | 0 | 0 | 0 | 0 | 31 | 1 | 2 | 3 | 1 | 1 | 0 | 0 | 1 |
| 137 | 137 | 1 | 26 | 2 | 3 | 2 | 28 | 4 | 0 | 2 | 0 | 0 | 0 | 0 | 0 | 0 | 0 | 0 | 0 | 0 | 0 | 0 | 0 | 0 | 0 | 0 | 0 | 53 | 1 | 3 | 2 | 1 | 2 | 1 | 1 | 1 |
| 138 | 138 | 2 | 27 | 2 | 6 | 4 | 35 | 2 | 1 | 0 | 0 | 0 | 0 | 0 | 0 | 0 | 0 | 0 | 0 | 0 | 0 | 0 | 0 | 0 | 0 | 0 | 0 | 19 | 1 | 3 | 3 | 1 | 2 | 1 | 1 | 0 |
| 139 | 139 | 2 | 40 | 2 | 2 | 1 | 13 | 4 | 1 | 1 | 0 | 0 | 1 | 1 | 3 | 0 | 0 | 0 | 0 | 0 | 0 | 0 | 0 | 0 | 0 | 0 | 0 | 0 | 0 | 3 | 2 | 1 | 2 | 1 | 0 | 1 |
| 140 | 140 | 2 | 33 | 3 | 2 | 2 | 15 | 4 | 0 | 0 | 0 | 0 | 0 | 0 | 0 | 0 | 0 | 0 | 0 | 0 | 0 | 0 | 0 | 0 | 0 | 0 | 0 | 0 | 0 | 2 | 2 | 1 | 1 | 1 | 0 | 0 |
| 141 | 141 | 1 | 31 | 1 | 3 | 1 | 13 | 3 | 0 | 0 | 0 | 0 | 0 | 0 | 1 | 0 | 0 | 0 | 0 | 0 | 0 | 0 | 0 | 0 | 0 | 0 | 0 | 0 | 0 | 2 | 2 | 1 | 1 | 1 | 0 | 0 |
| 142 | 142 | 2 | 31 | 1 | 3 | 1 | 45 | 2 | 0 | 0 | 0 | 0 | 0 | 0 | 0 | 0 | 0 | 0 | 0 | 0 | 0 | 0 | 0 | 0 | 0 | 0 | 0 | 0 | 0 | 2 | 2 | 1 | 2 | 0 | 1 | 1 |
| 143 | 143 | 2 | 19 | 2 | 3 | 1 | 20 | 4 | 0 | 0 | 0 | 0 | 0 | 0 | 0 | 3 | 0 | 0 | 0 | 0 | 0 | 0 | 0 | 0 | 0 | 0 | 0 | 33 | 1 | 2 | 3 | 1 | 2 | 1 | 1 | 1 |
| 144 | 144 | 1 | 31 | 2 | 5 | 1 | 28 | 2 | 1 | 0 | 0 | 0 | 0 | 0 | 0 | 0 | 0 | 0 | 0 | 0 | 0 | 0 | 0 | 0 | 0 | 0 | 0 | 20 | 1 | 2 | 3 | 1 | 1 | 0 | 0 | 1 |
| 145 | 145 | 2 | 53 | 3 | 3 | 5 | 35 | 2 | 1 | 1 | 1 | 0 | 0 | 0 | 0 | 0 | 0 | 0 | 0 | 0 | 0 | 0 | 0 | 0 | 0 | 0 | 0 | 13 | 0 | 2 | 3 | 1 | 2 | 1 | 1 | 0 |
| 146 | 146 | 1 | 33 | 2 | 6 | 1 | 11 | 2 | 0 | 0 | 2 | 0 | 0 | 0 | 0 | 0 | 0 | 0 | 0 | 0 | 0 | 0 | 0 | 0 | 0 | 0 | 0 | 12 | 0 | 2 | 2 | 1 | 1 | 0 | 1 | 0 |
| 147 | 147 | 2 | 19 | 2 | 3 | 1 | 11 | 1 | 1 | 2 | 0 | 1 | 0 | 0 | 3 | 3 | 1 | 0 | 2 | 0 | 1 | 0 | 0 | 2 | 0 | 0 | 0 | 19 | 1 | 2 | 3 | 1 | 1 | 1 | 1 | 1 |

**TABLE 3.3.** Depression Data (*Continued*)

| OBS | ID | SEX | AGE | MARITAL | EDUC | EMPLOY | INCOME | RELIG | C1 | C2 | C3 | C4 | C5 | C6 | C7 | C8 | C9 | C10 | C11 | C12 | C13 | C14 | C15 | C16 | C17 | C18 | C19 | C20 | CESD | CASES | DRINK | HEALTH | REGDOC | TREAT | BEDDAYS | ACUTEILL | CHRONILL |
|---|---|---|---|---|---|---|---|---|---|---|---|---|---|---|---|---|---|---|---|---|---|---|---|---|---|---|---|---|---|---|---|---|---|---|---|---|---|
| 148 | 148 | 2 | 23 | 1 | 4 | 1 | 9 | 2 | 0 | 0 | 0 | 0 | 0 | 0 | 0 | 1 | 1 | 0 | 0 | 0 | 1 | 0 | 0 | 1 | 0 | 0 | 1 | 0 | 5 | 0 | 1 | 1 | 1 | 2 | 0 | 1 | 0 |
| 149 | 149 | 1 | 27 | 3 | 3 | 1 | 19 | 1 | 0 | 0 | 0 | 0 | 1 | 0 | 0 | 0 | 1 | 0 | 1 | 1 | 1 | 0 | 2 | 2 | 0 | 0 | 0 | 0 | 9 | 0 | 1 | 2 | 1 | 2 | 0 | 0 | 0 |
| 150 | 150 | 1 | 57 | 3 | 3 | 1 | 28 | 2 | 0 | 0 | 0 | 0 | 0 | 0 | 0 | 0 | 0 | 1 | 0 | 0 | 0 | 1 | 0 | 0 | 0 | 0 | 0 | 0 | 2 | 0 | 1 | 2 | 1 | 2 | 0 | 0 | 1 |
| 151 | 151 | 2 | 26 | 4 | 4 | 2 | 7 | 1 | 2 | 2 | 2 | 0 | 1 | 0 | 0 | 0 | 2 | 1 | 1 | 1 | 2 | 0 | 2 | 1 | 0 | 1 | 0 | 0 | 18 | 1 | 1 | 3 | 1 | 2 | 0 | 0 | 1 |
| 152 | 152 | 1 | 31 | 3 | 3 | 1 | 13 | 1 | 0 | 1 | 2 | 0 | 0 | 0 | 0 | 0 | 1 | 0 | 0 | 1 | 2 | 1 | 1 | 1 | 1 | 1 | 0 | 0 | 11 | 0 | 1 | 2 | 2 | 2 | 0 | 1 | 1 |
| 153 | 153 | 2 | 74 | 5 | 3 | 4 | 5 | 2 | 0 | 1 | 0 | 0 | 0 | 0 | 0 | 0 | 2 | 1 | 0 | 0 | 1 | 1 | 0 | 1 | 1 | 0 | 0 | 0 | 10 | 0 | 2 | 4 | 2 | 2 | 1 | 0 | 1 |
| 154 | 154 | 2 | 24 | 1 | 5 | 1 | 8 | 2 | 0 | 0 | 1 | 0 | 1 | 0 | 0 | 0 | 0 | 1 | 0 | 0 | 0 | 0 | 1 | 0 | 0 | 0 | 0 | 0 | 4 | 0 | 1 | 1 | 1 | 2 | 0 | 0 | 0 |
| 155 | 155 | 2 | 21 | 1 | 5 | 1 | 5 | 2 | 0 | 0 | 0 | 0 | 0 | 0 | 0 | 1 | 0 | 0 | 0 | 0 | 0 | 1 | 0 | 0 | 0 | 0 | 0 | 0 | 2 | 0 | 1 | 2 | 1 | 1 | 0 | 1 | 1 |
| 156 | 156 | 2 | 22 | 1 | 3 | 1 | 11 | 1 | 0 | 0 | 0 | 0 | 1 | 0 | 0 | 0 | 0 | 0 | 0 | 1 | 0 | 0 | 0 | 1 | 0 | 0 | 0 | 0 | 3 | 0 | 1 | 2 | 1 | 1 | 0 | 1 | 1 |
| 157 | 157 | 2 | 28 | 1 | 3 | 1 | 19 | 1 | 0 | 0 | 0 | 0 | 0 | 0 | 0 | 0 | 0 | 0 | 0 | 2 | 0 | 0 | 0 | 0 | 0 | 0 | 0 | 0 | 4 | 0 | 1 | 2 | 1 | 1 | 0 | 0 | 1 |
| 158 | 158 | 1 | 49 | 1 | 3 | 1 | 16 | 2 | 0 | 0 | 0 | 0 | 0 | 0 | 0 | 0 | 1 | 0 | 0 | 1 | 3 | 0 | 0 | 0 | 0 | 0 | 0 | 0 | 5 | 0 | 1 | 2 | 2 | 2 | 0 | 0 | 1 |
| 159 | 159 | 2 | 47 | 3 | 5 | 1 | 32 | 1 | 1 | 0 | 0 | 0 | 0 | 0 | 0 | 0 | 0 | 0 | 0 | 1 | 0 | 0 | 0 | 0 | 0 | 0 | 0 | 0 | 1 | 0 | 2 | 1 | 1 | 2 | 0 | 1 | 1 |
| 160 | 160 | 2 | 32 | 2 | 5 | 1 | 32 | 1 | 0 | 1 | 0 | 0 | 0 | 0 | 0 | 0 | 0 | 0 | 0 | 0 | 0 | 0 | 0 | 0 | 1 | 0 | 0 | 0 | 1 | 0 | 1 | 1 | 1 | 1 | 0 | 1 | 1 |
| 161 | 161 | 2 | 36 | 2 | 6 | 1 | 35 | 1 | 0 | 0 | 0 | 0 | 1 | 0 | 0 | 0 | 0 | 0 | 0 | 0 | 0 | 0 | 0 | 0 | 0 | 0 | 0 | 0 | 1 | 0 | 1 | 1 | 2 | 2 | 0 | 0 | 0 |
| 162 | 162 | 1 | 45 | 2 | 5 | 1 | 65 | 1 | 0 | 0 | 0 | 0 | 1 | 0 | 0 | 0 | 0 | 0 | 0 | 0 | 0 | 0 | 1 | 0 | 1 | 0 | 0 | 0 | 9 | 0 | 1 | 1 | 2 | 2 | 0 | 0 | 0 |
| 163 | 163 | 2 | 43 | 2 | 7 | 2 | 42 | 2 | 0 | 2 | 0 | 0 | 2 | 0 | 0 | 1 | 0 | 0 | 0 | 2 | 0 | 0 | 1 | 0 | 0 | 1 | 0 | 0 | 9 | 0 | 1 | 2 | 1 | 2 | 0 | 0 | 0 |
| 164 | 164 | 2 | 22 | 1 | 3 | 1 | 5 | 4 | 0 | 0 | 0 | 0 | 0 | 0 | 0 | 0 | 0 | 0 | 0 | 0 | 0 | 0 | 0 | 0 | 0 | 0 | 0 | 0 | 0 | 0 | 1 | 2 | 1 | 2 | 0 | 0 | 0 |
| 165 | 165 | 2 | 20 | 1 | 4 | 1 | 36 | 4 | 0 | 0 | 0 | 0 | 0 | 0 | 0 | 0 | 0 | 0 | 0 | 0 | 0 | 0 | 0 | 0 | 0 | 0 | 0 | 0 | 0 | 0 | 1 | 1 | 1 | 2 | 0 | 0 | 0 |
| 166 | 166 | 1 | 36 | 2 | 7 | 1 | 65 | 3 | 0 | 0 | 0 | 0 | 0 | 0 | 0 | 0 | 0 | 0 | 0 | 0 | 0 | 0 | 0 | 0 | 0 | 0 | 0 | 0 | 0 | 0 | 1 | 1 | 1 | 1 | 0 | 0 | 0 |
| 167 | 167 | 1 | 59 | 2 | 6 | 2 | 45 | 1 | 0 | 0 | 0 | 0 | 0 | 0 | 0 | 0 | 1 | 0 | 0 | 0 | 0 | 0 | 1 | 1 | 0 | 0 | 0 | 0 | 3 | 0 | 1 | 1 | 1 | 2 | 0 | 0 | 1 |
| 168 | 168 | 2 | 57 | 2 | 5 | 2 | 55 | 1 | 0 | 0 | 0 | 0 | 0 | 0 | 0 | 0 | 0 | 0 | 0 | 0 | 0 | 0 | 0 | 0 | 0 | 0 | 0 | 0 | 0 | 0 | 2 | 1 | 1 | 2 | 0 | 0 | 1 |
| 169 | 169 | 2 | 23 | 2 | 3 | 1 | 19 | 1 | 0 | 0 | 0 | 0 | 0 | 0 | 0 | 0 | 1 | 0 | 0 | 0 | 0 | 0 | 0 | 0 | 0 | 0 | 0 | 0 | 1 | 0 | 1 | 1 | 1 | 2 | 0 | 0 | 1 |
| 170 | 170 | 2 | 74 | 5 | 2 | 4 | 8 | 1 | 0 | 1 | 2 | 0 | 1 | 0 | 0 | 0 | 0 | 0 | 0 | 0 | 2 | 0 | 0 | 0 | 0 | 0 | 0 | 0 | 6 | 0 | 1 | 1 | 1 | 2 | 0 | 0 | 1 |
| 171 | 171 | 1 | 29 | 2 | 3 | 1 | 19 | 1 | 0 | 0 | 0 | 0 | 0 | 0 | 0 | 0 | 0 | 0 | 0 | 0 | 0 | 0 | 0 | 0 | 0 | 0 | 0 | 0 | 1 | 0 | 2 | 3 | 1 | 1 | 0 | 0 | 1 |
| 172 | 172 | 2 | 65 | 2 | 2 | 4 | 9 | 1 | 0 | 0 | 0 | 0 | 2 | 0 | 0 | 0 | 2 | 0 | 0 | 0 | 0 | 0 | 0 | 0 | 0 | 0 | 0 | 0 | 12 | 0 | 1 | 1 | 1 | 2 | 0 | 0 | 0 |
| 173 | 173 | 1 | 50 | 3 | 1 | 2 | 5 | 1 | 0 | 2 | 1 | 0 | 1 | 0 | 0 | 0 | 1 | 2 | 1 | 0 | 2 | 1 | 0 | 0 | 1 | 0 | 3 | 0 | 20 | 1 | 1 | 1 | 1 | 2 | 0 | 0 | 1 |
| 174 | 174 | 1 | 31 | 2 | 2 | 1 | 19 | 1 | 0 | 0 | 0 | 0 | 2 | 0 | 0 | 0 | 2 | 1 | 0 | 3 | 0 | 3 | 3 | 2 | 0 | 1 | 0 | 0 | 30 | 1 | 1 | 3 | 1 | 1 | 0 | 0 | 1 |
| 175 | 175 | 2 | 57 | 3 | 3 | 1 | 6 | 1 | 0 | 0 | 0 | 0 | 0 | 2 | 0 | 0 | 1 | 0 | 0 | 0 | 0 | 0 | 0 | 0 | 0 | 0 | 0 | 0 | 5 | 0 | 1 | 3 | 1 | 1 | 0 | 0 | 1 |
| 176 | 176 | 1 | 70 | 1 | 3 | 4 | 4 | 1 | 0 | 0 | 0 | 2 | 0 | 0 | 0 | 0 | 2 | 0 | 0 | 2 | 0 | 0 | 0 | 0 | 0 | 0 | 0 | 2 | 21 | 1 | 1 | 2 | 2 | 2 | 0 | 0 | 0 |
| 177 | 177 | 2 | 25 | 1 | 3 | 3 | 6 | 4 | 1 | 1 | 2 | 2 | 2 | 2 | 0 | 1 | 1 | 1 | 0 | 0 | 3 | 3 | 0 | 3 | 0 | 0 | 0 | 0 | 7 | 0 | 1 | 3 | 1 | 1 | 0 | 0 | 0 |
| 178 | 178 | 2 | 45 | 2 | 4 | 2 | 55 | 1 | 0 | 0 | 1 | 0 | 1 | 0 | 0 | 0 | 0 | 0 | 1 | 0 | 1 | 0 | 1 | 0 | 1 | 0 | 0 | 0 | 7 | 0 | 1 | 1 | 1 | 2 | 0 | 0 | 0 |
| 179 | 179 | 1 | 36 | 2 | 5 | 1 | 23 | 4 | 0 | 0 | 1 | 0 | 1 | 0 | 0 | 0 | 1 | 0 | 0 | 0 | 0 | 0 | 0 | 0 | 0 | 0 | 0 | 0 | 15 | 0 | 1 | 1 | 1 | 2 | 0 | 0 | 1 |
| 180 | 180 | 1 | 25 | 2 | 3 | 1 | 23 | 1 | 0 | 0 | 1 | 0 | 0 | 0 | 0 | 1 | 1 | 0 | 0 | 0 | 1 | 1 | 0 | 2 | 1 | 2 | 1 | 0 | 1 | 0 | 1 | 2 | 1 | 1 | 0 | 0 | 1 |
| 181 | 181 | 1 | 58 | 1 | 3 | 1 | 15 | 2 | 0 | 0 | 1 | 0 | 0 | 0 | 0 | 0 | 0 | 0 | 0 | 0 | 0 | 0 | 0 | 0 | 0 | 0 | 0 | 0 | 1 | 0 | 1 | 2 | 1 | 1 | 0 | 0 | 1 |
| 182 | 182 | 2 | 59 | 5 | 3 | 1 | 13 | 1 | 2 | 3 | 2 | 2 | 3 | 3 | 1 | 0 | 1 | 2 | 3 | 0 | 3 | 2 | 1 | 3 | 1 | 0 | 0 | 0 | 33 | 1 | 1 | 3 | 1 | 1 | 0 | 0 | 1 |
| 183 | 183 | 1 | 26 | 1 | 3 | 1 | 9 | 1 | 0 | 1 | 0 | 0 | 0 | 1 | 0 | 0 | 0 | 0 | 0 | 2 | 0 | 1 | 1 | 1 | 0 | 0 | 0 | 1 | 8 | 0 | 1 | 1 | 1 | 2 | 0 | 1 | 1 |
| 184 | 184 | 1 | 21 | 1 | 3 | 1 | 19 | 1 | 1 | 0 | 0 | 0 | 0 | 0 | 0 | 0 | 3 | 0 | 0 | 0 | 0 | 0 | 0 | 0 | 0 | 0 | 0 | 0 | 5 | 0 | 1 | 1 | 2 | 2 | 0 | 0 | 1 |
| 185 | 185 | 2 | 83 | 5 | 2 | 4 | 6 | 4 | 1 | 1 | 0 | 0 | 0 | 0 | 0 | 0 | 3 | 0 | 0 | 0 | 0 | 0 | 0 | 0 | 0 | 0 | 0 | 0 | 7 | 0 | 2 | 4 | 1 | 1 | 0 | 0 | 1 |
| 186 | 186 | 2 | 34 | 3 | 2 | 1 | 7 | 4 | 1 | 1 | 1 | 1 | 1 | 0 | 1 | 1 | 0 | 2 | 0 | 1 | 1 | 2 | 1 | 1 | 0 | 0 | 0 | 0 | 16 | 1 | 1 | 1 | 1 | 1 | 0 | 0 | 0 |
| 187 | 187 | 1 | 42 | 2 | 3 | 1 | 23 | 2 | 0 | 0 | 0 | 0 | 0 | 0 | 0 | 0 | 2 | 0 | 0 | 0 | 0 | 2 | 0 | 1 | 0 | 0 | 0 | 0 | 16 | 1 | 1 | 2 | 1 | 1 | 1 | 1 | 1 |
| 188 | 188 | 2 | 42 | 3 | 3 | 2 | 7 | 1 | 2 | 1 | 2 | 0 | 0 | 0 | 1 | 2 | 1 | 1 | 0 | 0 | 0 | 2 | 1 | 0 | 2 | 0 | 3 | 0 | 39 | 1 | 1 | 2 | 1 | 1 | 0 | 1 | 1 |
| 189 | 189 | 2 | 24 | 1 | 3 | 1 | 13 | 4 | 3 | 3 | 3 | 3 | 3 | 0 | 2 | 3 | 3 | 3 | 1 | 1 | 0 | 2 | 0 | 1 | 1 | 0 | 0 | 0 | 9 | 0 | 1 | 2 | 2 | 1 | 0 | 0 | 0 |
| 190 | 190 | 2 | 80 | 5 | 2 | 4 | 5 | 1 | 1 | 0 | 0 | 0 | 0 | 0 | 2 | 1 | 0 | 0 | 1 | 2 | 0 | 1 | 0 | 0 | 0 | 0 | 0 | 0 | 8 | 0 | 1 | 2 | 2 | 1 | 0 | 0 | 0 |
| 191 | 191 | 2 | 18 | 1 | 2 | 1 | 15 | 2 | 1 | 1 | 1 | 0 | 0 | 0 | 1 | 1 | 1 | 0 | 0 | 1 | 0 | 0 | 0 | 0 | 0 | 0 | 0 | 0 | 3 | 0 | 1 | 1 | 1 | 1 | 0 | 0 | 0 |
| 192 | 192 | 2 | 58 | 5 | 2 | 1 | 9 | 1 | 0 | 0 | 1 | 0 | 0 | 0 | 0 | 0 | 0 | 0 | 0 | 1 | 0 | 0 | 0 | 0 | 0 | 0 | 0 | 0 | 2 | 0 | 1 | 2 | 1 | 1 | 0 | 0 | 0 |
| 193 | 193 | 2 | 54 | 2 | 3 | 2 | 7 | 2 | 0 | 0 | 0 | 0 | 0 | 0 | 0 | 0 | 0 | 0 | 1 | 0 | 0 | 0 | 0 | 0 | 1 | 0 | 0 | 0 | 3 | 0 | 1 | 2 | 1 | 1 | 0 | 0 | 0 |
| 194 | 194 | 2 | 47 | 3 | 4 | 5 | 45 | 1 | 0 | 0 | 0 | 0 | 0 | 0 | 0 | 1 | 0 | 0 | 0 | 1 | 0 | 0 | 0 | 0 | 0 | 0 | 0 | 0 | 3 | 0 | 1 | 2 | 1 | 2 | 0 | 0 | 1 |
| 195 | 195 | 1 | 61 | 2 | 2 | 6 | 24 | 1 | 0 | 0 | 0 | 0 | 0 | 0 | 0 | 0 | 0 | 0 | 1 | 0 | 0 | 1 | 0 | 0 | 0 | 0 | 0 | 0 | 5 | 0 | 1 | 1 | 1 | 2 | 1 | 1 | 1 |
| 196 | 196 | 1 | 18 | 1 | 2 | 6 | 24 | 2 | 0 | 1 | 0 | 0 | 0 | 0 | 0 | 0 | 0 | 0 | 1 | 0 | 0 | 1 | 0 | 0 | 0 | 0 | 0 | 0 | 5 | 0 | 1 | 1 | 1 | 2 | 1 | 1 | 0 |

**TABLE 3.3.** Depression Data (Continued)

| OBS | ID | SEX | AGE | MARITAL | EDUC | EMPLOY | INCOME | RELIG | C1 | C2 | C3 | C4 | C5 | C6 | C7 | C8 | C9 | C10 | C11 | C12 | C13 | C14 | C15 | C16 | C17 | C18 | C19 | C20 | CESD | CASES | DRINK | HEALTH | REGDOC | TREAT | BEDDAYS | ACUTEILL | CHRONILL |
|---|---|---|---|---|---|---|---|---|---|---|---|---|---|---|---|---|---|---|---|---|---|---|---|---|---|---|---|---|---|---|---|---|---|---|---|---|---|
| 197 | 197 | 2 | 51 | 2 | 5 | 1 | 45 | 1 | 0 | 0 | 0 | 0 | 0 | 0 | 0 | 0 | 0 | 0 | 0 | 0 | 0 | 0 | 0 | 0 | 0 | 0 | 1 | 0 | 4 | 0 | 1 | 1 | 1 | 2 | 0 | 0 | 0 |
| 198 | 198 | 1 | 28 | 2 | 2 | 5 | 26 | 2 | 0 | 0 | 0 | 0 | 0 | 0 | 0 | 0 | 0 | 0 | 0 | 0 | 0 | 0 | 0 | 0 | 0 | 0 | 0 | 0 | 0 | 0 | 1 | 1 | 1 | 2 | 0 | 0 | 0 |
| 199 | 199 | 2 | 44 | 2 | 3 | 1 | 9 | 1 | 0 | 0 | 0 | 0 | 0 | 0 | 0 | 0 | 0 | 0 | 0 | 0 | 0 | 0 | 0 | 0 | 0 | 0 | 0 | 0 | 0 | 0 | 2 | 1 | 1 | 2 | 0 | 1 | 1 |
| 200 | 200 | 2 | 31 | 3 | 3 | 1 | 23 | 1 | 2 | 0 | 0 | 0 | 0 | 2 | 2 | 0 | 2 | 3 | 0 | 2 | 0 | 0 | 0 | 0 | 0 | 0 | 0 | 0 | 8 | 1 | 1 | 2 | 1 | 2 | 1 | 0 | 1 |
| 201 | 201 | 2 | 50 | 3 | 3 | 7 | 19 | 1 | 0 | 0 | 0 | 0 | 0 | 0 | 0 | 0 | 0 | 0 | 0 | 0 | 0 | 0 | 0 | 0 | 0 | 0 | 0 | 0 | 3 | 0 | 1 | 3 | 1 | 2 | 0 | 0 | 1 |
| 202 | 202 | 1 | 59 | 3 | 2 | 1 | 7 | 1 | 0 | 0 | 0 | 0 | 0 | 0 | 0 | 0 | 0 | 0 | 0 | 0 | 0 | 0 | 0 | 0 | 0 | 0 | 0 | 0 | 3 | 0 | 1 | 2 | 1 | 1 | 0 | 0 | 1 |
| 203 | 203 | 2 | 33 | 3 | 4 | 1 | 19 | 1 | 0 | 0 | 0 | 0 | 0 | 0 | 0 | 0 | 0 | 0 | 0 | 0 | 0 | 0 | 0 | 0 | 0 | 0 | 0 | 0 | 2 | 0 | 1 | 2 | 1 | 2 | 0 | 1 | 1 |
| 204 | 204 | 2 | 26 | 5 | 3 | 1 | 19 | 1 | 0 | 0 | 0 | 0 | 0 | 0 | 0 | 0 | 0 | 0 | 0 | 0 | 0 | 0 | 0 | 0 | 0 | 0 | 0 | 0 | 2 | 0 | 1 | 2 | 1 | 1 | 0 | 1 | 1 |
| 205 | 205 | 2 | 58 | 2 | 3 | 6 | 6 | 1 | 0 | 0 | 0 | 0 | 0 | 0 | 0 | 0 | 0 | 0 | 0 | 0 | 0 | 0 | 0 | 0 | 0 | 0 | 0 | 0 | 0 | 0 | 2 | 2 | 1 | 2 | 0 | 0 | 0 |
| 206 | 206 | 1 | 67 | 5 | 2 | 5 | 8 | 1 | 0 | 0 | 0 | 2 | 0 | 0 | 0 | 0 | 0 | 0 | 0 | 0 | 0 | 0 | 0 | 0 | 0 | 0 | 0 | 0 | 3 | 0 | 2 | 3 | 1 | 2 | 0 | 0 | 0 |
| 207 | 207 | 2 | 49 | 2 | 3 | 1 | 28 | 1 | 0 | 0 | 0 | 0 | 0 | 0 | 0 | 0 | 0 | 0 | 0 | 0 | 0 | 0 | 0 | 0 | 0 | 0 | 0 | 0 | 0 | 0 | 2 | 2 | 1 | 2 | 0 | 0 | 0 |
| 208 | 208 | 1 | 42 | 2 | 3 | 5 | 28 | 1 | 0 | 0 | 0 | 0 | 0 | 0 | 0 | 0 | 0 | 0 | 0 | 0 | 0 | 0 | 0 | 0 | 0 | 0 | 0 | 0 | 0 | 0 | 1 | 1 | 1 | 2 | 0 | 0 | 0 |
| 209 | 209 | 2 | 61 | 2 | 2 | 7 | 23 | 2 | 0 | 0 | 0 | 0 | 0 | 0 | 0 | 0 | 0 | 0 | 0 | 0 | 0 | 0 | 0 | 0 | 0 | 0 | 0 | 0 | 0 | 0 | 1 | 3 | 1 | 1 | 0 | 0 | 0 |
| 210 | 210 | 2 | 32 | 5 | 3 | 1 | 23 | 2 | 0 | 0 | 0 | 2 | 0 | 2 | 0 | 0 | 0 | 0 | 0 | 0 | 0 | 0 | 0 | 0 | 0 | 3 | 0 | 1 | 13 | 1 | 1 | 3 | 1 | 1 | 1 | 0 | 1 |
| 211 | 211 | 2 | 47 | 2 | 4 | 1 | 28 | 1 | 0 | 0 | 0 | 0 | 0 | 0 | 0 | 0 | 0 | 0 | 0 | 0 | 0 | 0 | 0 | 0 | 0 | 0 | 0 | 0 | 0 | 0 | 4 | 2 | 1 | 2 | 0 | 0 | 0 |
| 212 | 212 | 1 | 59 | 2 | 3 | 5 | 6 | 2 | 0 | 0 | 0 | 0 | 3 | 0 | 0 | 0 | 0 | 0 | 0 | 0 | 0 | 3 | 0 | 0 | 0 | 0 | 0 | 0 | 9 | 0 | 4 | 4 | 2 | 1 | 0 | 0 | 0 |
| 213 | 213 | 2 | 25 | 5 | 3 | 1 | 8 | 1 | 1 | 0 | 0 | 0 | 0 | 0 | 0 | 0 | 0 | 0 | 0 | 0 | 0 | 0 | 0 | 0 | 0 | 0 | 0 | 0 | 4 | 0 | 3 | 3 | 1 | 2 | 0 | 0 | 0 |
| 214 | 214 | 1 | 74 | 2 | 3 | 5 | 15 | 1 | 0 | 0 | 0 | 0 | 0 | 0 | 0 | 0 | 0 | 0 | 0 | 0 | 0 | 0 | 0 | 0 | 0 | 0 | 0 | 0 | 0 | 0 | 3 | 1 | 1 | 1 | 0 | 0 | 0 |
| 215 | 215 | 2 | 30 | 5 | 4 | 1 | 11 | 1 | 0 | 0 | 0 | 0 | 0 | 0 | 0 | 0 | 0 | 0 | 0 | 0 | 0 | 0 | 0 | 0 | 0 | 0 | 0 | 0 | 0 | 0 | 3 | 1 | 1 | 2 | 0 | 0 | 0 |
| 216 | 216 | 2 | 59 | 5 | 4 | 7 | 19 | 2 | 0 | 0 | 0 | 3 | 3 | 3 | 0 | 0 | 3 | 0 | 0 | 0 | 0 | 0 | 0 | 0 | 0 | 0 | 0 | 0 | 5 | 1 | 1 | 2 | 1 | 2 | 0 | 0 | 0 |
| 217 | 217 | 2 | 46 | 2 | 3 | 5 | 19 | 1 | 0 | 0 | 0 | 0 | 0 | 0 | 0 | 0 | 0 | 0 | 0 | 0 | 0 | 0 | 0 | 0 | 0 | 0 | 0 | 0 | 3 | 0 | 1 | 3 | 1 | 1 | 0 | 0 | 0 |
| 218 | 218 | 2 | 51 | 3 | 4 | 7 | 35 | 2 | 0 | 0 | 0 | 0 | 0 | 2 | 0 | 0 | 0 | 0 | 0 | 0 | 0 | 0 | 0 | 0 | 0 | 0 | 0 | 0 | 9 | 1 | 2 | 4 | 1 | 2 | 0 | 0 | 1 |
| 219 | 219 | 2 | 25 | 2 | 5 | 1 | 11 | 2 | 1 | 0 | 0 | 0 | 0 | 0 | 0 | 0 | 0 | 0 | 0 | 0 | 0 | 0 | 0 | 0 | 0 | 0 | 0 | 0 | 3 | 0 | 1 | 2 | 1 | 1 | 0 | 1 | 0 |
| 220 | 220 | 2 | 18 | 5 | 4 | 5 | 65 | 2 | 0 | 0 | 0 | 2 | 0 | 2 | 0 | 0 | 0 | 0 | 0 | 0 | 0 | 0 | 0 | 0 | 0 | 0 | 0 | 0 | 7 | 0 | 1 | 3 | 1 | 2 | 0 | 0 | 0 |
| 221 | 221 | 2 | 52 | 2 | 4 | 7 | 55 | 2 | 0 | 0 | 0 | 2 | 2 | 0 | 2 | 0 | 0 | 0 | 0 | 0 | 0 | 0 | 0 | 0 | 0 | 0 | 0 | 0 | 9 | 1 | 2 | 4 | 1 | 2 | 0 | 0 | 1 |
| 222 | 222 | 2 | 37 | 2 | 5 | 1 | 28 | 1 | 0 | 0 | 0 | 0 | 0 | 0 | 0 | 0 | 0 | 0 | 0 | 0 | 0 | 0 | 0 | 0 | 0 | 0 | 0 | 0 | 3 | 0 | 1 | 3 | 1 | 2 | 0 | 0 | 0 |
| 223 | 223 | 2 | 42 | 2 | 4 | 1 | 13 | 1 | 0 | 0 | 0 | 2 | 0 | 0 | 0 | 0 | 0 | 0 | 0 | 0 | 0 | 0 | 0 | 0 | 0 | 0 | 0 | 0 | 1 | 0 | 1 | 3 | 1 | 2 | 0 | 0 | 0 |
| 224 | 224 | 2 | 43 | 2 | 4 | 1 | 37 | 2 | 0 | 0 | 0 | 0 | 0 | 0 | 0 | 0 | 0 | 0 | 0 | 0 | 0 | 0 | 0 | 0 | 0 | 0 | 0 | 0 | 3 | 0 | 2 | 2 | 1 | 2 | 0 | 0 | 0 |
| 225 | 225 | 2 | 63 | 5 | 5 | 7 | 31 | 1 | 0 | 0 | 0 | 0 | 0 | 0 | 0 | 0 | 0 | 0 | 0 | 0 | 0 | 0 | 0 | 0 | 0 | 0 | 0 | 0 | 7 | 0 | 2 | 3 | 1 | 2 | 0 | 0 | 0 |
| 226 | 226 | 2 | 89 | 5 | 5 | 7 | 45 | 3 | 0 | 3 | 0 | 2 | 0 | 0 | 2 | 0 | 0 | 0 | 0 | 0 | 0 | 0 | 0 | 0 | 0 | 0 | 0 | 0 | 11 | 1 | 2 | 2 | 1 | 2 | 0 | 0 | 1 |
| 227 | 227 | 2 | 23 | 3 | 4 | 5 | 15 | 1 | 0 | 0 | 0 | 2 | 0 | 0 | 0 | 0 | 0 | 0 | 0 | 0 | 0 | 0 | 0 | 0 | 0 | 0 | 0 | 0 | 0 | 0 | 1 | 2 | 1 | 2 | 0 | 0 | 0 |
| 228 | 228 | 2 | 60 | 2 | 5 | 7 | 8 | 1 | 0 | 0 | 0 | 0 | 0 | 0 | 0 | 0 | 0 | 0 | 0 | 0 | 0 | 0 | 0 | 0 | 0 | 0 | 0 | 0 | 0 | 0 | 1 | 2 | 1 | 2 | 0 | 0 | 0 |
| 229 | 229 | 1 | 22 | 5 | 5 | 1 | 55 | 2 | 0 | 0 | 0 | 0 | 2 | 0 | 0 | 0 | 0 | 0 | 0 | 0 | 0 | 0 | 0 | 0 | 0 | 0 | 0 | 0 | 5 | 0 | 1 | 2 | 1 | 2 | 0 | 0 | 0 |
| 230 | 230 | 2 | 42 | 2 | 3 | 1 | 45 | 2 | 0 | 0 | 0 | 0 | 0 | 0 | 0 | 0 | 0 | 0 | 0 | 0 | 0 | 0 | 0 | 0 | 0 | 0 | 0 | 0 | 0 | 0 | 1 | 2 | 1 | 2 | 0 | 0 | 0 |
| 231 | 231 | 2 | 19 | 1 | 3 | 2 | 65 | 2 | 2 | 2 | 0 | 0 | 0 | 0 | 0 | 0 | 2 | 0 | 0 | 0 | 0 | 0 | 0 | 0 | 0 | 0 | 0 | 0 | 28 | 1 | 1 | 2 | 1 | 2 | 0 | 0 | 0 |
| 232 | 232 | 2 | 30 | 2 | 4 | 1 | 55 | 4 | 0 | 0 | 0 | 0 | 0 | 2 | 0 | 0 | 0 | 0 | 0 | 0 | 0 | 0 | 0 | 0 | 0 | 0 | 0 | 0 | 0 | 0 | 1 | 2 | 1 | 2 | 0 | 0 | 0 |
| 233 | 233 | 2 | 68 | 5 | 3 | 7 | 13 | 1 | 0 | 0 | 0 | 0 | 0 | 0 | 0 | 0 | 0 | 0 | 0 | 0 | 0 | 0 | 0 | 0 | 0 | 0 | 0 | 0 | 1 | 0 | 3 | 2 | 1 | 2 | 0 | 0 | 0 |
| 234 | 234 | 2 | 19 | 2 | 4 | 5 | 37 | 6 | 0 | 0 | 0 | 0 | 3 | 0 | 0 | 2 | 0 | 0 | 0 | 0 | 0 | 0 | 0 | 0 | 0 | 0 | 0 | 0 | 5 | 0 | 1 | 2 | 1 | 2 | 0 | 0 | 0 |
| 235 | 235 | 1 | 63 | 5 | 3 | 1 | 23 | 1 | 3 | 3 | 0 | 3 | 3 | 2 | 0 | 0 | 2 | 3 | 2 | 3 | 0 | 0 | 0 | 0 | 0 | 0 | 1 | 0 | 13 | 1 | 1 | 2 | 1 | 2 | 1 | 0 | 1 |
| 236 | 236 | 2 | 32 | 2 | 4 | 1 | 19 | 1 | 0 | 0 | 0 | 0 | 0 | 0 | 0 | 0 | 0 | 0 | 0 | 0 | 0 | 0 | 0 | 0 | 0 | 0 | 0 | 0 | 5 | 0 | 1 | 3 | 1 | 2 | 0 | 0 | 0 |
| 237 | 237 | 2 | 57 | 2 | 3 | 7 | 45 | 2 | 0 | 0 | 0 | 0 | 0 | 0 | 0 | 0 | 0 | 0 | 0 | 0 | 0 | 0 | 0 | 0 | 0 | 0 | 0 | 0 | 3 | 0 | 3 | 2 | 1 | 2 | 0 | 0 | 0 |
| 238 | 238 | 2 | 36 | 2 | 3 | 1 | 37 | 3 | 0 | 0 | 0 | 0 | 0 | 0 | 0 | 2 | 0 | 0 | 0 | 0 | 0 | 0 | 0 | 0 | 0 | 0 | 0 | 0 | 1 | 0 | 1 | 2 | 1 | 2 | 0 | 1 | 0 |
| 239 | 239 | 2 | 23 | 2 | 4 | 5 | 23 | 1 | 3 | 3 | 0 | 3 | 0 | 0 | 0 | 0 | 0 | 0 | 0 | 0 | 0 | 0 | 0 | 0 | 0 | 0 | 0 | 0 | 3 | 0 | 1 | 2 | 1 | 2 | 0 | 1 | 1 |
| 240 | 240 | 2 | 36 | 2 | 3 | 1 | 17 | 1 | 0 | 0 | 0 | 2 | 0 | 0 | 0 | 0 | 0 | 0 | 0 | 0 | 0 | 0 | 0 | 0 | 0 | 0 | 0 | 0 | 1 | 0 | 1 | 2 | 1 | 2 | 0 | 0 | 1 |
| 241 | 241 | 2 | 40 | 2 | 3 | 1 | — | 1 | 0 | 0 | 0 | 0 | 0 | 0 | 0 | 0 | 0 | 0 | 0 | 0 | 0 | 0 | 0 | 0 | 0 | 0 | 0 | 0 | 7 | 0 | 2 | 2 | 1 | 2 | 1 | 1 | 1 |
| 242 | 242 | 2 | 54 | 2 | 2 | 5 | — | 1 | 0 | 0 | 0 | 0 | 0 | 0 | 0 | 0 | 0 | 0 | 0 | 0 | 0 | 0 | 0 | 0 | 0 | 0 | 0 | 0 | 0 | 0 | 1 | 1 | 1 | 1 | 0 | 0 | 1 |
| 243 | 243 | 2 | | 5 | 3 | 5 | | 1 | 0 | 0 | 0 | 0 | 0 | 0 | 0 | 0 | 0 | 0 | 0 | 0 | 0 | 0 | 0 | 0 | 0 | 0 | 0 | 0 | 0 | 0 | 1 | 1 | 1 | 1 | 0 | 0 | 1 |
| 244 | 244 | 2 | | 3 | 2 | 5 | | 1 | 0 | 0 | 0 | 0 | 0 | 0 | 0 | 0 | 0 | 0 | 0 | 0 | 0 | 0 | 0 | 0 | 0 | 0 | 0 | 0 | 0 | 0 | 2 | 1 | 1 | 1 | 0 | 0 | 1 |
| 245 | 245 | 2 | | 5 | 2 | 5 | | 1 | 0 | 0 | 0 | 0 | 0 | 0 | 0 | 0 | 0 | 0 | 0 | 0 | 0 | 0 | 0 | 0 | 0 | 0 | 0 | 0 | 0 | 0 | 2 | 2 | 1 | 1 | 0 | 0 | 1 |

**TABLE 3.3.** Depression Data (Concluded)

| OBS | ID | SEX | AGE | MARITAL | EDUC | EMPLOY | INCOME | RELIG | C1 | C2 | C3 | C4 | C5 | C6 | C7 | C8 | C9 | C10 | C11 | C12 | C13 | C14 | C15 | C16 | C17 | C18 | C19 | C20 | CESD | CASES | DRINK | HEALTH | REGDOC | TREAT | BEDDAYS | ACUTE-ILL | CHRON-ILL |
|---|---|---|---|---|---|---|---|---|---|---|---|---|---|---|---|---|---|---|---|---|---|---|---|---|---|---|---|---|---|---|---|---|---|---|---|---|---|
| 246 | 246 | 1 | 27 | 1 | 3 | 1 | 15 | 4 | 0 | 0 | 0 | 0 | 0 | 1 | 0 | 0 | 1 | 1 | 0 | 0 | 1 | 0 | 0 | 0 | 0 | 0 | 0 | 1 | 12 | 0 | 1 | 1 | 2 | 1 | 1 | 1 | 0 |
| 247 | 247 | 2 | 49 | 3 | 3 | 4 | 15 | 1 | 0 | 0 | 2 | 0 | 0 | 1 | 0 | 0 | 0 | 0 | 0 | 0 | 0 | 0 | 0 | 0 | 0 | 0 | 0 | 0 | 2 | 0 | 2 | 2 | 1 | 1 | 0 | 1 | 0 |
| 248 | 248 | 2 | 69 | 2 | 3 | 4 | 28 | 2 | 0 | 0 | 0 | 0 | 0 | 1 | 0 | 0 | 0 | 1 | 0 | 0 | 0 | 0 | 0 | 0 | 0 | 0 | 0 | 0 | 14 | 0 | 1 | 1 | 1 | 2 | 0 | 0 | 0 |
| 249 | 249 | 2 | 35 | 5 | 2 | 2 | 8 | 2 | 0 | 0 | 0 | 0 | 0 | 0 | 0 | 0 | 0 | 1 | 0 | 0 | 0 | 0 | 0 | 0 | 0 | 0 | 0 | 0 | 3 | 0 | 1 | 1 | 1 | 1 | 0 | 0 | 0 |
| 250 | 250 | 2 | 60 | 3 | 3 | 2 | 8 | 4 | 0 | 0 | 0 | 0 | 0 | 0 | 0 | 0 | 0 | 0 | 0 | 0 | 0 | 0 | 0 | 0 | 0 | 0 | 0 | 0 | 1 | 0 | 1 | 2 | 1 | 1 | 1 | 1 | 1 |
| 251 | 251 | 2 | 70 | 5 | 1 | 4 | 9 | 2 | 0 | 1 | 0 | 0 | 1 | 0 | 0 | 0 | 0 | 0 | 0 | 0 | 0 | 0 | 0 | 0 | 0 | 0 | 0 | 0 | 7 | 0 | 1 | 1 | 1 | 2 | 0 | 1 | 1 |
| 252 | 252 | 2 | 83 | 3 | 3 | 4 | 9 | 2 | 0 | 0 | 0 | 0 | 0 | 0 | 0 | 0 | 0 | 0 | 0 | 0 | 0 | 0 | 0 | 0 | 0 | 0 | 0 | 0 | 1 | 0 | 2 | 2 | 1 | 1 | 0 | 1 | 0 |
| 253 | 253 | 2 | 79 | 2 | 4 | 2 | 55 | 1 | 0 | 0 | 0 | 0 | 0 | 0 | 0 | 0 | 0 | 0 | 0 | 0 | 0 | 0 | 0 | 0 | 0 | 0 | 0 | 0 | 3 | 0 | 2 | 1 | 1 | 1 | 0 | 1 | 0 |
| 254 | 254 | 2 | 28 | 5 | 3 | 5 | 15 | 4 | 0 | 1 | 1 | 0 | 0 | 0 | 0 | 1 | 0 | 1 | 0 | 0 | 0 | 0 | 0 | 0 | 0 | 0 | 0 | 0 | 4 | 0 | 2 | 2 | 2 | 2 | 0 | 0 | 0 |
| 255 | 255 | 2 | 29 | 3 | 2 | 1 | 12 | 2 | 1 | 0 | 0 | 0 | 0 | 1 | 0 | 0 | 0 | 1 | 1 | 0 | 0 | 0 | 0 | 0 | 0 | 0 | 0 | 0 | 6 | 0 | 2 | 1 | 2 | 2 | 0 | 0 | 0 |
| 256 | 256 | 2 | 29 | 5 | 3 | 1 | 8 | 2 | 0 | 0 | 2 | 0 | 0 | 0 | 0 | 0 | 0 | 1 | 1 | 0 | 1 | 0 | 0 | 0 | 0 | 0 | 0 | 0 | 22 | 1 | 1 | 1 | 1 | 2 | 0 | 1 | 0 |
| 257 | 257 | 2 | 51 | 3 | 5 | 4 | 21 | 3 | 0 | 0 | 0 | 0 | 0 | 0 | 0 | 2 | 0 | 1 | 0 | 1 | 1 | 0 | 0 | 0 | 0 | 0 | 0 | 0 | 18 | 1 | 2 | 3 | 2 | 2 | 0 | 1 | 0 |
| 258 | 258 | 1 | 58 | 3 | 2 | 1 | 13 | 2 | 0 | 0 | 0 | 1 | 1 | 1 | 0 | 0 | 0 | 0 | 0 | 0 | 0 | 0 | 0 | 0 | 0 | 0 | 0 | 0 | 2 | 0 | 2 | 2 | 1 | 2 | 0 | 0 | 0 |
| 259 | 259 | 2 | 21 | 5 | 2 | 2 | 20 | 2 | 0 | 0 | 0 | 1 | 0 | 0 | 0 | 0 | 0 | 0 | 0 | 0 | 0 | 0 | 0 | 0 | 0 | 0 | 0 | 0 | 15 | 1 | 2 | 2 | 1 | 2 | 0 | 1 | 0 |
| 260 | 260 | 2 | 30 | 3 | 5 | 1 | 15 | 1 | 0 | 0 | 0 | 1 | 0 | 1 | 0 | 0 | 0 | 0 | 0 | 0 | 0 | 0 | 0 | 0 | 0 | 0 | 0 | 0 | 0 | 0 | 2 | 1 | 1 | 1 | 0 | 1 | 0 |
| 261 | 261 | 2 | 57 | 3 | 3 | 1 | 13 | 4 | 0 | 0 | 0 | 1 | 1 | 1 | 0 | 0 | 0 | 0 | 0 | 0 | 0 | 0 | 0 | 0 | 0 | 0 | 0 | 0 | 3 | 0 | 2 | 3 | 1 | 2 | 0 | 0 | 0 |
| 262 | 262 | 1 | 37 | 2 | 6 | 1 | 23 | 3 | 0 | 0 | 0 | 0 | 0 | 0 | 0 | 0 | 0 | 0 | 0 | 0 | 0 | 0 | 0 | 0 | 0 | 0 | 0 | 0 | 3 | 0 | 1 | 1 | 1 | 1 | 0 | 1 | 0 |
| 263 | 263 | 2 | 55 | 3 | 3 | 5 | 9 | 3 | 1 | 0 | 0 | 0 | 0 | 0 | 0 | 0 | 0 | 0 | 1 | 0 | 0 | 0 | 0 | 0 | 0 | 0 | 0 | 0 | 5 | 0 | 2 | 3 | 1 | 2 | 0 | 0 | 0 |
| 264 | 264 | 1 | 34 | 1 | 3 | 1 | 15 | 3 | 0 | 0 | 0 | 0 | 0 | 3 | 0 | 0 | 2 | 0 | 0 | 0 | 0 | 0 | 0 | 0 | 0 | 0 | 0 | 0 | 4 | 0 | 2 | 2 | 1 | 2 | 0 | 0 | 0 |
| 265 | 265 | 2 | 48 | 1 | 3 | 2 | 23 | 2 | 0 | 0 | 0 | 0 | 0 | 0 | 0 | 1 | 0 | 0 | 0 | 0 | 0 | 0 | 0 | 0 | 0 | 0 | 0 | 0 | 7 | 0 | 1 | 1 | 1 | 2 | 0 | 1 | 0 |
| 266 | 266 | 2 | 22 | 4 | 3 | 2 | 35 | 1 | 0 | 0 | 0 | 2 | 1 | 1 | 0 | 0 | 0 | 0 | 0 | 0 | 0 | 0 | 0 | 0 | 0 | 0 | 0 | 0 | 7 | 0 | 1 | 1 | 1 | 1 | 0 | 1 | 0 |
| 267 | 267 | 1 | 62 | 3 | 5 | 3 | 35 | 1 | 0 | 0 | 0 | 0 | 0 | 0 | 0 | 0 | 0 | 0 | 0 | 0 | 0 | 0 | 0 | 0 | 0 | 0 | 0 | 0 | 1 | 0 | 2 | 2 | 1 | 1 | 0 | 0 | 1 |
| 268 | 268 | 2 | 23 | 5 | 4 | 4 | 15 | 6 | 0 | 0 | 0 | 1 | 0 | 0 | 0 | 0 | 0 | 0 | 0 | 0 | 0 | 0 | 0 | 0 | 0 | 0 | 0 | 0 | 58 | 1 | 2 | 3 | 4 | 2 | 0 | 2 | 0 |
| 269 | 269 | 2 | 43 | 3 | 3 | 1 | 11 | 4 | 1 | 0 | 0 | 1 | 0 | 0 | 0 | 0 | 0 | 0 | 0 | 0 | 0 | 0 | 0 | 0 | 0 | 0 | 0 | 0 | 15 | 1 | 1 | 2 | 1 | 2 | 0 | 1 | 1 |
| 270 | 270 | 2 | 23 | 1 | 4 | 1 | 11 | 4 | 1 | 0 | 0 | 0 | 0 | 1 | 0 | 0 | 0 | 0 | 0 | 0 | 0 | 0 | 0 | 0 | 1 | 0 | 0 | 0 | 3 | 0 | 2 | 2 | 1 | 2 | 0 | 1 | 0 |
| 271 | 271 | 2 | 36 | 2 | 5 | 3 | 17 | 3 | 1 | 0 | 0 | 0 | 0 | 0 | 0 | 0 | 0 | 1 | 0 | 0 | 0 | 0 | 0 | 0 | 0 | 0 | 0 | 0 | 8 | 0 | 1 | 2 | 1 | 2 | 0 | 1 | 0 |
| 272 | 272 | 2 | 32 | 2 | 7 | 1 | 11 | 3 | 0 | 0 | 0 | 1 | 0 | 0 | 0 | 0 | 0 | 0 | 0 | 0 | 0 | 0 | 0 | 0 | 0 | 0 | 0 | 0 | 2 | 0 | 1 | 1 | 1 | 1 | 0 | 1 | 0 |
| 273 | 273 | 2 | 26 | 1 | 5 | 1 | 6 | 3 | 0 | 0 | 0 | 0 | 0 | 1 | 0 | 0 | 0 | 0 | 0 | 0 | 0 | 0 | 0 | 0 | 0 | 0 | 0 | 0 | 20 | 1 | 2 | 2 | 3 | 2 | 0 | 0 | 0 |
| 274 | 274 | 2 | 42 | 5 | 3 | 1 | 23 | 4 | 0 | 0 | 0 | 0 | 0 | 1 | 0 | 0 | 0 | 0 | 0 | 0 | 0 | 0 | 0 | 0 | 1 | 0 | 0 | 0 | 8 | 0 | 1 | 3 | 1 | 2 | 0 | 1 | 1 |
| 275 | 275 | 2 | 35 | 3 | 5 | 4 | 6 | 3 | 0 | 0 | 0 | 1 | 0 | 0 | 0 | 0 | 0 | 1 | 0 | 0 | 0 | 0 | 0 | 0 | 0 | 0 | 0 | 0 | 30 | 1 | 2 | 3 | 3 | 2 | 0 | 2 | 0 |
| 276 | 276 | 2 | 52 | 5 | 5 | 1 | 35 | 3 | 0 | 0 | 0 | 0 | 0 | 0 | 0 | 0 | 0 | 0 | 0 | 0 | 0 | 0 | 0 | 0 | 0 | 0 | 0 | 0 | 4 | 0 | 2 | 4 | 1 | 2 | 0 | 0 | 0 |
| 277 | 277 | 1 | 77 | 5 | 1 | 5 | 45 | 2 | 0 | 0 | 0 | 0 | 0 | 0 | 0 | 0 | 0 | 0 | 0 | 0 | 0 | 0 | 0 | 0 | 0 | 0 | 0 | 0 | 0 | 0 | 1 | 3 | 1 | 1 | 1 | 0 | 1 |
| 278 | 278 | 1 | 26 | 1 | 4 | 1 | 15 | 2 | 1 | 0 | 0 | 0 | 0 | 0 | 0 | 0 | 0 | 0 | 0 | 0 | 0 | 0 | 0 | 0 | 0 | 0 | 0 | 0 | 12 | 1 | 1 | 2 | 1 | 2 | 0 | 1 | 0 |
| 279 | 279 | 2 | 62 | 2 | 6 | 1 | 23 | 3 | 0 | 0 | 0 | 0 | 0 | 0 | 0 | 0 | 0 | 0 | 0 | 0 | 0 | 0 | 0 | 0 | 0 | 0 | 0 | 0 | 0 | 0 | 2 | 1 | 1 | 1 | 0 | 1 | 1 |
| 280 | 280 | 2 | 23 | 5 | 5 | 5 | 11 | 1 | 0 | 1 | 0 | 1 | 0 | 0 | 0 | 0 | 0 | 0 | 0 | 0 | 0 | 0 | 0 | 0 | 0 | 0 | 0 | 0 | 0 | 0 | 2 | 1 | 1 | 1 | 0 | 0 | 0 |
| 281 | 281 | 2 | 32 | 2 | 6 | 1 | 28 | 1 | 0 | 0 | 0 | 0 | 0 | 0 | 0 | 0 | 0 | 0 | 0 | 0 | 0 | 0 | 0 | 0 | 0 | 0 | 0 | 0 | 1 | 0 | 2 | 2 | 1 | 1 | 0 | 0 | 0 |
| 282 | 282 | 2 | 30 | 1 | 3 | 5 | 35 | 1 | 0 | 0 | 0 | 0 | 0 | 0 | 0 | 0 | 0 | 1 | 0 | 0 | 0 | 0 | 0 | 0 | 0 | 0 | 0 | 0 | 13 | 0 | 1 | 1 | 1 | 2 | 0 | 0 | 0 |
| 283 | 283 | 1 | 83 | 2 | 3 | 5 | 45 | 3 | 0 | 0 | 0 | 2 | 0 | 0 | 0 | 0 | 0 | 0 | 0 | 0 | 0 | 0 | 0 | 0 | 0 | 0 | 0 | 0 | 0 | 0 | 2 | 3 | 1 | 1 | 0 | 0 | 0 |
| 284 | 284 | 2 | 37 | 5 | 4 | 1 | 28 | 2 | 0 | 0 | 0 | 0 | 0 | 0 | 0 | 0 | 0 | 0 | 0 | 0 | 0 | 0 | 0 | 0 | 0 | 0 | 0 | 0 | 10 | 0 | 1 | 3 | 1 | 1 | 0 | 0 | 0 |
| 285 | 285 | 2 | 42 | 4 | 5 | 5 | 37 | 3 | 1 | 0 | 0 | 1 | 0 | 1 | 0 | 0 | 0 | 0 | 0 | 1 | 0 | 0 | 0 | 0 | 0 | 0 | 0 | 0 | 9 | 0 | 2 | 3 | 2 | 2 | 0 | 2 | 0 |
| 286 | 286 | 1 | 56 | 2 | 3 | 3 | 13 | 3 | 1 | 1 | 0 | 0 | 0 | 0 | 0 | 0 | 0 | 0 | 0 | 0 | 0 | 3 | 0 | 0 | 0 | 0 | 0 | 0 | 3 | 0 | 2 | 2 | 1 | 1 | 0 | 1 | 0 |
| 287 | 287 | 2 | 61 | 5 | 4 | 2 | 23 | 3 | 0 | 1 | 0 | 3 | 3 | 3 | 0 | 3 | 0 | 0 | 0 | 0 | 0 | 0 | 3 | 0 | 0 | 0 | 0 | 0 | 9 | 0 | 2 | 3 | 1 | 2 | 0 | 0 | 1 |
| 288 | 288 | 1 | 19 | 1 | 2 | 1 | 35 | 3 | 0 | 0 | 0 | 0 | 0 | 0 | 0 | 0 | 0 | 0 | 0 | 0 | 0 | 0 | 0 | 0 | 0 | 0 | 0 | 0 | 28 | 1 | 1 | 1 | 2 | 2 | 0 | 3 | 0 |
| 289 | 289 | 1 | 45 | 4 | 3 | 1 | 35 | 3 | 0 | 1 | 1 | 3 | 3 | 0 | 0 | 0 | 0 | 0 | 0 | 0 | 0 | 0 | 0 | 0 | 0 | 0 | 0 | 0 | 4 | 0 | 2 | 3 | 3 | 2 | 0 | 0 | 0 |
| 290 | 290 | 2 | 19 | 2 | 4 | 5 | 55 | 1 | 0 | 1 | 0 | 3 | 3 | 0 | 0 | 0 | 0 | 0 | 0 | 0 | 0 | 0 | 0 | 0 | 0 | 0 | 0 | 0 | 7 | 0 | 2 | 2 | 1 | 2 | 0 | 1 | 0 |
| 291 | 291 | 2 | 43 | 2 | 3 | 2 | 55 | 3 | 0 | 0 | 0 | 0 | 0 | 0 | 0 | 0 | 0 | 0 | 0 | 0 | 0 | 0 | 0 | 0 | 0 | 0 | 0 | 0 | 0 | 0 | 1 | 1 | 1 | 1 | 0 | 2 | 0 |
| 292 | 292 | 2 | 64 | 2 | 3 | 1 | 55 | 3 | 0 | 0 | 0 | 0 | 0 | 0 | 0 | 0 | 0 | 0 | 0 | 0 | 0 | 0 | 0 | 0 | 0 | 0 | 0 | 0 | 39 | 1 | 2 | 2 | 1 | 2 | 0 | 1 | 0 |
| 293 | 293 | 2 | 43 | 3 | 2 | 5 | 28 | 1 | 0 | 0 | 0 | 0 | 0 | 0 | 0 | 0 | 0 | 1 | 0 | 0 | 0 | 0 | 0 | 0 | 0 | 0 | 0 | 0 | 2 | 0 | 2 | 1 | 1 | 2 | 0 | 2 | 0 |
| 294 | 294 | 2 | 58 | 1 | 3 | 4 | 9 | 1 | 1 | 0 | 0 | 2 | 0 | 0 | 0 | 0 | 1 | 2 | 0 | 0 | 0 | 0 | 0 | 0 | 0 | 0 | 1 | 1 | 10 | 0 | 2 | 3 | 1 | 1 | 0 | 0 | 1 |

Note that investigators often alter the order of these operations. For example, some prefer to check for missing data and outliers and to make transformations prior to combining the data sets. This is particularly true in analyzing longitudinal data when the first data set may be available well before the others. This may also be an iterative process in that determining errors during data management may lead to entering new data to replace erroneous values.

Three statistical packages—BMDP, SAS, and SPSS—were noted as the packages used in this book. Other packages are given in Table 3.1. In evaluating a package it is sometimes helpful to examine what options it has for some of the operations given in this chapter, particularly the data management features. As noted, data can be entered using non-statistical packages, but the data management features are often essential for general statistical analysis.

## BIBLIOGRAPHY

Aneshensel, C. S., Frerichs, R. R., and Clark, V. A. 1981. Family roles and sex differences in depression. *Journal of Health and Social Behavior* 22:379–393.

Barnett, V., and Lewis, T. 1984. *Outliers in statistical data.* 2nd ed. New York: Wiley.

BMDP 1985. Using LOTUS 123 as a source of data for BMDPC, *BMDP Statistical Software Communications* 18:10–11.

Comstock, G. W., and Helsing, K. J. (1976). Symptoms of depression in two communities. *Psychological Medicine* 6:551–563.

Dunn, O. J., and Clark, V. A. 1987. *Applied statistics: Analysis of variance and regression.* 2nd ed. New York: Wiley.

Frerichs, R. R., Aneshensel, C. S., and Clark, V. A. 1981. Prevalence of depression in Los Angeles County. *American Journal of Epidemiology* 113:691–699.

Frerichs, R. R., Aneshensel, C. S., Clark, V. A., and Yokopenic, P. 1981. Smoking and depression: A community survey. *American Journal of Public Health* 71:637–640.

Radloff, L. S. 1977. The CES–D scale: A self-report depression scale for research in the general population. *Applied Psychological Measurement* 1:385–401.

Raskin, R. 1989. Statistical Software for the PC: Testing for Significance. *PC Magazine* Vol. 8, Num. 5:103–198.

Tukey, J. W. 1977. *Exploratory data analysis.* Reading, Mass: Addison-Wesley.

## PROBLEMS

3.1 Using a packaged program, compute a two-way table of income versus employment status from the depression data. From the data in this table, decide if there are any adults whose income is unusual considering their employment status. Are there any adults whose income is unusual considering their age or education?

3.2 List the programs from SAS that you would be likely to use to screen interval or ratio data. Repeat for SPSS and BMDP.

3.3 Describe the person in the depression data set who has the highest total CESD score.

3.4 Make a code book for a data set you have.

3.5 Using a packaged program, compute histograms for mothers' and fathers' heights and weights from the lung function data set given in Appendix B. Describe any cases that you consider potential outliers.

3.6 For the chronic respiratory disease study, use the information given in Section 1.2 and Appendix B to create a code book for the data set. If you were to create a more complete code book, what other information would you require?

3.7 For the data set given in Appendix B, use a packaged computer program to obtain a frequency count for the variables in columns 5, 19, and 33. Do you notice anything unusual?

3.8 For the data set given in Appendix B, use a packaged computer program to find out how many families have one child, two children, and three children between the ages of 7 and 18.

3.9 For the data set given in Appendix B, produce a two-way table of sex of child 1 versus sex of child 2 (for families with at least two children). Comment.

3.10 For the depression data set, determine which variables and observations do not fall within the ranges given in the code book.

3.11 For the program you intend to use, describe how you would add data from three more time periods to the depression data set.

3.12 For the data set in Appendix B, create a new variable; AGEDIFF = (age of child 1) − (age of child 2), for families with at least two children. Produce a frequency count of this variable. Are there any negative values? Comment.

3.13 Combine the results from the following two questions into a single variable:
   a. Have you been sick during the last two weeks?

   Yes, go to (b)_____
   No          _____

   b. How many days were you sick? _____

## Chapter Four

# DATA SCREENING AND DATA TRANSFORMATION

## 4.1 WHAT WILL YOU LEARN FROM THIS CHAPTER?

From this chapter you will learn:

- How to assess the effects of commonly used transformations (4.2).
- What a normal probability plot looks like for various distributions (4.3).
- How to assess whether data are normally distributed (4.3).
- Several methods for transforming data to achieve approximate normality (4.3).
- How to assess whether the data are independent (4.4).

Each computer package offers the users information to help decide if their data are normally distributed. They provide convenient methods for transforming the data to achieve approximate normality. They also include output for checking the independence of the observations. Hence the assumption of independent, normally distributed data that appears in many statistical tests can be, at least approximately, assessed. Note that most investigators will try to discard the most obvious outliers prior to assessing normality because such outliers can grossly distort the distribution.

## 4.2 COMMON TRANSFORMATIONS

In the analysis of data it is often useful to transform certain variables before performing the analyses. Examples of such uses are found in the next section and in Chapter 6. In this section we present some of the most commonly used transformations. If you are familiar with this subject, you may wish to skip to the next section.

To develop a feel for transformations, let us examine a plot of transformed values versus the original values of the variable. To begin with, a plot of values of a variable $X$ against itself produces a 45° diagonal line going through the origin, as shown in Figure 4.1a. One of the most commonly performed transformations is taking the *logarithm* (log) to the base 10. Recall that the logarithm $Y$ of a number $X$ satisfies the relationship $X = 10^Y$. That is, the logarithm of $X$ is the power $Y$ to which 10 must be raised in order to produce $X$. As shown in Figure 4.1b, the logarithm of 10 is 1 since $10 = 10^1$. Similarly, the logarithm of 1 is 0 since $1 = 10^0$, and the logarithm of 100 is 2 since $100 = 10^2$. Other values of logarithms can be obtained from tables of common logarithms or from a hand calculator with a log function. All statistical packages discussed in this book allow the user to make this transformation as well as others.

Note that an increase in $X$ from 1 to 10 increases the logarithm from 0 to 1, that is, an increase of one unit. Similarly, an increase in $X$ from 10 to 100 increases the logarithm also by one unit. For larger numbers it takes a great increase in $X$ to produce a small increase in log $X$. Thus the logarithmic transformation has the effect of stretching small values of $X$ and condensing large values of $X$. Note also that the logarithm of any number less than 1 is negative, and the logarithm of a value of $X$ that is less than or equal to 0 is not defined. In practice, if negative or zero values of $X$ are possible, the investigator may first add an appropriate constant to each value of $X$, thus making them all positive prior to taking the logarithms. The choice of the additive constant can have an important effect on the statistical properties of the transformed variable, as will be seen in the next section. The value added must be larger than the magnitude of the minimum value of $X$.

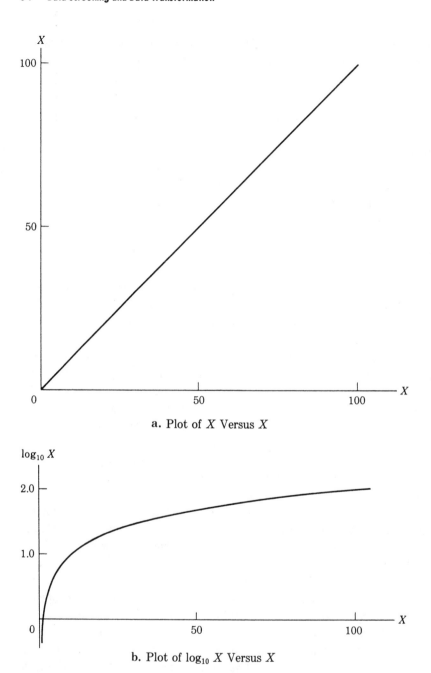

**a.** Plot of $X$ Versus $X$

**b.** Plot of $\log_{10} X$ Versus $X$

**FIGURE 4.1.** Plots of Selected Transformations of Variable $X$

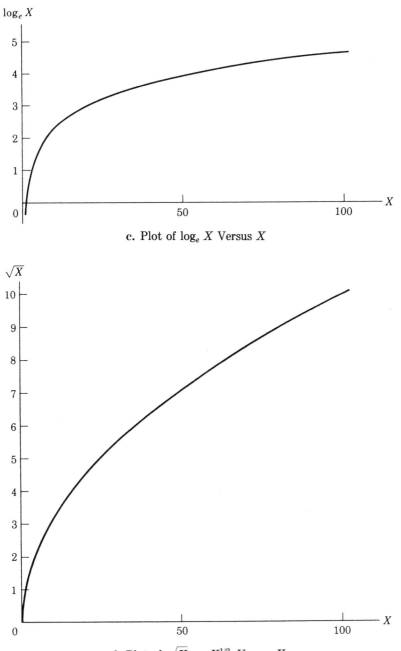

c. Plot of $\log_e X$ Versus $X$

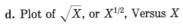

d. Plot of $\sqrt{X}$, or $X^{1/2}$, Versus $X$

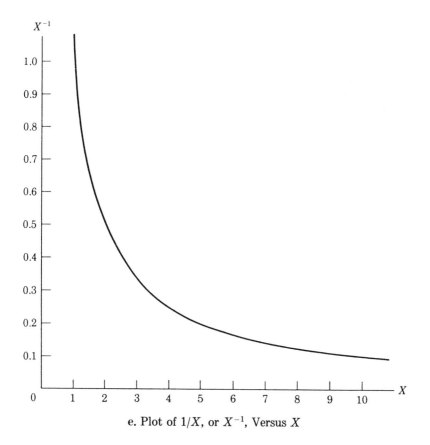

e. Plot of $1/X$, or $X^{-1}$, Versus $X$

Logarithms can be taken to any base. A familiar base is the number $e = 2.7183 \ldots$ Logarithms taken to the base $e$ are called *natural logarithms* and are denoted by $\log_e$ or ln. The natural logarithm of $X$ is the power to which $e$ must be raised to produce $X$. There is a simple relationship between the natural and common logarithms, namely,

$$\log_e X = 2.3026 \log_{10} X$$

$$\log_{10} X = 0.4343 \log_e X$$

Figure 4.1c shows that a graph of $\log_e X$ versus $X$ has the same shape as $\log_{10} X$, with the only difference being in the vertical scale; i.e., $\log_e X$ is larger than $\log_{10} X$ for $X > 1$ and smaller than $\log_{10} X$ for $X < 1$.

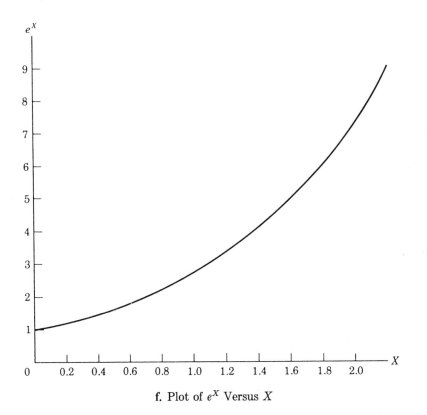

**f.** Plot of $e^X$ Versus $X$

The natural logarithm is used frequently in theoretical studies because of certain appealing mathematical properties.

Another class of transformations is known as *power transformations*. For example, the transformation $X^2$—i.e., $X$ raised to the power of 2—is used frequently in statistical formulas such as computing the variance. The most commonly used power transformations are the square root —i.e., $X$ raised to the power $\frac{1}{2}$—and the inverse $1/X$—i.e., $X$ raised to the power $-1$. Figure 4.1d shows a plot of the square root of $X$ versus $X$. Note that this function is also not defined for negative values of $X$. Compared with taking the logarithm, taking the square root also progressively condenses the values of $X$ as $X$ increases. However, the degree of condensation is not as severe as in the case of logarithms. That is, the square root of $X$ tapers off slower than log $X$, as can be seen by comparing Figure 4.1d with Figure 4.1b or 4.1c.

Unlike $\sqrt{X}$ and log $X$, the function $1/X$ decreases with increasing $X$, as shown in Figure 4.1e. To obtain an increasing function, you may use $-1/X$.

One way to characterize transformations is to note the power to which $X$ is raised. We denote that power by $p$. For the square root transformation $p = 1/2$ and for the inverse transformation $p = -1$. The logarithmic transformation can be thought of as corresponding to a value of $p = 0$ (see Tukey 1977). We can think of these three transformations as ranging from $p$ values of $-1$ to 0 to $1/2$. The effects of these transformations in reducing a long tailed distribution to the right are greater as the value of $p$ decreases from 1 (the $p$ value for no transformation) to $1/2$ to 0 to $-1$. Smaller values of $p$ result in a greater degree of transformation for $p < 1$. Thus, $p$ of $-2$ would result in a greater degree of transformation than a $p$ of $-1$. The logarithmic transformation reduces a long tail more than a square root transformation (see Figure 4.1d versus 4.1b or 4.1c) but not as much as the inverse transformation. The changes in the amount of transformation depending on the value of $p$ provide the background for one method of choosing an appropriate transformation. In the next section, after a discussion of how to assess whether you have a normal distribution, we give several strategies to assist in choosing appropriate $p$ values.

Finally, exponential functions are also sometimes used in data analysis. An *exponential function* of $X$ may be thought of as the antilogarithm of $X$. For example, the antilogarithm of $X$ to the base 10 is 10 raised to the power $X$. Similarly, the antilogarithm of $X$ to the base of $e$ is $e$ raised to the power $X$. The exponential function $e^X$ is illustrated in Figure 4.1f. The function $10^X$ has the same shape but increases faster than $e^X$. Both have the opposite effect of taking logarithms; i.e., they increasingly stretch the larger values of $X$. Additional discussion of these interrelationships can be found in Tukey (1977).

## 4.3 ASSESSING THE NEED FOR AND SELECTING A TRANSFORMATION

In the theoretical development of statistical methods some assumptions are usually made regarding the distribution of the variables being analyzed. Often the form of the distribution is specified. The most commonly assumed distribution for continuous observations is the *normal*, or Gaussian, *distribution*. Although the assumption is sometimes not crucial to the validity of the results, some check of normality is advisable in the early stages of analysis. In this section methods for choosing a transformation to induce normality are presented.

## Transformations to Induce Normality: Iterative Methods Using Normal Probability Plots

Iterative methods using normal probability plots present an appealing option for checking normality. Histograms for this purpose can be obtained from SAS, SPSS, and BMDP programs. If the population distribution is normal, the sample histogram should resemble the famous bell-shaped Gaussian curve. However, assessing the degree of normality or lack thereof is sometimes difficult to do by simply looking at the histograms, so a different graphical aid is required. A useful tool for this purpose is the *normal probability plot*. Such a plot can be obtained as the optional output of program BMDP5D, and from the SAS UNIVARIATE procedure using the PLOT option.

One axis of the probability plot shows values of $X$, and the other shows expected values, $Z$, of $X$ if its distribution were exactly normal. The computation of these expected $Z$ values is discussed in Johnson and Wichern (1988). Equivalently, this graph is a plot of the *cumulative distribution* found in the data set, with the vertical axis adjusted to produce a straight line if the data followed an exact normal distribution. Thus if the data were from a normal distribution, the normal probability plot should approximate a straight line, as shown in Figure 4.2a for a normal distribution with zero mean. In this graph values of the variable $X$ are shown on the horizontal axis, and values of the expected normal values appear on the vertical axis. The $Z$ values correspond to those usually listed in standard distribution tables. In computer program output, the $X$ and $Z$ axes are often interchanged (see SAS UNIVARIATE).

The remainder of Figure 4.2 illustrates various other distributions with their corresponding normal probability plots. The purpose of these probability plots is to help you associate the appearance of the probability plot with the shape of the histogram or frequency distribution. Figures 4.2b and 4.2f both show symmetric distributions with more

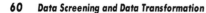

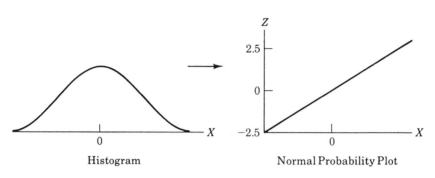

a. Normal Distribution

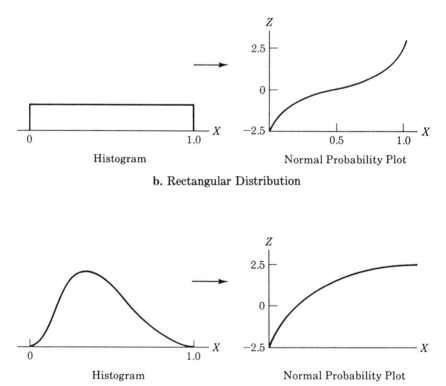

b. Rectangular Distribution

c. Lognormal (Skewed Right) Distribution

**FIGURE 4.2.** Plots of Hypothetical Histograms and the Resulting Normal Probability Plots from those Distributions

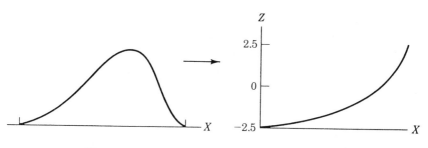

Histogram                    Normal Probability Plot

**d.** Negative Lognormal (Skewed Left) Distribution

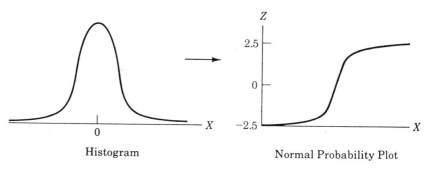

Histogram                    Normal Probability Plot

**e.** 1/Normal Distribution

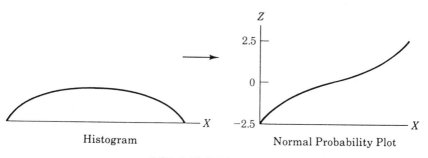

Histogram                    Normal Probability Plot

**f.** High-Tail Distribution

observations in the tails, i.e., heavier tails than those of the normal distribution. The corresponding normal probability plot is an inverted S. Taking the inverse of the data does not produce an approximately normal distribution in this case (Figure 4.2f) nor does it help in cases such as that shown in Figure 4.2b.

On the other hand, taking the logarithm of the data illustrated in Figure 4.2c would induce normality. The distribution before the transformation is called the *lognormal distribution*. This distribution is encountered in numerous real life applications. It can be recognized from its normal probability plot, whose curvature resembles a quarter of a circle. In this case we can create a new variable by taking logarithms to either base 10 or base $e$.

The investigator should request the normal probability plot of the transformed data. If the plot is still curved in the same direction, then subtracting a positive number that is smaller than the minimum $X$ value before taking logarithms may straighten out the remaining curvature. You are encouraged to experiment with various constants to find the one producing the nearest plot to a straight line. If, on the other hand, taking logarithms produces a plot with the curvature in the opposite direction, as shown in Figure 4.2d, then adding a positive constant prior to taking logarithms may be helpful. In some cases subtracting each $X$ value from a constant larger than the largest $X$ and then taking logarithms could also be helpful. For instance, subtracting each $X$ from 0 is the appropriate transformation in the case of Figure 4.2d, where all the data are negative.

As noted earlier, the inverse transformation is appropriate for data such as those illustrated in Figure 4.2e. In this case the histogram has tails lower than those for normal data, resulting in an S-shaped normal probability plot. Adding or subtracting an appropriate constant before taking the inverse may be helpful in this case as well.

Some transformations are commonly used in certain situations. Thus the square root transformation is used with Poisson-distributed variables. Such variables represent counts of events occurring randomly in space or time with a small probability such that the occurrence of one event does not affect another. Examples include the number of cells per unit area on a slide of biological material, the number of incoming phone calls per hour in a telephone exchange, and counts of radioactive material per unit of time. Similarly, the logarithmic transformation has been used on variables such as household income, the time elapsing before a specified number of Poisson events have occurred, systolic blood pressure of older individuals, and many other variables with long tails to the right. Finally, the inverse transformation has been used frequently on the length of time it takes a person or an animal to complete a certain task. Bennett and Franklin (1954) illustrate the use of transformations for counted data, and Tukey (1977) gives numerous examples of variables and appropriate transformations for them. Tukey (1957), Box and Cox (1964), Draper and Hunter (1969), and Bickel and Doksum (1981) discuss some systematic ways to find an appropriate transformation.

Another strategy for deciding what transformation to use is to progress up or down the values of $p$ depending on the shape of the normal probability plot. If it looks similar to Figure 4.2c, then a value of $p$ that is less than 1 is tried. Suppose $p = 1/2$ is tried. If this is not sufficient and the normal probability plot still is curved downward at the ends, then $p = 0$ or a logarithmic transformation can be tried. If the plot appears to be almost correct but a little more transformation is needed, a constant can be subtracted from each $X$ prior to the transformation. If the transformation appears to be a little too much, a constant can be added. Thus, transformations of the form $(X + C)^p$ are tried until the normal probability plot is as straight as possible. An example of the use of this type of transformation occurs when considering systolic blood pressure (SBP) for adult males. It has been found by several investigators that a transformation such as $\log(\text{SBP} - 75)$ results in data that appear to be normally distributed.

The investigator decreases or increases the value of $p$ until the resulting observations appear to be as close as possible to a normal distribution. Note that this does not mean that theoretically one has a normal distribution, since you are only working with a single sample.

Hines and Hines (1987) developed a method for reducing the amount of iteration needed by providing a graph which produces a suggested value of $p$ from information available in many computer programs. Using their graph one can directly estimate a reasonable value of $p$.

Their method is based on the use of quantiles that can either be obtained directly or approximated from computer output. A type of quantile that is widely available in computer output is the sample median. The median divides the observations of a variable into two equal parts. Suppose the observations are ordered from smallest to largest. The median is either the middle observation if the number of observations $N$

is odd, or the average of the middle two if $N$ is even. A quartile divides the values of the variable into four equal parts. A quantile is a value of the variable that divides the observations into a given number of equal parts. BMDP2D and SAS UNIVARIATE provide percentiles, which divide the output into 100 parts that can in turn be used to approximate the more general quantiles. The median is approximately the 50th percentile, and the first and third quartiles are approximately the 25th and 75th percentiles, respectively. What is needed in order to use their graph is the value of the median and the values of the symmetric quantiles (percentiles). By symmetric quantiles we mean, for example, the 10th and 90th percentiles or the 25th and 75th percentiles (the quartiles). These are called lower and upper quantiles.

If a distribution were symmetric about the median then the difference between the median and the lower quantile would equal the difference between the upper quantile and the median. By choosing a value of $p$ that results in these two differences being equal, one is approximately choosing a transformation that tends to make the data normally distributed (since symmetry is a property of the normal distribution). Using Figure 4.3 (kindly supplied by W. G. S. Hines and R. J. O'Hara Hines), one plots the ratio of the lower quantile of $X$ to the median of $X$ on the vertical axis versus the ratio of the median to the upper quantile on the horizontal axis for at least one quantile. The resulting value of $p$ is read off the closest curve.

For example, in the depression data set given in Chapter 3 a variable called INCOME is listed. This is family income in thousands of dollars. From BMDP2D, the median income is 15, the lower (first) quartile is 9, and the upper (third) quartile is 28. Hence the median is not halfway between the two quartiles, indicating a nonsymmetric or nonnormal distribution. If we plot $9/15 = 0.60$ on the vertical axis and $15/28 = 0.54$ on the horizontal axis of Figure 4.3, we get a value for $p$ of approximately $-1/3$. Such a value indicates the need for using $-1$ over the cubic root income if we wanted a symmetric variable that went from small to large incomes. Another possibility would be to subtract a constant from each $X$ and then replot the results. The minimum income was 2 so that 1.99 can safely be subtracted. Subtracting a constant simply shifts all quantiles by the same constant, so we have approximately $7/13 = 0.54$ on the vertical axis and $13/26 = 0.5$ on the horizontal axis. Now an appropriate value for $p$ is $-1/10$, which is close to 0, and a logarithmic

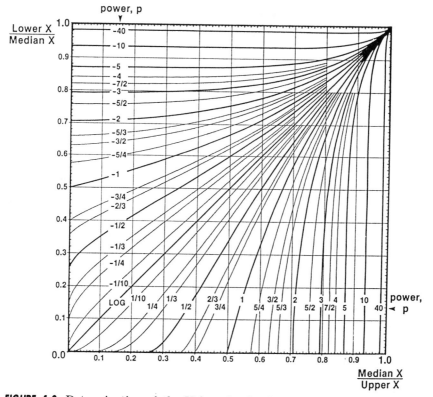

**FIGURE 4.3.** Determination of the Value of $p$ in the Power Transformation to Produce Approximate Normality

transformation should be considered. In other words, not as much of a transformation is needed. Additional quantiles could be estimated to better approximate $p$ from Figure 4.3.

It should be noted that not every distribution can be transformed to a normal distribution. For example, variable 29 in the depression data set is the sum of the results from the 20-item depression scale. The mode (the most commonly occurring score) is at zero, thus making it virtually impossible to transform these CESD scores to a normal distribution. However, if a distribution has a single mode in the interior of the distribution, then it is usually not too difficult to find a transformation producing a distribution that appears to be normal. Ideally, you should search for a transformation that has the proper scientific meaning in the context of your data.

## Statistical Tests for Normality

It is also possible to do formal statistical tests for normality. For example, the SAS UNIVARIATE procedure will compute a Shapiro-Wilks $W$ statistic if the number of observations is less than 2000. This test is well suited to small sample sizes (see Dunn and Clark 1987). The null hypothesis of normality is rejected for small values of $W$. The program prints out the value of $W$ and the associated probability ($p$ value) for testing the hypothesis that the data came from a normal distribution. If the $p$ value is small, then the data may not be normally distributed.

For larger $N$ the program prints a modified version of the Kolmogorov-Smirnov $D$ statistic and approximate $p$ values. Here the null hypothesis is rejected for large values of $D$ or a small $p$ value. Afifi and Azen (1979) discuss the characteristics of this test.

The BMDP2D program computes the skewness of a variable along with its standard error. *Skewness* is a measure of how nonsymmetric a distribution is. If the data are normally or symmetrically distributed, then the computed skewness will be close to zero. The numerical value of the ratio of the skewness to its standard error can be compared with tabled normal $Z$ values, and normality would be rejected if the value of the ratio is large. Positive skewness indicates that the distribution has a long tail to the right and probably looks like Figure 4.2c. (Note that it is important to remove outliers or erroneous observations prior to performing this test.)

Very little is known about what significance levels $\alpha$ should be chosen to compare with the $p$ value obtained from formal tests of normality. So the sense of increased preciseness gained by performing a formal test over examining a plot is somewhat of an illusion. From the normal probability plot you can both decide whether the data are normally distributed and get a suggestion about what transformation to use.

## Assessing the Need for a Transformation

In general, transformations are more effective in inducing normality when the standard deviation of the untransformed variable is large relative to the mean. If the standard deviation divided by the mean is less than $\frac{1}{4}$, then the transformation may not be necessary. In deciding whether to make a transformation, you may wish to perform the analysis with and without the proposed transformation. Examining the results will frequently convince you that the conclusions are not altered after making the transformation. In this case it is preferable to present the results in terms of the most easily interpretable units. And it is often helpful to conform to the customs of the particular field of investigation.

Sometimes, transformations are made to simplify later analyses rather than to approximate normal distributions. For example, it is known that FEV1 (forced expiratory volume in 1 second) and FVC (forced vital capacity) decrease in adults as they grow older. (See Section 1.3 for a discussion of these variables.) Some researchers will take the ratio FEV1/FVC and work with it because this ratio is less dependent on age. Using a variable that is independent of age can make analyses of a sample including adults of varying ages much simpler. In a practical sense, then, the researcher can use the transformation capabilities of the computer program packages to create new variables to be added to the set of initial variables rather than only modify and replace them.

If transformations alter the results, then you should select the transformation that makes the data conform as much as possible to the assumptions. If a particular transformation is selected, then all analyses should be performed on the transformed data, and the results should be presented in terms of the transformed values. Inferences and statements of results should reflect this fact.

## 4.4 ASSESSING INDEPENDENCE

Measurements on two or more variables collected from the *same* individual are not expected to be, nor are they assumed to be, independent of each other. On the other hand, independence of observations collected from *different* individuals or items is an assumption made in the theoretical derivation of most multivariate statistical analyses. This assumption is crucial to the validity of the results, and violating it may result in erroneous conclusions. Unfortunately, little is published about the quantitative effects of various degrees of nonindependence. Also, few practical methods exist for checking whether the assumption is valid.

In situations where the observations are collected from people, it is frequently safe to assume independence of observations collected from different individuals. Dependence could exist if a factor or factors exist to affect all of the individuals in a similar manner with respect to the variables being measured. For example, political attitudes of adult members of the same household cannot be expected to be independent. Inferences that assume independence are not valid when drawn from such responses. Similarly, biological data on twins or siblings are usually not independent.

Data collected in the form of a sequence either in time or in space can also be dependent. In those cases it is useful to plot the data in the appropriate sequence and search for trends or changes over time. This search may be done by using package scatter diagram programs such as BMDP6D, SAS PLOT, and SPSS-X SCATTERGRAM procedures. Values of the variable being considered appear on the vertical axis, and values of time, location, or observation identification number (ID) appear on the horizontal axis. If the plot resembles that shown in Figure 4.4a, little reason exists for suspecting lack of independence or non-randomness. The data in that plot were, in fact, obtained as a sequence of random standard normal numbers. In contrast, Figure 4.4b shows data that exhibit a positive trend. Figure 4.4c is an example of a temporary shift in the level of the observations, followed by a return to the initial level. This result may occur, for example, in laboratory data when equipment is temporarily out of calibration or when a substitute technician, following different procedures, is used temporarily. Finally, a common situation occurring in some business or industrial data is the existence of seasonal cycles, as shown in Figure 4.4d.

Some formal tests for the randomness of a sequence of observations are given in Brownlee (1965) or Bennett and Franklin (1954). One such test is presented in Chapter 6 of this book in the context of regression and correlation analysis. If you are dealing with series of observations you may also wish to study the area of forecasting and time series analysis. Some books on this subject are included in the bibliography at the end of this chapter and in the bibliography following Chapter 6.

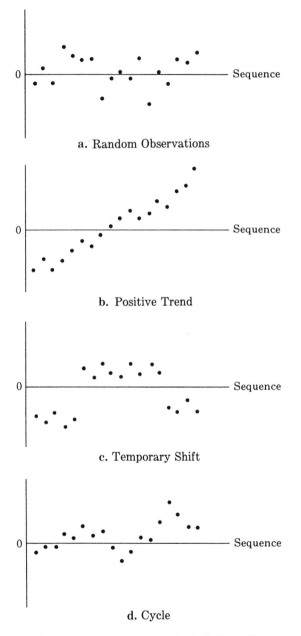

FIGURE 4.4. Graphs of Hypothetical Data Sequences Showing Lack of Independence

## SUMMARY

In this chapter we emphasized methods for determining if the data were normally distributed and for finding suitable transformations to make the data closer to a normal distribution. We also discussed methods for checking whether the data were independent.

No special order is best for all situations in data screening. However, most investigators would delete the obvious outliers first. They may then check the assumption of normality, attempt some transformations, and recheck the data for outliers. Other combinations of data-screening activities are usually dictated by the problem.

We wish to emphasize that data screening could be the most time-consuming and costly portion of data analysis. Investigators should not underestimate this aspect. If the data are not screened properly, much of the analysis may have to be repeated, resulting in an unnecessary waste of time and resources. Once the data have been carefully screened, the investigator will then be in a position to select the appropriate analysis to answer specific questions. In Chapter 5 we present a guide to the selection of the appropriate data analysis.

## BIBLIOGRAPHY

Afifi, A. A., and Azen, S. P. 1979. *Statistical analysis: A computer oriented approach.* 2nd ed. New York: Academic Press.

Bartlett, M. S. 1947. The use of transformations. *Biometrics* 3:39–52.

Bennett, C. A. and Franklin, N. L. 1954. *Statistical analysis in chemistry and the chemical industry.* New York: Wiley.

*Bickel, P. J., and Doksum, K. A. 1981. An analysis of transformations revisited. *Journal of the American Statistical Association* 76:296–311.

*Box, G. E. P., and Cox, D. R. 1964. An analysis of transformations. *Journal of the Royal Statistical Society* Series B, 26:211–252.

*Box, G. E. P., and Jenkins, G. M. 1976. *Time series analysis: forecasting and control.* Rev. ed. San Francisco: Holden-Day.

Brownlee, K. A. 1965. *Statistical theory and methodology in science and engineering.* 2nd ed. New York: Wiley.

Dixon, W. J., and Massey, F. J. 1983. *Introduction to statistical analysis.* 4th ed. New York: McGraw-Hill.

Draper, N. R., and Hunter, W. G. 1969. Transformations: Some examples revisited. *Technometrics* 11:23–40.

Dunn, O. J., and Clark, V. A. 1987. *Applied statistics: Analysis of variance and regression.* New York: Wiley.

Hines, W. G. S., and O'Hara Hines, R. J. 1987. Quick graphical power-law transformation selection. *The American Statistician* 41:21–24.

Hoaglin, D. C., Mosteller, F., and Tukey, J. W., eds. 1983. *Understanding robust and exploratory data analysis.* New York: Wiley.

Johnson, R. A., and Wichern, D. W. 1988. *Applied multivariate statistical analysis.* 2nd ed. Englewood Cliffs, N. J.: Prentice-Hall.

McCleary, R., and Hay, R. A. 1980. *Applied time series analysis for the social sciences.* Beverly Hills: Sage.

Mage, D. T. 1982. An objective graphical method for testing normal distribution assumptions using probability plots. *American Statistician* 36:116–120.

*Montgomery, D. C., and Johnson, L. A. 1976. *Forecasting and time series analysis.* New York: McGraw-Hill.

Ostrom, C. W. 1978. *Time series analysis: Regression techniques.* Beverly Hills: Sage.

*Tukey, J. W. 1957. On the comparative anatomy of transformations. *Annals of Mathematical Statistics* 28:602–632.

Tukey, J. W. 1977. *Exploratory data analysis.* Reading, Mass.: Addison-Wesley.

## PROBLEMS

4.1 Using the depression data set given in Table 3.3, create a variable equal to the negative of one divided by the cubic root of income. Display a normal probability plot of the new variable.

4.2 Take the logarithm to the base 10 of the income variable in the depression data set. Compare the histogram of income with the histogram of log(income). Also, compare the normal probability plots of income and log(income).

4.3 Repeat Problems 4.1 and 4.2, taking the square root of income.

4.4 Generate a set of 100 random normal deviates by using a computer program package. Display a histogram and normal probability plot of these values. Square these numbers by using transformations. Compare histograms and normal probability plots of the logarithms and the square roots of the set of squared normal deviates.

4.5 Take the logarithm of the CESD score plus 1 and compare the histograms of CESD and log(CESD + 1). (A small constant must be added to CESD because CESD can be zero.)

4.6 The accompanying data are from the New York Stock Exchange Composite Index for the August 9 through September 17, 1982, period. Run a program to plot these data in order to assess the lack of independence of

| Day | Month | Index |
|-----|-------|-------|
| 9 | Aug. | 59.3 |
| 10 | Aug. | 59.1 |
| 11 | Aug. | 60.0 |
| 12 | Aug. | 58.8 |
| 13 | Aug. | 59.5 |
| 16 | Aug. | 59.8 |
| 17 | Aug. | 62.4 |
| 18 | Aug. | 62.3 |
| 19 | Aug. | 62.6 |
| 20 | Aug. | 64.6 |
| 23 | Aug. | 66.4 |
| 24 | Aug. | 66.1 |
| 25 | Aug. | 67.4 |
| 26 | Aug. | 68.0 |
| 27 | Aug. | 67.2 |
| 30 | Aug. | 67.5 |
| 31 | Aug. | 68.5 |
| 1 | Sept. | 67.9 |
| 2 | Sept. | 69.0 |
| 3 | Sept. | 70.3 |
| 6 | Sept. | — |
| 7 | Sept. | 69.6 |
| 8 | Sept. | 70.0 |
| 9 | Sept. | 70.0 |
| 10 | Sept. | 69.4 |
| 13 | Sept. | 70.0 |
| 14 | Sept. | 70.6 |
| 15 | Sept. | 71.2 |
| 16 | Sept. | 71.0 |
| 17 | Sept. | 70.4 |

successive observations. (Note that the data cover five days per week, except for a holiday on Monday, September 6.) This time period saw an unusually rapid rise in stock prices (especially for August), coming after a protracted falling market. Compare this time series with prices for the current year.

4.7 Obtain a normal probability plot of the data set given in Problem 4.6. Suppose that you had been ignorant of the lack of independence of these data and had treated them as if they were independent samples. Assess whether they are normally distributed.

4.8 Obtain normal probability plots of mothers' and fathers' weights from the lung function data set in Appendix B. Discuss whether or not you consider weight to be normally distributed in the population from which this sample is taken.

4.9 Generate ten random normal deviates. Display a probability plot of these data. Suppose you didn't know the origin of these data. Would you conclude they were normally distributed? What is your conclusion based on the Shapiro-Wilks test? Do ten observations provide sufficient information to check normality?

4.10 Repeat Problem 4.8 with the weights expressed in ounces instead of pounds. How do your conclusions change? Obtain normal probability plots of the logarithm of mothers' weights expressed in pounds and then in ounces, and compare.

4.11 From the variables ACUTEILL and BEDDAYS described in Table 3.2, use arithmetic operations to create a single variable that takes on the value 1 if the person has been both bedridden and acutely ill in the last two months and that takes on the value 0 otherwise.

# Chapter Five

# SELECTING APPROPRIATE ANALYSES

## 5.1 WHAT WILL YOU LEARN FROM THIS CHAPTER?

From this chapter you will learn how to use the information on independent and dependent variables and Stevens's classification system to assist you in selecting analyses for your data. In particular, you will learn:

- How to decide what descriptive measures should be used (5.3).
- How to decide which measure of central tendency and dispersion should be used (5.3).
- How to determine which multivariate analyses presented in Chapters 7–16 fit your data (5.4).

## 5.2 WHY SELECTION OF ANALYSES IS OFTEN DIFFICULT

There are two reasons why deciding what descriptive measures or analyses to perform and report is often difficult for an investigator with real life data. First, in statistics textbooks statistical methods are presented in a logical order from the viewpoint of learning statistics but not from the viewpoint of doing data analysis by using statistics. Most texts are either mathematical statistics texts or are imitations of them with the mathematics simplified or left out. Also, when learning statistics for the first time, the student often finds mastering the techniques themselves tough enough without worrying about how to use them in the future. The second reason is that real life data often contain mixtures of types of data, which makes the choice of analysis somewhat arbitrary. Two trained statisticians presented with the same set of data will often opt for different ways of analyzing the set, depending on what assumptions they are willing to take into account in the interpretation of the analysis.

Acquiring a sense of when it is safe to ignore assumptions is difficult both to learn and to teach. Here, for the most part, an empirical approach will be suggested. For example, it is often a good idea to perform several different analyses, one where all the assumptions are met and one where some are not, and compare the results. The idea is to use statistics to obtain insights into the data and to determine how the system under study works.

One point to keep in mind is that the examples presented in many statistics books are often ones the authors have selected after a long period of working with a particular technique. Thus they usually are "ideal" examples, designed to suit the technique being discussed. This feature makes learning the technique simpler but does not provide insight into its use in typical real life situations. In this book we will attempt to be more flexible than standard textbooks so that you will gain experience with commonly encountered difficulties.

In the next section suggested graphical and descriptive statistics are given for several types of data collected for analysis. Note, however, that these suggestions should not be applied rigidly in all situations. They are meant to be a framework for assisting the investigator in analyzing and reporting the data.

## 5.3 APPROPRIATE STATISTICAL MEASURES UNDER STEVENS'S CLASSIFICATION

In Chapter 2 Stevens's system of classifying variables into nominal (naming results), ordinal (determination of greater than or less than),

**TABLE 5.1.** Descriptive Measures Depending Upon Stevens's Scale

| Classification | Graphical Measures | Measures of the Center of a Distribution | Measures of the Variability of a Distribution |
|---|---|---|---|
| Nominal | Bar graphs<br>Pie charts | Mode | Binomial or multinomial variance |
| Ordinal | Histogram | Median | Range<br>$P_{75} - P_{25}$ |
| Interval | Histogram with areas measurable | Mean = $\overline{X}$ | Standard deviation = $S$ |
| Ratio | Histogram with areas measurable | Geometric mean = $\left( \prod_{i=1}^{N} X_i \right)^{1/N}$<br>Harmonic mean = $N / \sum_{i=1}^{N} 1/X_i$ | Coefficient of variation = $S/\overline{X}$ |

interval (equal intervals between successive values of a variable), and ratio (equal intervals and a true zero point) was presented. In this section this system is used to obtain suggested descriptive measures. Table 5.1 shows appropriate graphical and computed measures for each type of variable. It is important to note that the descriptive measures are appropriate to that type of variable listed on the left *and to all below it*. Note also that $\Sigma$ signifies addition, $\Pi$ multiplication, and $P$ percentile in Table 5.1.

## Measures for Nominal Data

For nominal data, the order of the numbers has no meaning. For example, in the depression data set the respondent's religion was coded 1 = Protestant, 2 = Catholic, 3 = Jewish, and 4 = Other. Any other four distinct numbers could be chosen, and their order could be changed without changing the empirical operation of equality. The measures used to describe this data should *not* imply a sense of order.

Suitable graphical measures for nominal data are bar graphs or pie charts. Both of these graphical measures are available from SPSS–X (BARCHART and PIECHART commands) and SAS (CHART procedure). These graphical measures will show the proportion of respondents who have each of the four responses to the religion question. The length of each bar represents the proportion for bar graphs, and the size of the piece of the pie represents the proportion for the pie charts.

The *mode*, or outcome of a variable that occurs most frequently, is the only appropriate measure of the center of the distribution. For a variable such as sex, where only two outcomes are available, the variability, or variance, of the proportion of cases who are male (female) can be measured by the *binomial variance*, which is

$$\text{Estimated Variance of } p = \frac{p(1-p)}{N}$$

where $p$ is the proportion of respondents who are males (females). If more than two outcomes are possible, then the variance of the $i$th proportion is given by

$$\text{Estimated Variance of } p_i = \frac{p_i(1-p_i)}{N}$$

## Measures for Ordinal Data

For ordinal variables order or ranking does have relevance, and so more descriptive measures are available. In addition to the pie charts and bar graphs used for nominal data, histograms can now be used. The area under the histogram still has *no* meaning because the intervals between successive numbers are not necessarily equal. For example, in the depression data set a general health question was asked and later coded 1 = excellent, 2 = good, 3 = fair, and 4 = poor. The distance between 1 and 2 is not necessarily equal to the distance between 3 and 4 when these numbers are used to identify answers to the health question.

An appropriate measure of the center of the distribution is the median. Roughly speaking, the *median* is the value of the variable that half the respondents exceed and half do not. The *range*, or largest minus smallest value occurring, is a measure of how variable or disperse the distribution is. Another measure that is sometimes reported is the difference between two *percentiles*. For example, sometimes the 5th percentile ($P_5$) will be subtracted from the 95th percentile ($P_{95}$). The 5th percentile is the value of the variable that divides the total sample such that 5% of the respondents are below and 95% are above this value. Some investigators prefer to report the *interquartile range*, $P_{75} - P_{25}$, or the *quartile deviation*, $(P_{75} - P_{25})/2$.

Histograms are available from SPSS–X, SAS, or BMDP. The BMDP2D computes the median, the range, and the quartile deviation. The UNIVARIATE procedure of SAS provides the median, the range, and percentiles $P_{75}$, $P_{25}$, and $P_{75} - P_{25}$, as well as any other percentiles. The SPSS–X DESCRIPTIVES and FREQUENCIES procedures provide the median and the range. The percentiles for a particular outcome can be obtained from the SPSS–X FREQUENCIES procedure.

## Measures for Interval Data

For interval data the full range of descriptive statistics generally used are available to the investigator. This set includes graphical measures such as histograms, with the area under the histogram now having meaning. The well-known *mean* and *standard deviation* can now be used also. These descriptive statistics are part of the output in most programs and hence are easily obtainable.

## Measures for Ratio Data

The additional measures available for ratio data are seldom used. The *geometric mean* (GM) is sometimes used when the log transformation is used, since

$$\log \text{GM} = \frac{\sum\limits_{}^{N} \log X}{N}$$

It is also used when computing the mean of a process where there is a constant rate of change. For example, suppose a rapidly growing city has a population of 2500 in 1970 and a population of 5000 according to the 1980 census. An estimate of the 1975 population (or halfway between 1970 and 1980) can be estimated as

$$\text{GM} = \left( \prod^{2} X_i \right)^{1/2} = (2500 \times 5000)^{1/2} = 3525$$

The *harmonic mean* (HM) is the reciprocal of the arithmetic mean of the reciprocals of the data. It is used for obtaining a mean of rates when the quantity in the numerator is fixed. For example, if an investigator wishes to analyze distance per unit of time that $N$ cars require to run a fixed distance, then the harmonic mean should be used.

The *coefficient of variation* can be used to compare the variability of distributions that have different means. It is a unitless statistic.

BMDP1D prints out the coefficient of variation. In SAS the SUMMARY and the MEANS procedures, among others, provide the coefficient of variation. In SPSS–X, the harmonic mean is available in the ONEWAY program.

### Stretching Assumptions

In the data analyses given in the next section, it is the ordinal variables that often cause confusion. Some statisticians treat them as if they were nominal data, often splitting them into two categories if they are dependent variables or using the dummy variables described in Section 9.3 if they are independent variables. Other statisticians treat them as if they were interval data. It is usually possible to assume that the underlying scale is continuous, and that because of a lack of a sophisticated measuring instrument, the investigator is not measuring with an interval scale.

The question really is, How far off is the ordinal scale from an interval scale? If it is close, then using an interval scale makes sense; otherwise, not. Further discussion on this point and a method for converting ordinal or rank variables to interval variables is given in Abelson and Tukey (1959 and 1963).

Although assumptions are sometimes stretched so that ordinal variables can be treated as if they were interval, this stretching should *never* be done with nominal data, because complete nonsense is likely to result.

## 5.4 APPROPRIATE MULTIVARIATE ANALYSES UNDER STEVENS'S CLASSIFICATION

To decide on appropriate analyses, we must classify variables as follows:

**1.** Independent versus dependent
**2.** Nominal or ordinal versus interval or ratio

The classification of independent or dependent may differ from analysis to analysis, but the classification into Stevens's system should remain constant throughout the analysis phase of the study. Once these classifications are determined, it is possible to refer to Table 5.2 and decide what analysis should be considered.

In Table 5.2 nominal and ordinal variables have been combined because this book does not cover analyses appropriate only to nominal or ordinal data separately. An inexpensive summary of measures and tests for these types of variables is given in Reynolds (1977) and Hildebrand, Laing, and Rosenthal (1977). Interval and ratio variables have also been combined because the same analyses are used for both types of variables. There are many measures of association and many statistical methods not listed in the table. For further information on choosing analyses appropriate to various data types, see Gage (1963) or Andrews et al. (1981).

The first row of Table 5.2 includes analyses that can be done if there are no dependent variables. Note that if there is only one variable, it can be considered either dependent or independent. A single independent variable that is either interval or ratio can be screened by methods given in Chapters 3 and 4, and descriptive statistics can be obtained from many statistical programs. If there are several interval or ratio independent variables, then several techniques are listed in the table.

In Table 5.2 the numbers in the parentheses following some techniques refer to the chapters of this book where those techniques are described. For example, to determine the advantages of doing a principal components analysis and how to obtain results and interpret them, you would consult Chapter 11. A very brief description of this technique is also given in Chapter 1. If no number in parentheses is given, then that technique is not discussed in this book.

For interval or ratio dependent variables and nominal or ordinal independent variables, analysis of variance is the appropriate technique. Analysis of variance is not discussed in this book; for discussions of this topic, see Winer (1971); Dunn and Clark (1987); Box, Hunter, and Hunter (1978); or Miller (1986). Multivariate analysis of variance and Hotelling's $T^2$ are discussed in Afifi and Azen (1979) and Morrison (1976). Discussion of multiple-classification analysis can be found in Andrews et al. (1973). Structure models are discussed in Duncan (1975), and log-linear models are presented in Upton (1978). The term log-linear models used here applies to the analysis of nominal and ordinal data. The same term is used in Chapter 13 in connection with one method of survival analysis.

**TABLE 5.2.** Suggested Data Analysis Under Stevens's Classification

| | Independent Variable(s) | | | |
|---|---|---|---|---|
| | Nominal or Ordinal | | Interval or Ratio | |
| Dependent Variable(s) | 1 Variable | >1 Variable | 1 Variable | >1 Variable |
| No dependent variables | $\chi^2$ goodness of fit | Measures of association<br>Log-linear model<br>$\chi^2$ test of independence | Univariate statistics (e.g., one-sample $t$ tests)<br>Descriptive measures (5)<br>Tests for normality (4) | Correlation matrix (7)<br>Principal components (14)<br>Factor analysis (15)<br>Cluster analysis (16) |
| **Nominal or Ordinal** | | | | |
| 1 variable | $\chi^2$ test<br>Fisher's exact test | Log-linear model<br>Logistic regression (12) | Discriminant function (11)<br>Logistic regression (12)<br>Univariate statistics (e.g., two-sample $t$ tests) | Discriminant function (11)<br>Logistic regression (12) |
| >1 variable | Log-linear model | Log-linear model | Discriminant function (11) | Discriminant function (11) |
| **Interval or Ratio** | | | | |
| 1 variable | $t$ test<br>Analysis of variance<br>Survival analysis (13) | Analysis of variance<br>Multiple-classification analysis<br>Survival analysis (13) | Linear regression (6)<br>Correlation (6)<br>Survival analysis (13) | Multiple regression (7, 8, 9)<br>Survival analysis (13) |
| >1 variable | Multivariate analysis of variance<br>Analysis of variance on principal components<br>Hotelling's $T^2$<br>Profile analysis (16) | Multivariate analysis of variance<br>Analysis of variance on principal components | Canonical correlation (10) | Canonical correlation (10)<br>Path analysis<br>Structural models (LISREL, EQS) |

Table 5.2 provides a general guide for what analyses should be *considered*. We do not mean that other analyses couldn't be done but simply that the usual analyses are the ones that are listed. For example, methods of performing discriminant function analyses have been studied for noninterval variables, but this technique was originally derived with interval or ratio data; and the available programs in BMDP, SAS, and SPSS-X are written for interval or ratio data.

Judgment will be called for when the investigator has, for example, five independent variables, three of which are interval, while one is ordinal and one is nominal, with one dependent variable that is interval. Most investigators would use multiple regression, as indicated in Table 5.2. They might pretend that the one ordinal variable is interval and use dummy variables for the nominal variable (see Chapter 9). Another possibility is to categorize all the independent variables and to perform an analysis of variance on the data. That is, analyses that require fewer assumptions in terms of types of variables can always be done. Sometimes, both analyses are done and the results are compared. Because the packaged programs are so simple to run, multiple analyses are a realistic option.

In the examples given in Chapters 6 through 16, the data used will often not be ideal. In some chapters a data set has been created that fits all the usual assumptions to explain the technique, but then a nonideal, real life data set is also run and analyzed. It should be noted that when inappropriate variables are used in a statistical analysis, the association between the statistical results and the real life situation is weakened. However, the statistical models do not have to fit perfectly in order for the investigator to obtain useful information from them.

## SUMMARY

The Stevens system of classification of variables can be a helpful tool in deciding on the choice of descriptive measures as well as in sophisticated data analyses. In this chapter we presented a table to assist the investigator in each of these two areas. A beginning data analyst may benefit from practicing the advice given in this chapter and from consulting more experienced researchers.

The recommended analyses are intended as general guidelines and are by no means exclusive. It is a good idea to try more than one way of analyzing the data whenever possible. Also, special situations may require specific analyses, perhaps ones not covered thoroughly in this book. Some investigators may wish to consult the detailed recommendations given in Andrews et al. (1981).

## BIBLIOGRAPHY

Abelson, R. P., and Tukey, J. W. 1959. Efficient conversion of nonmetric information into metric information. *Proceedings of the Social Statistics Section, American Statistical Association*: 226–230.

*————. 1963. Efficient utilization of non-numerical information in quantitative analysis: General theory and the case of simple order. *Annals of Mathematical Statistics* 34:1347–1369.

Afifi, A. A., and Azen, S. P. 1979. *Statistical analysis: A computer oriented approach*. 2nd ed. New York: Academic Press.

Andrews, F. M., Klem, L., Davidson, T. N., O'Malley, P. M., and Rodgers, W. L. 1981. *A guide for selecting statistical techniques for analyzing social sciences data*. 2nd ed. Ann Arbor: Institute for Social Research, University of Michigan.

Andrews, F. M., Morgan, J., Sonquist, J., and Klem, L. 1973. *Multiple classification analysis*. 2nd ed. Ann Arbor: Institute for Social Research, University of Michigan.

Box, G. E. P., Hunter W. G., and Hunter, J. S. 1978. *Statistics for experimenters*. New York: Wiley.

*Duncan, O. D. 1975. *Introduction to structural equation models*. New York: Academic Press.

Dunn, O. J., and Clark, V. A. 1987. *Applied statistics: Analysis of variance and regression*. New York: Wiley.

Gage, N. L., ed. 1963. *Handbook of research on teaching*. Chicago: Rand McNally.

Hildebrand, D. K., Laing, J. D., and Rosenthal, H. 1977. *Analysis of ordinal data*. Beverly Hills: Sage.

*Joreskog, K. G., and Sorbom, D. 1986. *LISREL VI: Analysis of linear structural relationships by the method of maximum likelihood instrumental variables, and least square methods*. Mooresville, Ind.: Scientific Software.

Miller, R. G., Jr. 1986. *Beyond ANOVA: Basics of applied statistics*. New York: Wiley.

*Morrison, D. F. 1975. *Multivariate statistical methods*. New York: McGraw-Hill.

Reynolds, H. T. 1977. *Analysis of nominal data*. Beverly Hills: Sage.

Stevens, S. S. 1951. Mathematics, measurement and psychophysics. In Stevens, S. S., ed. *Handbook of experimental psychology.* New York: Wiley.

Upton, G. J. G. 1978. *The analysis of cross-tabulated data.* New York: Wiley.

Winer, B. J. 1971. *Statistical principles in experimental design.* 2nd ed. New York: McGraw-Hill.

## PROBLEMS

5.1 Compute an appropriate measure of the center of the distribution for the following variables from the depression data set: MARITAL, INCOME, AGE, and HEALTH.

5.2 An investigator is attempting to determine the health effects on families of living in crowded urban apartments. Several characteristics of the apartment have been measured, including square feet of living area per person, cleanliness, and age of the apartment. Several illness characteristics for the families have been measured also, such as number of infectious diseases and number of bed days per month for each child, and overall health rating for the mother. Suggest an analysis to use with these data.

5.3 A coach has made numerous measurements on successful basketball players, such as height, weight, and strength. He also knows which position each player is successful at. He would like to obtain a function from these data that would predict which position a new player would be best at. Suggest an analysis to use with these data.

5.4 A college admissions committee wishes to predict which prospective student will successfully graduate. To do so, the committee intends to obtain the college grade point averages for a sample of college seniors and compare these with their high school grade point averages and Scholastic Aptitude Test scores. Which analysis should the committee use?

5.5 Data on men and women who have died have been obtained from health maintenance organization records. These data include age at death, height, weight, and several physiological and life-style measurements such as blood pressure, smoking status, dietary intake, and usual amount of exercise. The immediate and underlying causes of death are also available. From these data we would like to find out which variables predict death due to various underlying causes. (This procedure is known as *risk factor analysis.*) Suggest possible analyses.

5.6 Large amounts of data are available from the United Nations and other international organizations on each country and sovereign state of the world, including health, education, and commercial data. An economist would like to invent a descriptive system for the degree of development of each country on the basis of these data. Suggest possible analyses.

5.7 For the data described in Problem 5.6 we wish to put together similar countries into groups. Suggest possible analyses.

5.8   For the data described in Problem 5.6 we wish to relate health data such as infant mortality (the proportion of children dying before the age of one year) and life expectancy (the expected age at death of a person born today if the death rates remain unchanged) to other data such as gross national product per capita, percentage of people older than 15 who can read and write (literacy), average daily caloric intake per capita, average energy consumption per year per capita, and number of persons per practicing physician. Suggest possible analyses. What other variables would you include in your analysis?

5.9   A member of the admissions committee notices that there are several women with high high school grade point averages but low SAT scores. He wonders if this pattern holds for both men and women in general, only for women in general, or only in a few cases. Suggest ways to analyze this problem.

5.10  Two methods are currently used to treat a particular type of cancer. It is suspected that one of the treatments is twice as effective as the other in prolonging survival regardless of the severity of the disease at diagnosis. A study is carried out. After the data are collected, what analysis should the investigators use?

5.11  A psychologist would like to predict whether or not a respondent in the depression study described in Chapter 3 is depressed. To do this, she would like to use the information contained in the following variables: MARITAL, INCOME, and AGE. Suggest analyses.

5.12  Using the data described in Appendix Table B.1, an investigator would like to predict a child's lung function based on that of the parents and the area they live in. What analyses would be appropriate to use?

5.13  In the depression study, information was obtained on the respondent's religion (see Chapter 3). Describe why you think it is incorrect to obtain an average score for religion across the 294 respondents.

# Part Two

# *APPLIED REGRESSION ANALYSIS*

## Chapter Six

# SIMPLE LINEAR REGRESSION AND CORRELATION

## 6.1 WHAT WILL YOU LEARN FROM THIS CHAPTER?

From this chapter you will learn how to examine the relationship between two variables based on a sample. In particular, you will learn:

- Why and where regression and correlation are used (6.2, 6.3).
- The correct meaning of the words *regression* and *correlation* (6.4, 6.5).
- How to obtain the appropriate regression equation and statistics that explain the relationship between the variables (6.4, 6.5).
- How to predict the value of one variable from the value of another (6.4, 6.5).
- How to test appropriate hypotheses regarding the strength of the relationship (6.6, 6.7).
- How to perform a residual analysis in order to validate or improve the derived equation (6.8).
- How and when to use transformations to improve the equation (6.9).

- How and when to use weighted regression (6.10).
- How to correctly use the technique of regression in special contexts such as calibration (6.11).
- How to use regression analysis to analyze and explain paired data (6.11).
- Which are the appropriate computational methods or programs to obtain the numerical results (6.12).
- What to watch out for in performing the analysis (6.13).

If you are reading about simple linear regression for the first time, skip Sections 6.9, 6.10, and 6.11 in your first reading. If this chapter is a review for you, you can skim most of it, but read the above-mentioned sections in detail.

## 6.2 WHEN ARE REGRESSION AND CORRELATION USED?

The methods described in this chapter are appropriate for studying the relationship between two variables $X$ and $Y$. By convention, $X$ is called the *independent variable* and is plotted on the horizontal axis. The variable $Y$ is called the *dependent variable* and is plotted on the vertical axis.

The data for regression analysis can arise in two forms:

1. *Fixed-X case*: The values of $X$ are selected by the researchers or forced on them by the nature of the situation. For example, in the problem of predicting the sales for a company, the total sales are given for each year. Year is the fixed-$X$ variable, and its values are imposed on the investigator by nature. In an experiment to determine the growth of a plant as a function of temperature, a researcher could randomly assign plants to three different preset temperatures that are maintained in three greenhouses. The three temperature values then become the fixed values for $X$.

2. *Variable-X case*: The values of $X$ and $Y$ are both random variables. In this situation, cases are selected randomly from the population, and both $X$ and $Y$ are measured. All survey data are of this type, whereby individuals are chosen and various characteristics are measured on each.

Regression and correlation analysis can be used for either of two main purposes:

**1.** *Descriptive*: The kind of relationship and its strength are examined. This examination can be done graphically or by the use of descriptive equations. Tests of hypotheses and confidence intervals can serve to draw inferences regarding the relationship.
**2.** *Predictive*: The equation relating $Y$ and $X$ can be used to predict the value of $Y$ for a given value of $X$. Prediction intervals can also be used to indicate a likely range of the predicted value of $Y$.

## 6.3 DATA EXAMPLE

In this section we present an example that we use in the remainder of the chapter to illustrate the methods of regression and correlation.

Data were obtained from a sample of factory workers in a particular industry. (Since these are proprietary data, no further identification will be made of their source. The data set will be referred to as the factory workers data.) In these factories management was concerned about possible effects of the working environment on respiratory function.

One of the major early indicators of reduced respiratory function is FEV1 or forced expiration volume in the first second (amount of air exhaled in 1 second). It is known that taller males tend to have a higher FEV1, and we wished to determine the relationship between height and FEV1. A particular age group was chosen (20–24 years old) since it is also known that FEV1 decreases with age in adults. We also excluded females since different relationships exist for women and most of the workers were males. We thus obtained a sample of 87 cases. These data belong in the variable-$X$ case, where $X$ is height (in inches) and $Y$ is FEV1 (in liters). Here we may be concerned with describing the relationship between FEV1 and height, a descriptive purpose. We may also use the resulting equation to determine expected or normal FEV1 for a given height, a predictive use.

In Figure 6.1 a scatter diagram of the data is reproduced as printed by the computer program BMDP6D. In this graph the horizontal axis is divided into 70 units, and the vertical axis into 45 units. The actual values of the observations are rounded off to fit into this grid. For that reason

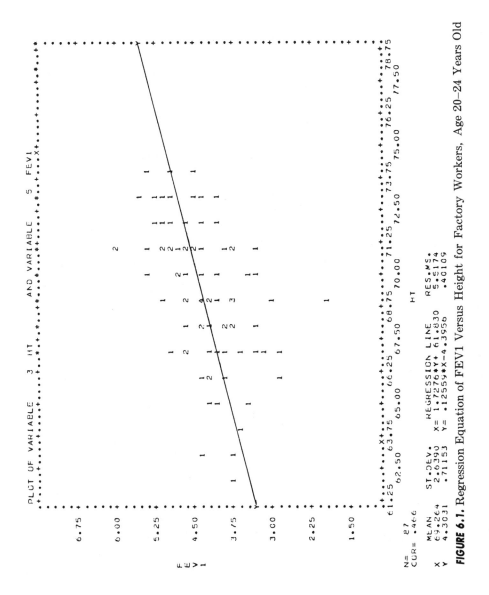

**FIGURE 6.1.** Regression Equation of FEV1 Versus Height for Factory Workers, Age 20–24 Years Old

more observations appear to coincide than is actually the case. The digits on the graph represent the number of points at that grid location. (If more than nine points occur at one location, some programs will use letters of the alphabet to indicate numbers.) A more detailed scatter diagram could be done by a computer-driven plotter or by hand.

Under the scatter plot are printed various statistics. These include $N = 87$ persons, the mean for each variable ($\bar{X} = 69.264$ in. height and $\bar{Y} = 4.303$ L), and the standard deviation ($S_X = 2.639$ in. and $S_Y = 0.7115$ L). In addition, COR $= 0.466$ is the estimated *correlation coefficient* (which will be discussed in Section 6.5).

The *regression line*, which will be defined in Section 6.4, is

$$Y = -4.396 + 0.126X$$

A visual representation of the regression line is obtained by connecting the two $Y$ symbols appearing on the outside border of the grid. Note that FEV1 is an increasing function of height, and the rate of increase (the coefficient of $X$) is 0.126 L/in.

For predictive purposes we would expect a male factory worker, age 20–24 years old, whose height is 5 ft 10 in., or 70 in., to have an FEV1 value of

$$\text{FEV1} = -4.394 + (0.126)(70) = 4.3957$$

or

$$\text{FEV1} = 4.40 \text{ (rounded off)}$$

To take an extreme example, suppose a person is 2 ft tall. Then this equation would predict a negative value of FEV1. This example illustrates the danger of using the regression equation outside the range of concern. A safe policy is to restrict the use of the equation to the range of $X$ observed in the sample.

The *residual mean square* (RES. MS.) in the bottom line of output in Figure 6.1 is an estimate of the variance around the regression line. The square root of this quantity $[S = (0.40109)^{1/2} = 0.63]$ is the estimated standard deviation of the regression line. As will be shown in Section 6.4, this standard deviation can be used to compute confidence intervals and prediction intervals.

The output of Figure 6.1 also includes the estimated regression equation of height on FEV1 (height as the dependent variable and FEV1 as the independent variable). Although the regression of $X$ and $Y$ is not meaningful in this example, it may be useful in other applications such as calibration (discussed in Section 6.11).

## 6.4 DESCRIPTION OF METHODS OF REGRESSION: FIXED-X CASE

In this section we present the background assumptions, models, and formulas necessary for understanding simple linear regression. The theoretical background is simpler for the fixed-$X$ case than for the variable-$X$ case, so we will begin with it.

### Assumptions and Background

For each value of $X$ we conceptualize a distribution of values of $Y$. This distribution is described partly by its mean and variance at each fixed $X$ value:

$$\text{mean of } Y \text{ values at a given } X = \alpha + \beta X$$

and

$$\text{variance of } Y \text{ values at a given } X = \sigma^2$$

The basic idea of *simple linear regression* is that the means of $Y$ lie on a straight line when plotted against $X$. Secondly, the variance of $Y$ at a given $X$ is assumed to be the same for all values of $X$. The latter assumption is called *homoscedasticity*, or homogeneity of variance. Figure 6.2 illustrates the distribution of $Y$ at three values of $X$. Note from Figure 6.2 that $\sigma^2$ is not the variance of all the $Y$'s from their mean but is, instead, the variance of $Y$ at a given $X$. It is clear that the means lie on a straight line in the range of concern and that the variance or the degree of variation is the same at the different values of $X$. Outside the range of concern it is immaterial to our analysis what the curve looks like, and in most practical situations linearity will hold over a limited range of $X$. This figure illustrates that extrapolation of a linear relationship beyond the range of concern can be dangerous.

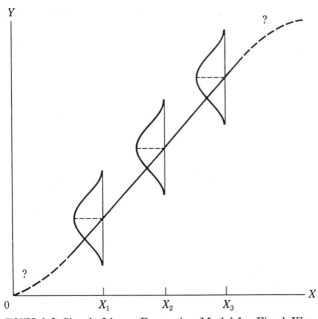

**FIGURE 6.2.** Simple Linear Regression Model for Fixed $X$'s

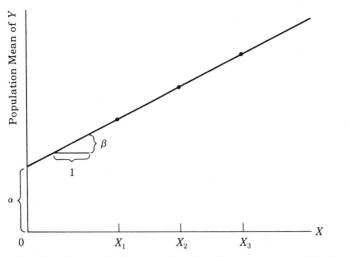

**FIGURE 6.3.** Theoretical Regression Line Illustrating $\alpha$ and $\beta$ Geometrically

The expression $\alpha + \beta X$, relating the mean of $Y$ to $X$, is called the *population least squares regression equation*. Figure 6.3 illustrates the meaning of the parameters $\alpha$ and $\beta$. The parameter $\alpha$ is the *intercept* of this line. That is, it is the mean of $Y$ when $X = 0$. The *slope* $\beta$ is the amount of change in the mean of $Y$ when the value of $X$ is increased by one unit. A negative value of $\beta$ signifies that the mean of $Y$ decreases as $X$ increases.

## Least Squares Method

The parameters $\alpha$ and $\beta$ are estimated from a sample collected according to the fixed-$X$ model. The *sample estimates* of $\alpha$ and $\beta$ are denoted by $A$ and $B$, respectively, and the resulting regression line is called the *sample least squares regression equation*.

To illustrate the method, we consider a hypothetical sample of four points, where $X$ is fixed at $X_1 = 5$, $X_2 = 5$, $X_3 = 10$, and $X_4 = 10$. The sample values of $Y$ are $Y_1 = 14$, $Y_2 = 17$, $Y_3 = 27$, and $Y_4 = 22$. These points are plotted in Figure 6.4.

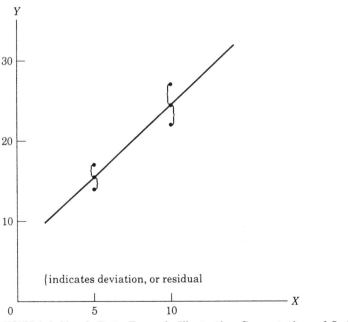

**FIGURE 6.4.** Simple Data Example Illustrating Computations of Output Given in Figure 6.1

The output from the computer would include the following information:

|   | MEAN | ST.DEV. | REGRESSION LINE | RES.MS. |
|---|------|---------|-----------------|---------|
| X | 7.5  | 2.8868  |                 |         |
| Y | 20.0 | 5.7155  | Y = 6.5 + 1.8X  | 8.5     |

The *least squares method* finds the line that minimizes the sum of squared vertical deviations from each point in the sample to the point *on* the line corresponding to the $X$ value. It can be shown mathematically that the least squares line is

$$\hat{Y} = A + BX$$

where

$$B = \frac{\sum (X - \bar{X})(Y - \bar{Y})}{\sum (X - \bar{X})^2}$$

and

$$A = \bar{Y} - B\bar{X}$$

Here $\bar{X}$ and $\bar{Y}$ denote the sample means of $X$ and $Y$, and $\hat{Y}$ denotes the predicted value of $Y$ for a given $X$.

The *deviation* (or *residual*) for the first point is computed as follows:

$$Y(1) - \hat{Y}(1) = 14 - [6.5 + 1.8(5)] = -1.5$$

Similarly, the other residuals are $+1.5$, $-2.5$, and $+2.5$, respectively. The sum of the squares of these deviations is 17.0. No other line can be fitted to produce a smaller sum of squared deviations than 17.0.

The estimate of $\sigma^2$ is called the *residual mean square* (RES. MS.) and is computed as

$$S^2 = \text{RES. MS.} = \frac{\sum (Y - \hat{Y})^2}{N - 2}$$

The number $N - 2$, called the *residual degrees of freedom*, is the sample size minus the number of parameters in the line (in this case, $\alpha$ and $\beta$).

Using $N - 2$ as a divisor in computing $S^2$ produces an *unbiased estimate* of $\sigma^2$. In the example,

$$S^2 = \text{RES. MS.} = \frac{17}{4 - 2} = 8.5$$

The square root of the residual mean square is called the *standard deviation of the estimate* and is denoted by $S$.

Packaged regression programs will also produce the *standard errors* of $A$ and $B$. These statistics are computed as

$$\text{SE}(A) = S\left[\frac{1}{N} + \frac{\bar{X}^2}{\sum(X - \bar{X})^2}\right]^{1/2}$$

and

$$\text{SE}(B) = \frac{S}{[\sum(X - \bar{X})^2]^{1/2}}$$

## Confidence and Prediction Intervals

For each value of $X$ under consideration a population of $Y$ values is assumed to exist. Confidence intervals and tests of hypotheses concerning the intercept, slope, and line may be made with assurance when three assumptions hold: (1) The $Y$ values are assumed to be normally distributed; (2) their means lie on a straight line; and (3) their variances are all equal.

For example, confidence for the slope $B$ can be computed by using the standard error of $B$. The *confidence interval* (CI) for $B$ is

$$\text{CI} = B \pm t \cdot \text{SE}(B)$$

where $t$ is the $100(1 - \alpha/2)$ percentile of the $t$ distribution with $N - 2$ df (*degrees of freedom*); see Appendix Table A.2. Similarly, the confidence interval for $A$ is

$$\text{CI} = A \pm t \cdot \text{SE}(A)$$

where the same degrees of freedom are used for $t$.

The value $\hat{Y}$ computed for a particular $X$ can be interpreted in two ways:

**1.** $\hat{Y}$ is the *point estimate of the mean* of $Y$ at that value of $X$.
**2.** $\hat{Y}$ is the *estimate of the $Y$ value* for any individual with the given value of $X$.

The investigator may supplement these point estimates with interval estimates. The *confidence interval* (CI) *for the mean of $Y$* at a given value of $X$, say $X^*$, is

$$CI = \hat{Y} \pm t \cdot S \left[ \frac{1}{N} + \frac{(X^* - \bar{X})^2}{\sum (X - \bar{X})^2} \right]^{1/2}$$

where $t$ is the $100(1 - \alpha/2)$ percentile of the $t$ distribution with $N - 2$ df (see Appendix Table A.2).

For an individual $Y$ value the confidence interval is called the *prediction interval* (PI). The prediction interval (PI) for an individual $Y$ at $X^*$ is computed as

$$PI = Y \pm tS \left[ 1 + \frac{1}{N} + \frac{(X^* - \bar{X})^2}{\sum (X - \bar{X})^2} \right]^{1/2}$$

where $t$ is the same as for the confidence interval for the mean of $Y$.

In summary, for the fixed-$X$ case we presented the model for simple regression analysis and methods for estimating the parameters of the model. Later in the chapter we will return to this model and present special cases and other uses. Next, we present the variable-$X$ model.

## 6.5 DESCRIPTION OF METHODS OF REGRESSION AND CORRELATION: VARIABLE-X CASE

In this section we present, for the variable-$X$ case, material similar to that given in the previous section.

For this model both $X$ and $Y$ are random variables measured on cases that are randomly selected from a population. One example is the factory workers data set, where FEV1 was predicted from height. The fixed-$X$ regression model applies in this case when we treat the $X$ values as if they were preselected. (This technique is justifiable theoretically by *conditioning* on the $X$ values that happened to be obtained in the sample.) Therefore all the previous discussion and formulas are precisely the same for this case as for the fixed-$X$ case. In addition, since both $X$ and $Y$ are considered random variables, other parameters can be useful for describing the model. These include the means and variances for $X$ and $Y$ over the entire population ($\mu_X$, $\mu_Y$, $\sigma_X^2$, and $\sigma_Y^2$). The sample estimates for these parameters are usually included in computer output. For example, in the analysis of the factory workers data these estimates were printed at the bottom of Figure 6.1.

As a measure of how the variables $X$ and $Y$ vary together, a parameter called the *population covariance* is often estimated. The population covariance of $X$ and $Y$ is defined as the average of the product $(X - \mu_X)(Y - \mu_Y)$ over the entire population. This parameter is denoted by $\sigma_{XY}$. If $X$ and $Y$ tend to increase together, $\sigma_{XY}$ will be positive. If, on the other hand, one tends to increase as the other decreases, $\sigma_{XY}$ will be negative.

So that the magnitude of $\sigma_{XY}$ is standardized, its value is divided by the product $\sigma_X \sigma_Y$. The resulting parameter, denoted by $\rho$, is called the *product moment correlation coefficient*, or simply the *correlation coefficient*. The value of

$$\rho = \frac{\sigma_{XY}}{\sigma_X \sigma_Y}$$

lies between $-1$ and $+1$, inclusive. The sample estimate for $\rho$ is the *sample correlation coefficient r*, or

$$r = \frac{S_{XY}}{S_X S_Y}$$

where

$$S_{XY} = \frac{\sum (X - \bar{X})(Y - \bar{Y})}{N - 1}$$

The sample statistic $r$ also lies between $-1$ and $+1$, inclusive. Further interpretation of the correlation coefficient is given in Thorndike (1978).

Tests of hypotheses and confidence intervals for the variable-$X$ case require that $X$ and $Y$ be jointly normally distributed. Formally, this requirement is that $X$ and $Y$ follow a bivariate normal distribution. Examples of the appearance of bivariate normal distributions are given in Section 6.7. If this condition is true, it can be shown that $Y$ also satisfies the three conditions for the fixed-$X$ case.

## 6.6 INTERPRETATION OF RESULTS: FIXED-X CASE

In this section we present methods for interpreting the results of a regression output.

First, the type of the sample must be determined. If it is a fixed-$X$ sample, the statistics of interest are the intercept and slope of the line and the standard error of the estimate; point and interval estimates for $\alpha$ and $\beta$ have already been discussed.

The investigator may also be interested in *testing hypotheses* concerning the parameters. A commonly used test is for the null hypothesis

$$H_0: \beta = \beta_0$$

The test statistic is

$$t = \frac{(B - \beta_0)[\sum (X - \bar{X})^2]^{1/2}}{S}$$

where $S$ is the square root of the residual mean square and the computed value of $t$ is compared with the tabled $t$ percentiles (Appendix Table A.2) with $N - 2$ degrees of freedom to obtain the $P$ value. Many computer programs will print the standard error of $B$. Then the $t$ statistic is simply

$$t = \frac{B - \beta_0}{\text{SE}(B)}$$

A common value of $\beta_0$ is $\beta_0 = 0$, indicating independence of $X$ and $Y$; i.e., the mean value of $Y$ does not change as $X$ changes.

Tests concerning $\alpha$ can also be performed for the null hypothesis $H_0$: $\alpha = \alpha_0$, using

$$t = \frac{A - \alpha_0}{S\{(1/N) + [\bar{X}^2/\sum(X - \bar{X})^2]\}^{1/2}}$$

Values of this statistic can also be compared with the tabled $t$ percentiles (Appendix Table A.2) with $N - 2$ degrees of freedom to obtain the $t$ value. If the standard error of $A$ is printed by the program, the test statistic can be computed simply as

$$t = \frac{A - \alpha_0}{\text{SE}(A)}$$

For example, to test whether the line passes through the origin, the investigator would test the hypothesis of $\alpha_0 = 0$.

It should be noted that rejecting the null hypothesis $H_0$: $\beta = 0$ is no indication in and of itself of the magnitude of the slope. An observed $B = 0.1$, for instance, might be found significantly different from zero, while a slope of $B = 1.0$ might be considered inconsequential in a particular application. The importance and strength of a relationship between $Y$ and $X$ is a separate question from the question of whether certain parameters are significantly different from zero. The test of the hypothesis $\beta = 0$ is a preliminary step to determine whether the magnitude of $B$ should be further examined. If the null hypothesis is rejected, then the magnitude of the effect of $X$ on $Y$ should be investigated.

One way of investigating the *magnitude of the effect* of a typical $X$ value on $Y$ is to multiply $B$ by $\bar{X}$ and to contrast this result with $\bar{Y}$. If $B\bar{X}$ is small relative to $\bar{Y}$, then the magnitude of the effect of $B$ in predicting $Y$ is small. Another interpretation of $B$ can be obtained by first deciding on two typical values of $X$, say $X_1$ and $X_2$, and then calculating the difference $B(X_2 - X_1)$. This difference measures the change in $Y$ when $X$ goes from $X_1$ to $X_2$.

To infer *causality*, we must justify that all other factors possibly affecting $Y$ have been controlled in the study. One way of accomplishing this control is to design an experiment in which such intervening factors are held fixed while only the variable $X$ is set at various levels. Standard statistical wisdom also requires randomization in the assignment to the various $X$ levels in the hope of controlling for any other factors not accounted for (see Box 1966).

## 6.7 INTERPRETATION OF RESULTS: VARIABLE-X CASE

In this section we present methods for interpreting the results of a regression and correlation output. In particular, we will look at the ellipse of concentration and the coefficient of correlation.

For the variable-$X$ model the regression line and its interpretation remain valid. Strictly speaking, however, causality cannot be inferred from this model. Here we are concerned with the bivariate distribution of $X$ and $Y$. We can safely estimate the means, variances, covariance, and correlation of $X$ and $Y$, i.e., the distribution of pairs of values of $X$ and $Y$ measured on the same individual. (Although these parameter estimates are printed by the computer, they are meaningless in the fixed-$X$ model.) For the variable-$X$ model the interpretations of the means and variances of $X$ and $Y$ are the usual measures of location (center of the distribution) and variability. We will now concentrate on how the *correlation coefficient* should be interpreted.

### Ellipse of Concentration

The *bivariate distribution* of $X$ and $Y$ is best interpreted by a look at the scatter diagram from a random sample. If the sample comes from a bivariate normal distribution, the data will tend to cluster around the means of $X$ and $Y$ and will approximate an ellipse called the *ellipse of concentration*. Note in Figure 6.1 that the points representing the data could be enclosed by an ellipse of concentration.

An ellipse can be characterized by the following:

1. The center.
2. The major axis, i.e., the line going from the center to the farthest point on the ellipse.
3. The minor axis, i.e., the line going from the center to the nearest point on the ellipse (the minor axis is always perpendicular to the major axis).

**4.** The ratio of the length of the minor axis to the length of the major axis. If this ratio is small, the ellipse is thin and elongated; otherwise, the ellipse is fat and round-shaped.

## Interpreting the Correlation Coefficient

For ellipses of concentration the center is at the point defined by the means of $X$ and $Y$. The directions and lengths of the major and minor axes are determined by the two variances and the correlation coefficient. For fixed values of the variances the ratio of the length of the minor axis to that of the major axis, and hence the shape of the ellipse, is determined by the correlation coefficient $\rho$.

In Figure 6.5 we represent ellipses of concentration for various bivariate normal distributions in which the means of $X$ and $Y$ are both zero and the variances are both one (standardized $X$'s and $Y$'s). The case $\rho = 0$, Figure 6.5a, represents independence of $X$ and $Y$. That is, the value of one variable has no effect on the value of the other, and the ellipse is a perfect circle. Higher values of $\rho$ correspond to more elongated ellipses, as indicated in Figures 6.5b through 6.5e.

We see that for very high values of $\rho$ one variable conveys a lot of information about the other. That is, if we are given a value of $X$, we can guess the corresponding $Y$ value quite accurately. We can do so because the range of the possible values of $Y$ for a given $X$ is determined by the width of the ellipse at that value of $X$. This width is small for a large value of $\rho$. For negative values of $\rho$ similar ellipses could be drawn where the major axis (long axis) has a negative slope, i.e., in the northwest/southeast direction.

Another interpretation of $\rho$ stems from the concept of the *conditional distribution*. For a specific value of $X$ the distribution of the $Y$ value is called the conditional distribution of $Y$ given that value of $X$. The word *given* is translated symbolically by a vertical line, so $Y$ given $X$ is written as $Y|X$. The variance of the conditional distribution of $Y$, variance $(Y|X)$, can be expressed as

$$\text{variance}\,(Y|X) = \sigma_Y^2(1 - \rho^2)$$

or

$$\sigma_{Y|X}^2 = \sigma_Y^2(1 - \rho^2)$$

Note that $\sigma_{Y|X} = \sigma$ as given in Section 6.4.

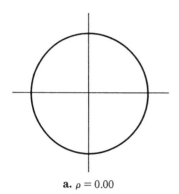

**a.** $\rho = 0.00$

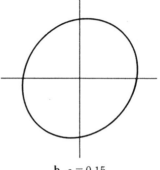

**b.** $\rho = 0.15$

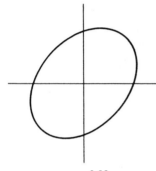

**c.** $\rho = 0.33$

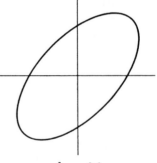

**d.** $\rho = 0.6$

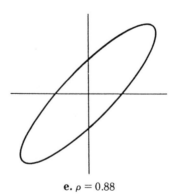

**e.** $\rho = 0.88$

**FIGURE 6.5.** Hypothetical Ellipses of Concentration for Various $\rho$ Values

This equation can be written in another form, as

$$\rho^2 = \frac{\sigma_Y^2 - \sigma_{Y|X}^2}{\sigma_Y^2}$$

The term $\sigma_{Y|X}^2$ measures the variance of $Y$ when $X$ has a specific fixed value. Therefore this equation states that $\rho^2$ is the proportion of variance of $Y$ reduced because of knowledge of $X$. This result is often loosely expressed by saying that $\rho^2$ is the proportion of the variance of $Y$ "explained" by $X$.

A better interpretation of $\rho$ is to note that

$$\frac{\sigma_{Y|X}}{\sigma_Y} = (1 - \rho^2)^{1/2}$$

This value is a measure of the proportion of the standard deviation of $Y$ not explained by $X$. For example, if $\rho = \pm 0.8$, then 64% of the variance of $Y$ is explained by $X$. However, $(1 - 0.8^2)^{1/2} = 0.6$, saying that 60% of the standard deviation of $Y$ is not explained by $X$. Since the standard deviation is a better measure of variability than the variance, it is seen that when $\rho = 0.8$, more than half of the variability of $Y$ is still not explained by $X$. If instead of using $\rho^2$ from the population we use $r^2$ from a sample, then

$$S_{Y|X}^2 = S^2 = \left(\frac{N-1}{N-2}\right)S_Y^2(1 - r^2)$$

and the results must be adjusted for sample size.

An important property of the correlation coefficient is that its value is not affected by the units of $X$ or $Y$ or any linear transformation of $X$ or $Y$. For instance, $X$ was measured in inches in the example shown in Figure 6.1, but the correlation between height and FEV1 is the same if we change the units of height to centimeters or the units of FEV1 to milliliters. In general, adding (subtracting) a constant to either variable or multiplying either variable by a constant will not alter the value of the correlation. Since $\hat{Y} = A + BX$, it follows that the correlation between $Y$ and $\hat{Y}$ is the same as that between $Y$ and $X$.

If we make the additional assumption that $X$ and $Y$ have a bivariate normal distribution, then it is possible to test the null hypothesis $H_0: \rho = 0$

by computing the test statistic

$$t = \frac{r(N-2)^{1/2}}{(1-r^2)^{1/2}}$$

with $N-2$ degrees of freedom. For the factory workers example of Figure 6.1, $r = 0.466$. To test the hypothesis $H_0: \rho = 0$ versus the alternative $H_1: \rho > 0$, we compute

$$t = \frac{0.466(87-2)^{1/2}}{(1-0.466^2)^{1/2}} = 4.86$$

with 85 df. This statistic results in $P < 0.001$, and the observed $r$ is significantly greater than zero.

Tests of null hypotheses other than $\rho = 0$ and confidence intervals for $\rho$ can be found in many textbooks (see Brownlee 1965; Dunn and Clark 1987; Afifi and Azen 1979). As before, a test of $\rho = 0$ should be made before attempting to interpret the magnitude of the sample correlation coeffi-

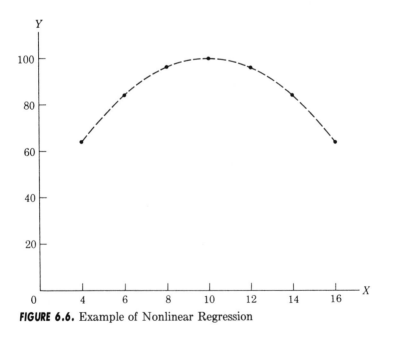

**FIGURE 6.6.** Example of Nonlinear Regression

cient. Note that the test of $\rho = 0$ is equivalent to the test of $\beta = 0$ given earlier.

All of the above interpretations were made with the assumption that the data follow a bivariate normal distribution, which implies that the mean of $Y$ is related to $X$ in a linear fashion. If the regression of $Y$ on $X$ is nonlinear, it is conceivable that the sample correlation coefficient is near zero when $Y$ is, in fact, strongly related to $X$. For example, in Figure 6.6 we can quite accurately predict $Y$ from $X$ [in fact, the points fit a curve $Y = 100 - (X - 10)^2$ exactly]. However, the sample correlation coefficient is $r = 0.0$. (An appropriate regression equation can be fitted by the techniques of polynomial regression; see Section 7.8. Also, in Section 6.9 we discuss the role of transformations in reducing nonlinearities.)

## 6.8 FURTHER EXAMINATION OF COMPUTER OUTPUT

Most packaged regression programs include other useful statistics in the printed output. To obtain some of this output, the investigator usually has to run one of the multiple regression programs discussed in Chapter 7. In this section we introduce some of these statistics in the context of simple linear regression.

### Standardized Regression Coefficient

The *standardized regression coefficient* is the slope in the regression equation if $X$ and $Y$ are standardized. Standardization of $X$ and $Y$ is achieved by subtracting the respective means from each set of observations and dividing the differences by the respective standard deviations. The resulting set of standardized sample values will have a mean of zero and a standard deviation of one for both $X$ and $Y$. After standardization the intercept in the regression equation will be zero, and for simple linear regression (one $X$ variable) the standardized slope will be equal to the correlation coefficient $r$. In multiple regression, where several $X$ variables are used, the standardized regression coefficients help quantify the relative contribution of each $X$ variable (see Chapter 7).

### Analysis of Variance Table

The test for $H_0$: $\beta = 0$ was discussed in Section 6.6 using the $t$ statistic. This test allows one-sided or two-sided alternatives. When the two-sided

**TABLE 6.1.** ANOVA Table for Simple Linear Regression

| Source of Variation | Sums of Squares | df | Mean Square | F |
|---|---|---|---|---|
| Regression | $\Sigma\,(\hat{Y} - \overline{Y})^2$ | 1 | $SS_{reg}/1$ | $MS_{reg}/MS_{res}$ |
| Residual | $\Sigma\,(Y - \hat{Y})^2$ | $N - 2$ | $SS_{res}/(N - 2)$ | |
| Total | $\Sigma\,(Y - \overline{Y})^2$ | $N - 1$ | | |

**TABLE 6.2.** ANOVA Example from Figure 6.1

| Source of Variation | Sums of Squares | df | Mean Square | F |
|---|---|---|---|---|
| Regression | 9.4469 | 1 | 9.4469 | 23.55 |
| Residual | 34.0927 | 85 | 0.40109 | |
| Total | 43.5396 | 86 | | |

alternative is chosen, it is possible to represent the test in the form of an *analysis of variance (ANOVA) table.* A typical ANOVA table is represented in Table 6.1. If $X$ were useless in predicting $Y$, our best guess of the $Y$ value would be $\overline{Y}$ regardless of the value of $X$. To measure how different our fitted line $\hat{Y}$ is from $\overline{Y}$, we calculate the sums of squares for regression as $\Sigma\,(\hat{Y} - \overline{Y})^2$, summed over each data point. (Note that $\overline{Y}$ is the average of all the $\hat{Y}$ values.) The residual mean square is a measure of how poorly or how well the regression line fits the actual data points. A large residual mean square indicates a poor fit. The $F$ ratio is, in fact, the squared value of the $t$ statistic described in Section 6.6 for testing $H_0$: $\beta = 0$.

Table 6.2 shows the ANOVA table for the factory workers data example. Note that the $F$ ratio of 23.55 is the square of the $t$ value of 4.86 computed previously. Also, the residual mean square (0.401) is the same as that given in Figure 6.1.

## Data Screening in Simple Linear Regression

For simple linear regression, a scatter diagram such as that shown in Figure 6.1 is one of the best tools for determining whether or not the data fit the basic model. Most researchers find it simplest to examine the plot of $Y$ against $X$. Alternatively, the residuals $e = Y - \hat{Y}$ can be plotted against $X$. Table 6.3 shows the data for the hypothetical example presented in Figure 6.4. Also shown are the predicted values $\hat{Y}$ and the residuals $e$. Note that, as expected, the mean of the $\hat{Y}$ values is equal to $\bar{Y}$. Also, it will always be the case that the mean of the residuals is zero. The variance of the residuals from the same regression line will be discussed later in this section.

Examples of three possible scatter diagrams and residual plots are illustrated in Figure 6.7. In Figure 6.7a, the idealized bivariate $(X, Y)$ normal distribution model is illustrated, using contours similar to those in Figure 6.5. In the accompanying residual plot, the residuals plotted against $X$ would also approximate an ellipse. An investigator could make several conclusions from Figure 6.7a. One important conclusion is that no evidence exists against the *linearity of the regression* of $Y$ or $X$. Also, there is no evidence for the existence of outliers (discussed later in this section). In addition, the normality assumption used when confidence intervals are calculated or statistical tests are performed is not obviously violated.

In Figure 6.7b the ellipse is replaced by a fan-shaped figure. This shape suggests that as $X$ increases, the standard deviation of $Y$ also increases. Note that the assumption of linearity is not obviously violated but that the assumption of homoscedasticity does not hold. In this case, the use of weighted least squares is recommended (see Section 6.10).

In Figure 6.7c the ellipse is replaced by a crescent-shaped form, indicating that the regression of $Y$ is not linear in $X$. One possibility for solving this problem is to fit a quadratic equation as discussed in Chapter 7. Another possibility is to transform $X$ into log $X$ or some other function

**TABLE 6.3.** Hypothetical Data Example from Figure 6.4

| i | X | Y | $\hat{Y} = 6.5 + 1.8X$ | $e = Y - \hat{Y} =$ Residual |
|---|---|---|---|---|
| 1 | 5 | 14 | 15.5 | −1.5 |
| 2 | 5 | 17 | 15.5 | 1.5 |
| 3 | 10 | 27 | 24.5 | 2.5 |
| 4 | 10 | 22 | 24.5 | −2.5 |
| Mean | 7.5 | 20 | 20 | 0 |

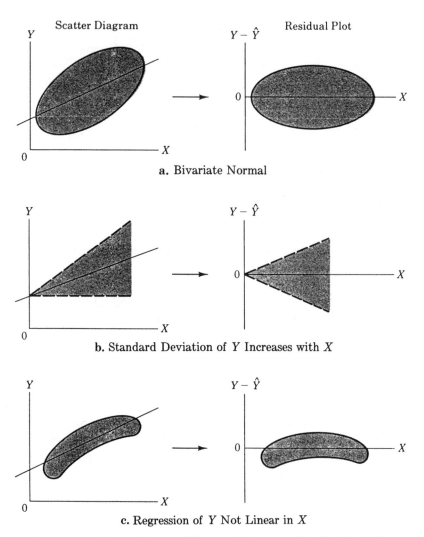

**FIGURE 6.7.** Hypothetical Scatter Plots and Corresponding Residual Plots

of $X$, and then fit a straight line to the transformed $X$ values and $Y$. This concept will be discussed in Section 6.9.

Formal tests exist for testing the linearity of the simple regression equation when multiple values of $Y$ are available for at least some of the $X$ values. However, most investigators assess linearity by plotting either $Y$ against $X$ or $Y - \hat{Y}$ against $X$. The scatter diagram with $Y$ plotted against

$X$ is often simpler to interpret. Lack of fit to the model may be easier to assess from the residual plots as shown on the right-hand side of Figure 6.7. Since the residuals always have a zero mean, it is useful to draw a horizontal line through the zero point on the vertical axis as has been done in Figure 6.7. This will aid in checking whether the residuals are symmetric around their mean (which is expected if the residuals are normally distributed). Unusual clusters of points can alert the investigator to possible anomalies in the data.

The distribution of the residuals about zero is made up of two components. One is called a random component and reflects incomplete prediction of $Y$ from $X$ and/or imprecision in measuring $Y$. But if the linearity assumption is not correct, then a second component will be mixed with the first, reflecting lack of fit to the model. Most formal analyses of residuals only assume that the first component is present. In the following discussion it will be assumed that there is no appreciable effect of non-linearity in the analysis of residuals. It is important that the investigator assess the linearity assumption if further detailed analysis of residuals is performed.

Most multiple regression programs provide lists and plots of the residuals. The investigator can either use these programs, even though a simple linear regression is being performed, or obtain the residuals by using the transformation $Y - (A + BX) = Y - \hat{Y} = e$ and then proceed with the plots.

The raw sample regression residuals $e$ have unequal variances and are slightly correlated. Their magnitude depends on the variation of $Y$ about the regression line. Most multiple regression programs provide numerous adjusted residuals that have been found useful in regression analysis. For simple linear regression, the various forms of residuals have been most useful in drawing attention to important *outliers*. The detection of outliers was discussed in Chapter 3 and the simplest of the techniques discussed there, plotting of histograms, can be applied to residuals. That is, histograms of the residual values can be plotted and examined for extremes.

In recent years, procedures for the detection of outliers in regression programs have focused on three types of outliers: (1) outliers in $Y$ from the regression line, (2) outliers in $X$, and (3) outliers that have a large influence on the estimate of the slope coefficient (see Chatterjee and Hadi 1988). The programs include numerous types of residuals and other statistics so that the user can detect these three types of outliers. The more commonly used ones will be discussed here (see Belsley, Kuh, and Welsh 1980; Cook and Weisberg 1982; or Chatterjee and Hadi 1988 for a more detailed discussion). A listing of the options in packaged programs is deferred to Section 7.10 since they are mostly available in the multiple regression programs. The discussion is presented in this chapter since plots and formulas are easier to understand for the simple linear regression model.

## Outliers in Y

Since the sample residuals do not all have the same variance and their magnitude depends on how closely the points lie to the straight line, they are often simpler to interpret if they are standardized. If we analyze only a single variable, $Y$, then it can be standardized by computing $(Y - \bar{Y})/S_Y$ which will have a mean zero and a standard deviation of one. As discussed in Chapter 3, formal tests for detection of outliers are available but often researchers simply investigate all standardized residuals that are larger than a given magnitude, say three. This general rule is based on the fact that, if the data are normally distributed, the chances of getting a value greater in magnitude than three is very small. This simple rule does not take sample size into consideration but is still widely used. A comparable standardization for a residual in regression analysis would be $(Y - \hat{Y})/S$ or $e/S$, but this does not take the unequal variance into account. The adjusted residual that accomplishes this is usually called a *studentized residual*:

$$\text{Studentized residual} = \frac{e}{S\sqrt{1 - h}},$$

where $h$ is a quantity called *leverage* (to be defined later when outliers in $X$ are discussed). Note that this residual is called a standardized residual in BMDP programs. (Standard nomenclature in this area is yet to be finalized and it is safest to read the description in the computer program used to be sure of the definition.) In addition, since a single outlier can greatly affect the regression line (particularly in small samples), what are often called *deleted* studentized residuals can be obtained. These residuals are computed in the same manner as studentized residuals, with the exception that the $i$th deleted studentized residual is computed from a regression line fitted to all but the $i$th observation. Deleted residuals have the advantage of removing the effect of an extreme outlier in assessing the magnitude of that outlier. Such residuals are called RSTUDENT in REG procedure in SAS, DRESID in SPSS–X, and DSTRESID (deleted standardized residuals) in the BMDP2R program. In simple linear regression, if the residuals are normally distributed, the deleted studentized residual follows the student-$t$ distribution with $N - 2$ degrees of freedom. These and other types of residuals can either be obtained in the form of plots of the desired residual against $\hat{Y}$ or $X$, or in lists. The plots are useful for large samples or quick scanning. The lists of the residuals (sometimes accompanied by a simple plot of their distance from zero for each observation) are useful for identifying the actual observation that has a large residual in $Y$.

Once the observations that have large residuals are identified, the researcher can examine those cases more carefully to decide whether to leave them in, correct them, or declare them missing values.

### Outliers in X

Possible outliers in $X$ are measured by statistics called *leverage* statistics. One measure of leverage for simple linear regression is called $h$, where

$$h = \frac{1}{N} + \frac{(X - \bar{X})^2}{\sum (X - \bar{X})^2}$$

When $X$ is far from $\bar{X}$, then leverage is large and vice versa. The size of $h$ is limited to the range

$$\frac{1}{N} < h < 1$$

The leverage $h$ for the $i$th observation tells how much $Y$ for that observation contributes to $\hat{Y}$. If we change $Y$ by a quantity $\Delta Y$, then $h \Delta Y$ is the resulting change in $\hat{Y}$. Observations with large leverages possess the potential for having a large effect on the slope of the line. Figure 6.8 includes some observations that illustrate the difference between points that are outliers in $X$ and in $Y$. Point 1 is an outlier in $Y$ (large residual), but has low leverage since $X$ is close to $\bar{X}$. It will affect the estimate of the intercept but not the slope. It will tend to increase the estimate of $S$ and

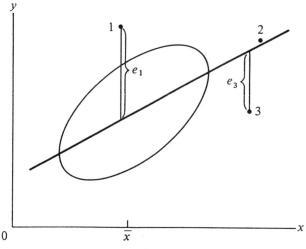

**FIGURE 6.8.** Illustration of the Effect of Outliers

hence the standard error of the slope of $B$. Point 2 has high leverage, but will not affect the estimate of the slope much because it is not an outlier in $Y$. Point 3 is both a high leverage point and an outlier in $Y$, so it will tend to reduce the value of the slope coefficient $B$ and to tip the regression line downward. It will also affect the estimate of $A$ and $S$, and thus have a large effect on the statements made concerning the regression line. Note that this is true even though the residual $e$ is less for point 3 than it is for point 1. Thus, looking solely at residuals in $Y$ may not tell the whole story and leverage statistics are important to examine if outliers are a concern.

## Influential Observations

A direct approach to the detection of outliers is to determine the *influence* of each observation on the slope coefficient $B$. Cook (1977) derived a function called Cook's distance, which provides a scaled distance between the value of $B$ when all observations are used and $B(-i)$, the slope when the $i$th observation is omitted. This distance is computed for each of the $N$ observations. Observations resulting in large values of Cook's distance should be examined as possible influential points or outliers. Cook (1977) suggests comparing the Cook's distance with the percentiles of the $F$ distribution with $P + 1$ and $N - P - 1$ degrees of freedom. Here, $P$ is the number of independent variables. Cook's distances exceeding the 95th percentile are recommended for careful examination and possible removal from the data set.

Other distance measures have been proposed (see Chatterjee and Hadi 1988). The modified Cook's distance measures the effect of an observation on both the slope $B$ and on the variance $S^2$. Welsch–Kuh's distance measure (also called DFFITS) is the scaled distance between $\hat{Y}$ and $\hat{Y}(-i)$, i.e., $\hat{Y}$ derived with the $i$th observation deleted. Large values of DFFITS also indicate an influential observation. DFFITS tends to measure the influence on $B$ and $S^2$, simultaneously.

A general lesson from the research work on outliers in regression analysis is that, when one examines either the scatter diagram of $Y$ versus $X$ or the plot of the residuals versus $X$, more attention should be given to the points that are outliers in both $X$ and $Y$ than to those that are only outliers in $Y$.

## Lack of Independence Among Residuals

When the observations can be ordered in time or place, plots of the residuals similar to those given in Figure 4.3 can be made and the discussion in Section 4.4 applies here directly. If the observations are independent, then successive residuals should not be appreciably correlated. The *serial* correlation, which is simply the correlation between successive residuals, can be used to assess lack of independence. For a sufficiently large $N$, the significance levels for the usual test $\rho = 0$ apply approximately to the serial correlation. Another test statistic available in some packaged programs is the Durbin–Watson statistic. The Durbin–Watson statistic is approximately equal to $2(1 -$ serial correlation$)$. Thus when the serial correlation is 0, the Durbin–Watson statistic is close to 2. Tables of significance levels can be found in Thiel (1971). The Durbin–Watson statistic is used to test whether the serial correlation is zero when it is assumed that the correlation between successive residuals is restricted to a correlation between immediately adjacent residuals.

## Normality of Residuals

Some regression programs provide normal probability plots of the residuals to enable the user to decide whether the data approximate a normal distribution. If the residuals are not normally distributed, then the distribution of $Y$ at each value of $X$ is not normal.

For simple linear regression with the variable $X$ model, many researchers assess bivariate normality by examining the scatter diagram of $Y$ versus $X$ to see if the points approximately fall within an ellipse.

## 6.9 ROBUSTNESS AND TRANSFORMATIONS FOR REGRESSION ANALYSIS

In this section we define the concept of robustness in statistical analysis, and we discuss the role of transformations in regression and correlation.

### Robustness and Assumptions

Regression and correlation analysis make certain assumptions about the population from which the data were obtained. A *robust analysis* is one that is useful even though all the assumptions are not met. For the purpose of fitting a straight line, we assume that the $Y$ values are normally distributed, the population regression equation is linear in the range of concern, and the variance of $Y$ is the same for all values of $X$. Linearity can be checked graphically, and transformations can help straighten out a nonlinear regression line.

The *assumption of homogeneity of variance* is not crucial for the resulting least squares line. In fact, the least squares estimates of $\alpha$ and $\beta$ are unbiased whether or not the assumption is valid. However, if glaring irregularities of variance exist, weighted least squares can improve the fit. In this case the weights are chosen to be proportional to the inverse of the variance. For example, if the variance is a linear function of $X$, then the weight is $1/X$.

The *assumption of normality* of the $Y$ values of each value of $X$ is made only when tests of hypotheses are performed or confidence intervals are calculated. It is generally agreed in the statistical literature that slight departures from this assumption do not appreciably alter our inferences if the sample size is sufficiently large.

The *lack of randomness* in the sample can seriously invalidate our inferences. Confidence intervals are often optimistically narrow because the sample is not truly a random one from the whole population to which we wish to generalize.

In all of the preceding analyses *linearity* of the relationship between $X$ and $Y$ was assumed. Thus careful examination of the scatter diagram should be the first step in any regression analysis. It is advisable to explore various transformations of $Y$ and/or $X$ if nonlinearity of the original measurements is apparent.

## Transformations

The subject of *transformations* has been discussed in detail in the literature; for examples, see Hald (1952) and Mosteller and Tukey (1977). In this subsection we present some typical graphs of the relationship between $Y$ and $X$ and some practical transformations.

In Chapter 4 we discussed the effects of transformations on the frequency distribution. There it was shown that taking the logarithm or square root of a number condensed the magnitude of larger numbers and stretched the magnitude of values less than one. Conversely, raising a number to a power greater than one stretches the large values and condenses the values less than one. These properties are useful in selecting the appropriate transformation to straighten out a nonlinear graph of one variable as a function of another.

Mosteller and Tukey (1977) present typical regression curves that are not linear in $X$ as belonging to one of the quadrants of Figure 6.9a. A very common case is illustrated in Figure 6.9b, which is represented by the fourth quadrant of the circle in Figure 6.9a. For example, the curve in Figure 6.9b might be made linear by transforming $X$ to $\log X$, to $-1/X$, or to $X^{1/2}$. Another possibility would be to transform $Y$ to $Y^2$. The other three cases are also indicated in Figure 6.9a. The remaining quadrants are interpreted in a similar fashion.

Other transformations could also be attempted, such as powers other than those indicated. It may also be useful to first add or subtract a constant from all values of $X$ or $Y$ and then take a power or logarithms. For example, sometimes taking $\log X$ does not straighten out the curve sufficiently. Subtracting a constant $C$ (which must be smaller than the smallest $X$ value) and then taking the logarithm has a greater effect.

The availability of packaged programs greatly facilitates the choice of an appropriate transformation. New variables can be created that are functions of the original variables, and scatter diagrams can be obtained of the new transformed variables. Visual inspection will often indicate the best transformation. Also, the magnitude of the correlation coefficient $r$ will indicate the best linear fit since it is a measure of linear association.

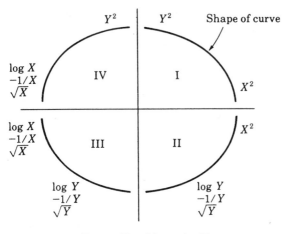

**a.** Curves Not Linear in $X$

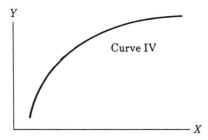

**b.** Detail of Fourth Quadrant

**FIGURE 6.9.** Choice of Transformation: Typical Curves and Appropriate Transformation

Attention should also be paid to transformations that are commonly used in the field of application and that have a particular scientific basis or physical rationale.

Once the transformation is selected, all subsequent estimates and tests are performed in terms of the transformed values. Since the variable to be predicted is usually the dependent variable $Y$, transforming $Y$ can complicate the interpretation of the resulting regression equation more than if $X$ is transformed.

For example, if log $X$ is used instead of $X$, the resulting equation is

$$Y = A + B \log_{10} X$$

This equation presents no problems in interpreting the predicted values of $Y$, and most investigators accept the transformation of $\log_{10} X$ as reasonable in certain situations.

However, if log $Y$ is used instead of $Y$, the resulting equation is

$$\log_{10} Y = A + BX$$

Then the predicted value of $Y$, say $Y^*$, must be detransformed, that is,

$$Y^* = 10^{A+BX}$$

Thus slight biases in fitting log $Y$ could be detransformed into large biases in predicting $Y$. For this reason most investigators look for transformations of $X$ first.

## 6.10 OTHER OPTIONS IN COMPUTER PROGRAMS

In this section we discuss two options available from computer programs: regression through the origin and weighted regression.

### Regression Through the Origin

Sometimes an investigator is convinced that the *regression line* should pass *through the origin*. In this case the appropriate model is simply the mean of

$$Y = \beta X$$

That is, the intercept is forced to be zero. The programs usually give the option of using this model and estimate $\beta$ as

$$B = \frac{\sum XY}{\sum X^2}$$

To test $H_0$: $\beta = \beta_0$, the test statistic is

$$t = \frac{B - \beta_0}{S/(\sum X^2)^{1/2}}$$

where

$$S = \sqrt{\frac{\sum (Y - BX)^2}{N - 1}}$$

and $t$ has $N - 1$ degrees of freedom.

## Weighted Least Squares Regression

The investigator may also request a *weighted least squares regression line*. In weighted least squares each observation is given an individual weight reflecting its importance or degree of variability. There are three common situations in which a weighted regression line is appropriate:

1. The variance of the distribution at a given $X$ is a function of the $X$ value. An example of this situation was shown in Figure 6.7b.

2. Each $Y$ observation is, in fact, the mean of several determinations, and that number varies from one value of $X$ to another.

3. The investigator wishes to assign different levels of importance to different points. For example, data from different countries could be weighted either by the size of the population or by the perceived accuracy of the data.

Formulas for weighted least squares are discussed in Dunn and Clark (1987) and Brownlee (1965). In case 1 the weights are the inverse of the variances of the point. In case 2 the weights are the number of determinations at each $X$ point. In case 3 the investigator must make up numerical weights to reflect the perception of importance or accuracy of the data points.

In weighted least squares regression the estimates of $\alpha$ and $\beta$ and other statistics are adjusted to reflect these special characteristics of the observations. In most situations the weights will not affect the results appreciably unless they are quite different from each other. Since it is considerably more work to compute a weighted least squares regression equation, it is recommended that one of the computer programs listed in Section 6.12 be used, rather than hand calculations.

## 6.11 SPECIAL APPLICATIONS OF REGRESSION

In this section we discuss some important applications of regression analysis that require some caution when the investigator is using regression techniques.

### Calibration

A common situation in laboratories or industry is one in which an instrument that is designed to measure a certain characteristic needs calibration. For example, the concentration of a certain chemical could be measured by an instrument. For calibration of the instrument several compounds with known concentrations, denoted by $X$, could be used, and the measurements, denoted by $Y$, could be determined by the instrument for each compound. As a second example, a costly or destructive method of measurement that is very accurate, denoted by $X$, could be compared with another method of measurement, denoted by $Y$, that is a less costly or nondestructive method. In either situation more than one determination of the level of $Y$ could be made for each value of $X$.

The object of the analysis is to derive, from the calibration data, a formula or a graph that can be used to predict the unknown value of $X$ from the measured $Y$ value in future determinations. We will present two methods (indirect and direct) that have been developed and a technique for choosing between them.

1. *Indirect method*: Since $X$ is assumed known with little error and $Y$ is the random variable, this classical method begins with finding the usual regression of $Y$ on $X$,

$$\hat{Y} = A + BX$$

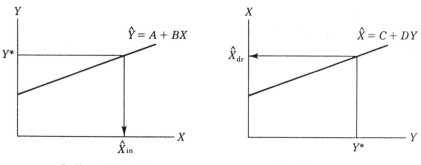

a. Indirect Method          b. Direct Method

**FIGURE 6.10.** Illustration of Indirect and Direct Calibration Methods

Then for a determination $Y^*$ the investigator obtains the indirect estimate of $X$, $\hat{X}_{in}$, as

$$\hat{X}_{in} = \frac{Y^* - A}{B}$$

This method is illustrated in Figure 6.10a.

**2.** *Direct method*: Here we pretend that $X$ is the dependent variable and $Y$ is the independent variable and fit the regression of $X$ on $Y$ as illustrated in Figure 6.10b. We denote the estimate of $X$ by $\hat{X}_{dr}$; thus

$$\hat{X}_{dr} = C + DY^*$$

To compare the two methods, we compute the quantities $\hat{X}_{in}$ and $\hat{X}_{dr}$ for all of the data in the sample. Then we compute the correlation between $\hat{X}_{in}$ and $X$, denoted by $r(in)$, and the correlation between $\hat{X}_{dr}$ and $X$, denoted by $r(dr)$. We then use the method that results in the higher correlation. Further discussion of this problem is found in Krutchkoff (1967), Berkson (1969), Shukla (1972), and Lwin and Maritz (1982).

It is advisable, before using either of these methods, that the investigator test whether the slope $B$ is significantly different from zero. This test can be done by the usual $t$ test. If the hypothesis

$H_0: \beta = 0$ is not rejected, this result is an indication that the instrument should not be used, and the investigator should not use either equation.

## Forecasting

*Forecasting* is the technique of analyzing historical data in order to provide estimates of the future values of certain variables of interest. If data taken over time $(X)$ approximate a straight line, forecasters might assume that this relationship would continue in the future and would use simple linear regression to fit this relationship. This line is extended to some point in the future. The difficulty in using this method lies in the fact that we are never sure that the linear trend will continue in the future. At any rate, it is advisable not to make long-range forecasts by using this method. Further discussion and additional methods are found in Montgomery and Johnson (1976), Levenbach and Cleary (1981), and Chatfield (1984).

## Paired Data

Another situation where regression analysis may be appropriate is the *paired-sample case*. A paired sample consists of observations taken at two time points (pre- and post-) on the same individual or consists of matched pairs, where one member of the pair is subject to one treatment and the other is a control or is subject to another treatment.

Many investigators apply only the paired $t$ test to such data and as a result may miss an important relationship that could exist in the data. The investigator should, at least, plot a scatter diagram of the data and look for possible linear and nonlinear relationships. If a *linear relationship* seems appropriate, then the techniques of regression and correlation described in this chapter are useful in describing this relationship. If a *nonlinear relationship* exists, transformations should be considered.

## Focus on Residuals

Sometimes, performing the regression analysis may be only a preliminary step to obtaining the residuals, which are the quantities of interest to be saved for future analysis. For example, in the bone density study described in Chapter 1, the variable age has a major effect on predicting bone density of elderly women. One method of taking age into account in future analyses would be to fit a simple regression equation predicting bone density, with age as the only independent variable. Then future analyses of bone density could be performed using the residuals from this equation instead of the original bone density measurements. These residuals would have a mean of zero. Examination of subgroup means is then easy to interpret since their difference from zero can be easily seen.

## Adjusted Estimates of Y

An alternative method to using residuals is to use the slope coefficient $B$ from the regression equation to obtain an age-adjusted bone density for each individual, which is an estimate of their bone density at the average age for the entire sample. This adjusted bone density is computed for each individual from the following equation,

$$\text{adjusted } Y = \bar{Y} + B(\bar{X} - X)$$

where $X$ and $Y$ are the values for each individual. The adjusted $Y$ has a mean $\bar{Y}$ and thus allows the researcher to work with numbers of the usual magnitude.

## 6.12 DISCUSSION OF COMPUTER PROGRAMS

In this section only the programs and commands needed for simple linear regression are discussed. Some of the statistics mentioned in this chapter are not available in output from certain of these programs. These statistics can be obtained by using multiple linear regression programs, described in Section 7.10, which can be used to perform simple linear regression as well. In many applications of simple linear regression, however, the programs described here are not only sufficient but actually preferable to the ones listed in Chapter 7 since they are easier to peruse. Table 6.4 summarizes the features of the computer programs discussed in this section.

One illustration of simple linear regression output was given in Figure 6.1. The BMDP6D scatterplot program provides the most commonly used statistics and a convenient method for drawing the regression line on

**TABLE 6.4.** Summary of Computer Output from BMDP, SAS, and SPSS–X for Simple Linear Regression

| Output | BMDP | SAS | SPSS–X |
|---|---|---|---|
| Means and Standard Deviations | 6D | UNIVARIATE, CORR | DESCRIPTIVES |
| Regression line | 6D | REG | PLOT |
| Test $\beta = 0$ | 6D* | REG | PLOT |
| Standard Error of $B$ | 2R | REG | PLOT |
| Correlation Coefficient | 6D | REG, CORR | PLOT |
| $S$ or Mean Square Residual | 6D | REG | REGRESSION |
| Plotted Regression Line | 6D | PLOT | PLOT |
| Plots by Groups | 6D | PLOT | PLOT |
| Control of Plot Size | 6D | PLOT | PLOT |
| Control of Tick Marks | | PLOT | |

* Includes test of $\rho = 0$ which is equivalent to test of $\beta = 0$.

the scatter diagram (drawing a line between the two $Y$ values given on the border). The plot size and scale can be controlled by specifying minimum and maximum limits. The points in the plot can be identified by group and different symbols used for the various groups. For example, if the factory workers had worked in three different buildings in the factory, letters can be used to identify workers from the three buildings in the same scatter diagram. This program is available for both mainframe and personal computers.

The SAS PLOT procedure produces scatter diagrams and offers a wide variety of options. One especially useful feature is the ease of putting tick marks on the $X$ and $Y$ axes. Identification of groups is also straightforward. A scatter diagram of the observations and a graph of the fitted regression line can be done in a single plot. The SAS Procedures Guide 6.03 provides an example of the statements needed to accomplish this. The line is not as easy to see as that drawn with a ruler, but a line could be drawn through the plotted $\hat{Y}$ points. The CORR procedure can be used to obtain the correlation coefficient and the significance levels of the test of the null hypothesis $\rho = 0$. Sample means and standard deviations are also included in this output. The most straightforward way to obtain the equation of the regression line is to use the multiple regression REG procedure. The above procedures are all available for use on mainframe computers and for the PC.

The PLOT procedure in SPSS, found in both the mainframe SPSS–X and personal computer PC+ versions, performs a simple linear regression by specifying FORMAT = REGRESSION. In addition to a scatterplot of $Y$ versus $X$, this command calculates the regression estimates of the intercept and slope along with their standard errors, the correlation coefficient, and a test statistic for the hypothesis $\rho = 0$. It is straightforward to plot points by groups and to control the plot size. Means and standard deviations can be calculated using the DESCRIPTIVES procedure. Alternatively, the multiple regression procedure REGRESSION calculates the means and standard deviations as well as the regression estimates and statistics. It can also optionally produce a plot. The user has less control over the plot, however. This procedure is discussed in greater detail in Section 7.10.

## 6.13 WHAT TO WATCH OUT FOR

In Sections 6.8 and 6.9, methods for checking for outliers, normality, homogeneity of variance, and independence were presented along with a brief discussion of the importance of including checks in the analysis. In this section, other cautions will be emphasized.

1. *Sampling process*: In the development of the theory for linear regression, the sample is assumed to be obtained randomly in such a way that it represents the whole population. Often, convenience samples, which are samples of easily available cases, are taken for economic or other reasons. In surveys of the general population, more complex sampling methods are often employed. The net effect of this lack of simple random sampling from a defined population is likely to be an underestimate of the variance and possibly bias in the regression line. Researchers should describe their sampling procedures carefully so their results can be assessed with this in mind.

2. *Errors in variables*: Random errors in measuring $Y$ do not bias the slope and intercept estimates but do affect the estimate of the variance. Errors in measuring $X$ produce estimates of $\beta$ that are biased towards zero. Methods for estimating regression equations that are resistant to outliers are given in Hoaglin, Mosteller, and Tukey (1983) and Fuller (1987). Such methods should be considered if measurement error is likely to be large.

**3.** The use of nominal or ordinal, rather than interval or ratio, data in regression analysis has several possible consequences. First, often the number of distinct values can be limited. For example, in the depression data set, the self-reported health scale has only four possible outcomes: excellent, good, fair, or poor. It has been shown that when interval data are grouped into seven or eight categories, later analyses are little affected; but coarser grouping may lead to biased results. Secondly, the intervals between the numbers assigned to the scale values (1, 2, 3, and 4 for the health scale, for example) may not be properly chosen. Perhaps poor health should be given a value of 5 as it is an extreme answer and is more than one unit from fair health (see Abelson and Tukey 1963).

**4.** *Choice of model*: Particularly for the variable-$X$ model, it is essential not only that the model be in the correct analytic form, i.e., linear if a linear model is assumed, but also that the appropriate $X$ variables be chosen. This will be discussed in more detail in Chapter 8. Here, since only one $X$ variable is used, it is possible that an important confounding variable (one correlated with both $X$ and $Y$) may be omitted. This could result in a high correlation coefficient between $X$ and $Y$ and a well-fitting line that is caused by the omission of the confounding variable. For example, FEV1 decreases with age for adults as does bone density. Hence, we would expect to see spurious correlation and regression between bone density and FEV1. In Section 6.11, the use of residuals to assist in solving this problem was discussed and further results will be given in the following chapters on multiple regression analysis.

**5.** In using the computed regression equation to predict $Y$ from $X$, it is advisable to restrict the prediction to $X$ values within the range observed in the sample, unless the investigator is certain that the same linear relationship between $Y$ and $X$ is valid outside of this range.

**6.** An observed correlation between $Y$ and $X$ should not be interpreted to mean a causal relationship between $X$ and $Y$ regardless of the magnitude of the correlation. The correlation may be due to a causal effect of $X$ on $Y$, of $Y$ on $X$, or of other variables on both $X$ and $Y$. Causation should be inferred only after a careful examination of a variety of theoretical and experimental evidence, not merely from statistical studies based on correlation.

**7.** In most multivariate studies, $Y$ is regressed on more than an $X$ variable. This subject will be treated in the next three chapters.

## SUMMARY

In this chapter we gave a conventional presentation of simple linear regression, similar to the presentations found in many statistical text-books. Thus the reader familiar with the subject should be able to make the transition to the mode we follow in the remainder of the book.

We made a clear distinction between random- and fixed-$X$ variable regression. Whereas most of the statistics computed apply to both cases, certain statistics apply only to the random-$X$ case. Computer programs are written to provide more output than is sometimes needed or appropriate. You should be aware of which model is being assumed so that you can make the proper interpretation of the results.

If packaged programs are used, it may be sufficient to run one of the simple plotting programs in order to obtain both the plot and the desired statistics. It is good practice to plot many variables against each other in the preliminary stages of data analysis. This practice allows you to examine the data in a straightforward fashion. However, it is also advisable to use the more sophisticated programs as warm-ups for the more complex data analyses discussed in later chapters.

The concepts of simple linear regression and correlation presented in this chapter can be extended in several directions. Chapter 7 treats the fitting of linear regression equations to more than one independent variable. Chapter 8 gives methods for selecting independent variables. Additional topics in regression analysis are covered in Chapter 9. These topics include missing values, dummy variables, segmented regression, and ridge regression.

## BIBLIOGRAPHY

### General References

*Abelson, R. P., and Tukey, J. W. 1963. Efficient utilization of non-numerical information in quantitative analysis: General theory and the case of simple order. *Annals of Mathematical Statistics* 34:1347–1369

Acton, F. S. 1959. *The analysis of straight-line data*, New York: Wiley.

Afifi, A. A., and Azen, S. P. 1979. *Statistical analysis: A computer oriented approach.* 2nd ed. New York: Academic Press.

Allen, D. M., and Cady, F. B. 1982. *Analyzing experimental data by regression.* Belmont, Calif.: Lifetime Learning.

Box, G. E. P. 1966. Use and abuse of regression. *Technometrics* 8:625–629.

Brownlee, K. A. 1965. *Statistical theory and methodology in science and engineering.* 2nd ed. New York: Wiley.

Dixon, W. J., and Massey, F. J. 1983. *Introduction to statistical analysis.* 4th ed. New York: McGraw-Hill.

Dunn, O. J., and Clark, V. A. 1987. *Applied statistics: Analysis of variance and regression.* 2nd ed. New York: Wiley.

Fuller, W. A. 1987. *Measurement error models.* New York: Wiley.

Hald, A. 1952. *Statistical theory with engineering applications.* New York: Wiley.

Hoaglin, D. C., Mosteller, F., and Tukey, J. W., eds. 1983. *Understanding robust and exploratory data analysis.* New York: Wiley.

Klitgaard, R. 1985. *Data analysis for development.* New York: Oxford University Press.

Lewis-Beck, M. S. 1980. *Applied regression: An introduction.* Beverly Hills: Sage.

Mosteller, F., and Tukey, J. W. 1977. *Data analysis and regression.* Reading, Mass.: Addison-Wesley.

Theil, H. 1971. *Principles of econometrics.* New York: Wiley.

Thorndike, R. M. 1978. *Correlational procedures for research.* New York: Gardner Press.

Williams, E. J. 1959. *Regression analysis.* New York: Wiley.

## Calibration

Berkson, J. 1969. Estimation of a linear function for a calibration line. *Technometrics* 11:649–660.

*Hunter, W. G., and Lamboy, W. F. 1981. A Bayesian analysis of the linear calibration problem. *Technometrics* 23:323–343.

Krutchkoff, R. G. 1967. Classical and inverse regression methods of calibration. *Technometrics* 9:425–440.

Lwin, T., and Maritz, J. S. 1982. An analysis of the linear-calibration controversy from the perspective of compound estimation. *Technometrics* 24:235–242.

Naszodi, L. J. 1978. Elimination of bias in the course of calibration. *Technometrics* 20:201–206.

Shukla, G. K. 1972. On the problem of calibration. *Technometrics* 14:547–554.

## Influential Observations and Outlier Detection

Barnett, V., and Lewis, T. 1984. *Outliers in statistical data.* 2nd ed. New York: Wiley.

Belsley, D. A., Kuh, E., and Welsch, R. E. 1980. *Regression diagnostics: Identifying influential data and sources of colinearity.* New York: Wiley.

Chatterjee, S., and Hadi, A. S. 1988. *Sensitivity analysis in linear regression.* New York: Wiley.

Cook, R. D. 1977. Detection of influential observations in linear regression. *Technometrics* 19:15–18.

*————. 1979. Influential observations in linear regression. *Journal of the American Statistical Association* 74:169–174.

*Cook, R. D., and Weisberg, S. 1980. Characterization of an empirical influence function for detecting influential cases in regression. *Technometrics* 22:495–508.

Cook, R. D., and Weisberg, S. 1982. *Residuals and influence in regression.* London: Chapman and Hall.

Velleman, P. F., and Welsch, R. E. 1981. Efficient computing of regression diagnostics. *The American Statistician* 35:234–242.

## Transformations

Atkinson, A. C. 1985. *Plots, transformations and regressions.* New York: Oxford University Press.

Bennett, C. A., and Franklin, N. L. 1954. *Statistical analysis in chemistry and the chemical industry.* New York: Wiley.

*Bickel, P. J., and Doksum, K. A. 1981. An analysis of transformations revisited. *Journal of the American Statistical Association* 76:296–311.

*Box, G. E. P., and Cox, D. R. 1964. An analysis of transformations. *Journal of the Royal Statistical Society* Series B, 26:211–252.

Carroll, R. J., and Ruppert, D. 1988. *Transformation and weighting in regression.* London: Chapman and Hall.

Draper, N. R., and Hunter, W. G. 1969. Transformations: Some examples revisited. *Technometrics* 11:23–40.

Hald, A. 1952. *Statistical theory with engineering applications.* New York: Wiley.

Kowalski, C. J. 1970. The performance of some rough tests for bivariate normality before and after coordinate transformations to normality. *Technometrics* 12:517–544.

Mosteller, F., and Tukey, J. W. 1977. *Data analysis and regression.* Reading, Mass.: Addison-Wesley.

*Tukey, J. W. 1957. On the comparative anatomy of transformations. *Annals of Mathematical Statistics* 28:602–632.

## Time Series and Forecasting

*Box, G. E. P., and Jenkins, G. M. 1976. *Time series analysis: Forecasting and control.* Rev. ed. San Francisco: Holden-Day.

Chatfield, C. 1984. *The analysis of time series: An introduction.* 3rd ed. London: Chapman and Hall.

Levenbach, H., and Cleary, J. P. 1981. *The beginning forecaster: The forecasting process through data analysis.* Belmont, Calif.: Lifetime Learning.

————. 1982. *The professional forecaster: The forecasting process through data analysis,* Belmont, Calif.: Lifetime Learning.

McCleary, R., and Hay, R. A. 1980. *Applied time series analysis for the social sciences.* Beverly Hills: Sage.

Montgomery, D. C., and Johnson, L. A. 1976. *Forecasting and time series analysis.* New York: McGraw-Hill.

## PROBLEMS

6.1 In Table 8.1, financial performance data of 30 chemical companies are presented. Use growth in earnings per share, labeled EPS5, as the dependent variable and growth in sales, labeled SALESGR5, as the independent variable. (A description of these variables is given in Section 8.3.) Plot the data, compute a regression line, and test that $\beta = 0$ and $\alpha = 0$. Are earnings affected by sales growth for these chemical companies? Which company's earnings were highest, considering its growth in sales?

6.2 From the family lung function data set in Appendix B, perform a regression analysis of weight on height for fathers. Repeat for mothers. Determine the correlation coefficient and the regression equation for fathers and mothers. Test that the coefficients are significantly different from zero for both sexes. Also, find the standardized regression equation and report it. Would you suggest removing the woman who weighs 267 lb from the data set? Discuss why the correlation for fathers appears higher than that for mothers.

6.3 In Problem 4.6, the New York Stock Exchange Composite Index for August 9 through September 17, 1982, was presented. Data for the daily volume of transactions, in millions of shares, for all of the stocks traded on the exchange and included in the Composite Index are given below, together with the Composite Index values, for the same period as that in Problem 4.6. Describe how volume appears to be affected by the price index, using regression analysis. Describe whether or not the residuals from your regression analysis are serially correlated. Plot the index versus time and volume versus time, and describe the relationships you see in these plots.

| Day | Month | Index | Volume |
|-----|-------|-------|--------|
| 9 | Aug. | 59.3 | 63 |
| 10 | Aug. | 59.1 | 63 |
| 11 | Aug. | 60.0 | 59 |
| 12 | Aug. | 58.8 | 59 |
| 13 | Aug. | 59.5 | 53 |
| 16 | Aug. | 59.8 | 66 |
| 17 | Aug. | 62.4 | 106 |
| 18 | Aug. | 62.3 | 150 |
| 19 | Aug. | 62.6 | 93 |
| 20 | Aug. | 64.6 | 113 |
| 23 | Aug. | 66.4 | 129 |
| 24 | Aug. | 66.1 | 143 |
| 25 | Aug. | 67.4 | 123 |
| 26 | Aug. | 68.0 | 160 |
| 27 | Aug. | 67.2 | 87 |
| 30 | Aug. | 67.5 | 70 |
| 31 | Aug. | 68.5 | 100 |
| 1 | Sept. | 67.9 | 98 |
| 2 | Sept. | 69.0 | 87 |
| 3 | Sept. | 70.3 | 150 |
| 6 | Sept. | – | – |
| 7 | Sept. | 69.6 | 81 |
| 8 | Sept. | 70.0 | 91 |
| 9 | Sept. | 70.0 | 87 |
| 10 | Sept. | 69.4 | 82 |
| 13 | Sept. | 70.0 | 71 |
| 14 | Sept. | 70.6 | 98 |
| 15 | Sept. | 71.2 | 83 |
| 16 | Sept. | 71.0 | 93 |
| 17 | Sept. | 70.4 | 77 |

6.4   For the data in Problem 6.3, pretend that the index increases linearly in time and use linear regression to obtain an equation to forecast the index value as a function of time. Using "volume" as a weight variable, obtain a weighted least squares forecasting equation. Does weighted least squares help the fit? Obtain a recent value of the index (from a newspaper). Does either forecasting equation predict the true value correctly (or at least somewhere near it)? Explain.

6.5   Repeat Problem 6.2, using log(weight) and log(height) in place of the original variables. Using graphical and numerical devices, decide whether the transformations help.

6.6   Examine the plot you produced in Problem 6.1 and choose some transformation for $X$ and/or $Y$ and repeat the analysis described there. Compare the correlation coefficients for the original and transformed variables, and decide whether the transformation helped. If so, which transformation was helpful?

6.7   Using the depression data set described in Table 3.2, perform a regression analysis of depression, as measured by CESD, on income. Plot the residuals.

Does the normality assumption appear to be met? Repeat using the logarithm of CESD instead of CESD. Is the fit improved?

6.8   Continuation of Problem 6.7: Calculate the variance of CESD for observations in each of the groups defined by income as follows: INCOME $< 30$, INCOME between 30 and 39 inclusive, INCOME between 40 and 49 inclusive, INCOME between 50 and 59 inclusive, INCOME $> 59$. For each observation, define a variable WEIGHT equal to 1 divided by the variance of a CESD within the income group to which it belongs. Obtain a weighted least squares regression of the untransformed variable CESD on income, using the values of WEIGHT as weights. Compare the results with the unweighted regression analysis. Is the fit improved by weighting?

6.9   From the depression data set described in Table 3.2, create a data set containing only the variables AGE and INCOME.

a) Find the regression of income on age.

b) Successively add and then delete each of the following points:

| AGE | INCOME |
|-----|--------|
| 42  | 120    |
| 80  | 150    |
| 180 | 15     |

and repeat the regression each time with the single extra point. How does the regression equation change? Which of the new points are outliers? Which are influential?

Use the family lung function data given in Appendix B for the next four problems.

6.10  For the oldest child, perform the following regression analyses: FEV1 on weight, FEV1 on height, FVC on weight, and FVC on height. Note the values of the slope and correlation coefficient for each regression and test whether they are equal to zero. Discuss whether height or weight is more strongly associated with lung function in the oldest child.

6.11  What is the correlation between height and weight in the oldest child? How would your answer to the last part of Problem 6.10 change if $\rho = 1$? $\rho = -1$? $\rho = 0$?

6.12  Examine the residual plot from the regression of FEV1 on height for the oldest child. Choose an appropriate transformation, perform the regression with the transformed variable, and compare the results (statistics, plots) with the original regression analysis.

6.13  For the mother, perform a regression of FEV1 on weight. Test whether the coefficients are zero. Plot the regression line on a scatter diagram of MFEV1 versus MWEI. On this plot, identify the following groups of points:

Group 1: ID $= 12, 33, 45, 42, 94, 144$;
Group 2: ID $= 7, 94, 105, 107, 115, 141, 144$

Remove the observations in group 1 and repeat the regression analysis. How does the line change? Repeat for group 2.

## Chapter Seven

# MULTIPLE REGRESSION AND CORRELATION

## 7.1 WHAT WILL YOU LEARN FROM THIS CHAPTER?

From this chapter you will learn how to examine the relationship between one dependent or response variable $Y$ and a set of independent predictor variables $X_1$ to $X_P$. In particular you will learn:

- When multiple regression is used (7.2, 7.3).
- The correct meaning of the phrases multiple linear regression, multiple correlation, and partial correlation (7.4, 7.5).
- How to obtain appropriate equations and statistics to describe the interrelationships (7.4, 7.5, 7.6, 7.7).
- How to predict the value of one variable from a set of values of other variables (7.4).
- How to read and interpret a correlation or covariance matrix (7.5, 7.9).
- How to perform a residual analysis in order to improve the derived equations (7.8).

- How and when to use transformations to improve the equations (7.8).
- About the role of interactions in regression analysis (7.8).
- How to compute and compare regression equations for subgroups of data (7.9).
- Which are the appropriate computer programs (7.10).
- What to watch out for in multiple regression analysis (7.11).

The methods described in this chapter for examining the interrelationships among a set of variables are fundamental to an understanding of all of the multivariate techniques discussed in the remainder of the book. On a first reading Sections 7.8 and 7.9 may be skipped.

## 7.2 WHEN ARE MULTIPLE REGRESSION AND CORRELATION USED?

The methods described in this chapter are appropriate for studying the relationship between several $X$ variables and one $Y$ variable. By convention, the $X$ variables are called *independent variables*, although they do not have to be statistically independent and are permitted to be intercorrelated. The $Y$ variable is called the *dependent variable*.

As in Chapter 6, the data for multiple regression analysis can come from one of two situations:

1. *Fixed-X case*: The levels of the various $X$'s are selected by the researcher or dictated by the nature of the situation. For example, in a chemical process an investigator can set the temperature, the pressure, and the length of time that a vat of material is agitated and then measure the concentration of a certain chemical. A regression analysis can be performed to *describe* or *explain* the relationship between the independent variables and the dependent variable (the concentration of a certain chemical) or to *predict* the dependent variable.

2. *Variable-X case*: A random sample of individuals is taken from a population, and the $X$ variables and the $Y$ variable are measured on each individual. For example, a sample of adults might be taken

from Los Angeles and information obtained on their age, income, and education in order to see whether these variables predict their attitudes toward air pollution. Regression and correlation analyses can be performed in the variable-$X$ case. Both *descriptive* and *predictive information* can be obtained from the results.

Multiple regression is one of the most commonly used statistical methods, and an understanding of its use will help you comprehend other multivariate techniques.

## 7.3 DATA EXAMPLE

In Chapter 6 the data from a sample of factory workers, age 20–24, were analyzed. These data fit the variable-$X$ case. Height was used as the $X$ variable in order to predict FEV1, and the following equation was obtained:

$$FEV1 = -4.39 + .126(\text{height, in inches})$$

Several such equations could have been obtained, one for each age interval. However, in analyzing such data, investigators usually want to obtain a single equation that is useful with a wider age range. Because FEV1 decreases with age in adults, we need to fit a multiple regression with both age and height as independent variables and FEV1 as the dependent variable. We expect the slope coefficient for age to be negative and the slope coefficient for height to be positive. Also, since prolonged smoking is known to decrease FEV1, separate regression equations should be fitted for smokers and nonsmokers.

A geometric representation of the simple regression of FEV1 on age and height, respectively, is shown in Figures 7.1a and 7.1b. The multiple regression equation is represented by a plane, as shown in Figure 7.1c. Note that the plane slopes upward as a function of height and downward as a function of age. A hypothetical individual whose FEV1 is large relative to his age and height appears above both simple regression lines as well as above the multiple regression plane.

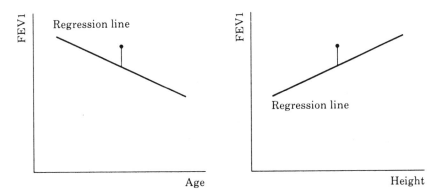

**a.** Simple Regression of FEV1 on Age    **b.** Simple Regression of FEV1 on Height

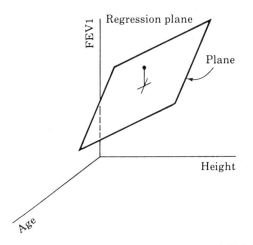

**c.** Multiple Regression of FEV1 on Age and Height

**FIGURE 7.1.** Hypothetical Representation of Simple and Multiple Regression Equations of FEV1 on Age and Height

For illustrative purposes Figure 7.2 shows how a plane can be constructed. Constructing such a regression plane involves the following steps:

**1.** Draw lines on the $X_1$, $Y$ wall, setting $X_2 = 0$.

**2.** Draw lines on the $X_2$, $Y$ wall, setting $X_1 = 0$.

**3.** Drive nails in the walls at the lines drawn in steps 1 and 2.

**4.** Connect the pairs of nails by strings and tighten the strings.

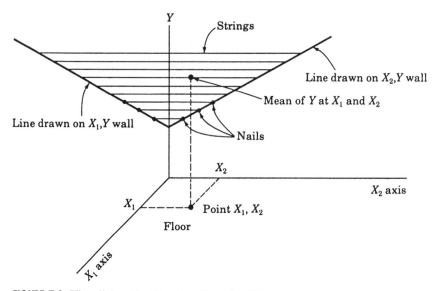

**FIGURE 7.2.** Visualizing the Construction of a Plane

The resulting strings in step 4 form a plane. This plane is the regression plane of $Y$ on $X_1$ and $X_2$. The mean of $Y$ at a given $X_1$ and $X_2$ is the point on the plane vertically above the point $X_1$, $X_2$.

Data for 448 smokers were analyzed by BMDP1R, and the following descriptive statistics were printed:

| VARIABLE | MEAN | STD DEVIATION |
|---|---|---|
| AGE | 39.43 | 11.87 |
| HEIGHT | 68.76 | 2.63 |
| FEV1 | 3.62 | 0.82 |

The program also produced the following estimated regression equation:

$$\hat{FEV}1 = -1.40 - 0.036(\text{age}) + 0.093(\text{height})$$

As expected, there is a positive slope associated with height and a negative slope associated with age. For predictive purposes, we would expect a 22-year-old male whose height is 70 in., for example, to have an FEV1 value of $-1.40 - 0.036(22) + 0.093(70) = 4.32\,\text{L}$.

Note that a value of $-4.39 + 0.126(70) = 4.4$ L will be obtained for a person in the age group 20–24 for the same height when we are using the *first* equation given in this section. In the single-predictor equation the coefficient for height is 0.126. This value is the rate of change of FEV1 for young males as a function of height when no other variables are taken into account. With two predictors the coefficient for height is 0.093, which is interpreted as the rate of change of FEV1 as a function of height *after adjusting for age*. The latter slope is also called the *partial regression coefficient* of FEV1 on height after adjusting for age. Even when both equations are derived from the same sample, the simple and partial regression coefficients may not be comparable.

The output of this program includes other items that will be discussed later in this chapter.

## 7.4 DESCRIPTION OF TECHNIQUES: FIXED-X CASE

In this section we present the background assumptions, model, and formulas necessary for an understanding of multiple linear regression for the fixed-$X$ case. Computations can quickly become tedious when there are several independent variables, so we assume that you will obtain output from a packaged computer program. Therefore we present a minimum of formulas and place the main emphasis on the techniques and interpretation of results. This section is slow reading and requires concentration.

Since there is more than one $X$ variable, we use the notation $X_1, X_2, \ldots,$ $X_P$ to represent $P$ possible variables. In packaged programs these variables may appear in the output as $X(1)$, X1, VAR 1, etc. For the fixed-$X$ case values of the $X$ variables are assumed to be fixed in advance in the sample. At each combination of levels of the $X$ variables, we conceptualize a distribution of the values of $Y$. This distribution of $Y$ values is assumed to have a mean value equal to $\alpha + \beta_1 X_1 + \beta_2 X_2 + \cdots + \beta_P X_P$ and a variance equal to $\sigma^2$ at given levels of $X_1, X_2, \ldots, X_P$.

When $P = 2$, the surface is a plane, as depicted in Figure 7.1c or 7.2. The parameter $\beta_1$ is the rate of change of the mean of $Y$ as a function of $X_1$, where the value of $X_2$ is held fixed. Similarly, $\beta_2$ is the rate of change of the mean of $Y$ as a function of $X_2$ when $X_1$ is fixed. Thus $\beta_1$ and $\beta_2$ are the slopes of the regression plane with regard to $X_1$ and $X_2$.

When $P > 2$, the regression plane generalizes to a so-called *hyperplane*, which cannot be represented geometrically on two-dimensional paper. Some people conceive the vertical axis as always representing the $Y$ variable, and they think of all of the $X$ variables as being represented by the horizontal plane. In this situation the hyperplane can still be imagined as in Figure 7.1c. The parameters $\beta_1, \beta_2, \ldots, \beta_P$ represent the slope of the regression hyperplane with respect to $X_1, X_2, \ldots, X_P$, respectively. The betas are called *partial regression coefficients*. For example, $\beta_1$ is the rate of change of the mean of $Y$ as a function of $X_1$ when the levels of $X_2, \ldots, X_P$ are held fixed. In this sense it represents the change of the mean of $Y$ as a function of $X_1$ after adjusting for $X_2, \ldots, X_P$.

Again we assume that the variance of $Y$ is homogeneous over the range of concern of the $X$ variables. Usually, for the fixed-$X$ case, $P$ is rarely larger than three or four.

## Least Squares Method

As for simple linear regression, the method of least squares is used to obtain estimates of the parameters. These estimates, denoted by $A$, $B_1$, $B_2, \ldots, B_P$, are printed in the output of any multiple regression program. The estimate $A$ is usually labeled "intercept," and $B_1, B_2, \ldots, B_P$ are usually given in tabular form under the label "coefficient" or "regression coefficient" by variable name. The formulas for these estimates are mathematically derived by minimizing the sums of the squared vertical deviations. Formulas may be found in many of the books listed in the Bibliography at the end of this chapter.

The *predicted value* $\hat{Y}$ for a given set of values $X_1^*, X_2^*, \ldots, X_P^*$ is then calculated as

$$\hat{Y} = A + B_1 X_1^* + B_2 X_2^* + \cdots + B_P X_P^*$$

The estimate of $\sigma^2$ is the *residual mean square*, obtained as

$$S^2 = \text{RES. MS.} = \frac{\sum (Y - \hat{Y})^2}{N - P - 1}$$

The square root of the residual mean square is called the *standard error of the estimate*. It represents the variation unexplained by the regression plane.

Packaged programs usually print the value of $S^2$ and the standard errors for the regression coefficients (and sometimes for the intercept). These quantities are useful in computing confidence intervals and prediction intervals around the computed $\hat{Y}$. These intervals can be computed from the output in the following manner.

## Prediction and Confidence Intervals

To obtain *prediction* and *confidence intervals*, we need access to the estimated covariances or correlations among the regression coefficients. For example, BMDP1R prints these correlations as an optional output. When these estimates are available, we compute the estimated variance of $\hat{Y}$ as

$$\text{Var } \hat{Y} = \frac{S^2}{N} + [(X_1^* - \bar{X}_1)^2 \text{ Var } B_1 + (X_2^* - \bar{X}_2)^2 \text{ Var } B_2 + \cdots$$
$$+ (X_P^* - \bar{X}_P)^2 \text{ Var } B_P] + [2(X_1^* - \bar{X}_1)(X_2^* - \bar{X}_2) \text{ Cov}(B_1, B_2)$$
$$+ 2(X_1^* - \bar{X}_1)(X_3^* - \bar{X}_3) \text{ Cov}(B_1, B_3) + \cdots]$$

The variances (Var) of the various $B_i$ are computed as the squares of the standard errors of the $B_i$, $i$ going from 1 to $P$. The covariances (Cov) are computed from the standard errors and the correlations among the regression coefficients. For example,

$$\text{Cov}(B_1, B_2) = (\text{standard error } B_1)(\text{standard error } B_2)[\text{Corr}(B_1, B_2)]$$

If $\hat{Y}$ is interpreted as an estimate of the mean of $Y$ at $X_1^*, X_2^*, \ldots, X_P^*$, then the confidence interval for this mean is computed as

$$\text{CI}(\text{mean } Y \text{ at } X_1^*, X_2^*, \ldots, X_P^*) = \hat{Y} \pm t\sqrt{\text{Var } \hat{Y}}$$

where $t$ is the $100(1 - \alpha/2)$ percentile of the $t$ distribution with $N - P - 1$ degrees of freedom.

When $\hat{Y}$ is interpreted as an estimate of the $Y$ value for an individual, then the variance of $\hat{Y}$ is increased by $S^2$, similar to what is done in simple linear regression. Then the prediction interval is

$$\text{PI}(\text{individual } Y \text{ at } X_1^*, X_2^*, \ldots, X_P^*) = \hat{Y} \pm t\sqrt{S^2 + \text{Var } \hat{Y}}$$

where $t$ is the same as it is for the confidence interval above. Note that these intervals require the additional assumption that for any set of levels of $X$ the values of $Y$ are normally distributed.

In summary, for the fixed-$X$ case we presented the model for multiple linear regression analysis and discussed estimates of the parameters. Next, we present the variable-$X$ model.

## 7.5 DESCRIPTION OF TECHNIQUES: VARIABLE-X CASE

For this model the $X$'s and the $Y$ variable are random variables measured on cases that are randomly selected from a population. An example is the factory worker data, where FEV1, height, age, and smoking status were measured on a sample of workers. As in Chapter 6, the previous discussion and formulas given for the fixed-$X$ case apply to the variable-$X$ case. (This result is justifiable theoretically by conditioning on the $X$ variable values that happen to be obtained in the sample.) Furthermore, since the $X$ variables are also random variables, the joint distribution of $Y$, $X_1$, $X_2, \ldots, X_P$ is of interest.

When there is only one $X$ variable, it was shown in Chapter 6 that the bivariate distribution of $X$ and $Y$ can be characterized by its region of concentration. These regions are ellipses if the data come from a bivariate normal distribution. Two such ellipses and their corresponding regression lines are illustrated in Figures 7.3a and 7.3b.

When two $X$ variables exist, the regions of concentration become three-dimensional forms. These forms take the shape of ellipsoids if the joint distribution is multivariate normal. In Figure 7.3c such an ellipsoid with the corresponding regression plane of $Y$ on $X_1$ and $X_2$ is illustrated. The regression plane passes through the point representing the population means of the three variables and intersects the vertical axis at a distance $\alpha$ from the origin. Its position is determined by the slope coefficients $\beta_1$ and $\beta_2$. When $P > 2$, we can imagine the horizontal plane representing all the $X$ variables and the vertical plane representing $Y$. The ellipsoid of concentration becomes the so-called *hyperellipsoid*.

### Estimation of Parameters

For the variable-$X$ case several additional parameters are needed to characterize the joint distribution. These include the population means,

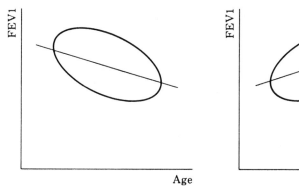

**a.** Regression of FEV1 on Age

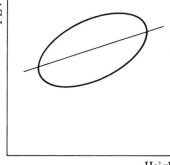

**b.** Regression of FEV1 on Height

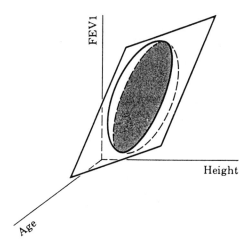

**c.** Regression of FEV1 on Age and Height

**FIGURE 7.3.** Hypothetical Regions of Concentration and Corresponding Regression Lines and Planes for the Population Variable-$X$ Model

$\mu_1, \mu_2, \ldots, \mu_P$ and $\mu_Y$, the population standard deviations $\sigma_1, \sigma_2, \ldots, \sigma_P$ and $\sigma_Y$, and the covariances of the $X$ and $Y$ variables.

The variances and covariances are usually arranged in the form of a square array called the *covariance matrix*. For example, if $P = 3$, the covariance matrix is a four-by-four array of the form

|  |  | $\overline{\quad X \quad}$ | | | |
|---|---|---|---|---|---|
|  |  | 1 | 2 | 3 | Y |
| | 1 | $\sigma_1^2$ | $\sigma_{12}$ | $\sigma_{13}$ | $\sigma_{1Y}$ |
| $X$ | 2 | $\sigma_{12}$ | $\sigma_2^2$ | $\sigma_{23}$ | $\sigma_{2Y}$ |
| | 3 | $\sigma_{13}$ | $\sigma_{23}$ | $\sigma_3^2$ | $\sigma_{3Y}$ |
| | Y | $\sigma_{1Y}$ | $\sigma_{2Y}$ | $\sigma_{3Y}$ | $\sigma_Y^2$ |

A dashed line is included to separate the dependent variable $Y$ from the independent variables. The estimates of the means, variances, and covariances are available in the output of most regression programs. The estimated variances and covariances are denoted by $S$ instead of $\sigma$.

In addition to the estimated covariance matrix, an estimated *correlation matrix* is also available, which is given in the same format:

|  |  | $\overline{\quad X \quad}$ | | | |
|---|---|---|---|---|---|
|  |  | 1 | 2 | 3 | Y |
| | 1 | 1 | $r_{12}$ | $r_{13}$ | $r_{1Y}$ |
| $X$ | 2 | $r_{12}$ | 1 | $r_{23}$ | $r_{2Y}$ |
| | 3 | $r_{13}$ | $r_{23}$ | 1 | $r_{3Y}$ |
| | Y | $r_{1Y}$ | $r_{2Y}$ | $r_{3Y}$ | 1 |

Since the correlation of a variable with itself must be equal to 1, the diagonal elements of the correlation matrix are equal to 1. The off-diagonal elements are the simple correlation coefficients described in Section 6.5. As before, their numerical values always lie between $+1$ and $-1$.

As an example, the following covariance matrix is obtained from data from the 448 factory workers who are smokers. For the sake of illustration we include a third $X$ variable, FVC, defined as the forced vital capacity, which is the total volume of air exhaled by the workers in one breath:

|   |        | —— X —— |        |       | Y     |
|---|--------|---------|--------|-------|-------|
|   |        | Age     | Height | FVC   | FEV1  |
| X | Age    | 140.80  | −3.62  | −5.41 | −5.35 |
|   | Height | −3.62   | 6.94   | 1.14  | 0.78  |
|   | FVC    | −5.41   | 1.14   | 0.85  | 0.65  |
| Y | FEV1   | −5.35   | 0.78   | 0.65  | 0.67  |

Note that this matrix is a symmetric matrix. That is, if a mirror were placed along the diagonal, then the elements in the lower triangle would be mirror images of those in the upper triangle. For example, the covariance of age and height is $-3.62$. Also, the covariance between height and age is $-3.62$.

Symmetry holds also for the correlation matrix that follows:

|   |        | —— X —— |        |       | Y     |
|---|--------|---------|--------|-------|-------|
|   |        | Age     | Height | FVC   | FEV1  |
| X | Age    | 1       | −0.12  | −0.49 | −0.55 |
|   | Height | −0.12   | 1      | 0.47  | 0.36  |
|   | FVC    | −0.49   | 0.47   | 1     | 0.87  |
| Y | FEV1   | −0.55   | 0.36   | 0.87  | 1     |

All of the correlations are computed from the appropriate elements of the covariance matrix. For example, the correlation between age and height is

$$r_{12} = \frac{S_{12}}{\sqrt{S_1^2 S_2^2}} = \frac{S_{12}}{S_1 S_2}$$

or

$$-0.12 = \frac{-3.62}{\sqrt{140.8}\sqrt{6.94}}$$

Note that the correlations are computed between $Y$ and each $X$ as well as among the $X$ variables. We will return to interpreting these simple correlations in Section 7.7.

## Multiple Correlation

So far, we have discussed the concept of correlation between two variables. It is also possible to describe the strength of the linear relationship between $Y$ and a set of $X$ variables by using the *multiple correlation coefficient*. In the population we will denote this multiple correlation by $\mathcal{R}$. It represents the simple correlation between $Y$ and the corresponding point on the regression plane for all possible combinations of the $X$ variables. Each individual in the population has a $Y$ value and a corresponding point on the plane computed as

$$Y' = \alpha + \beta_1 X_1 + \beta_2 X_2 + \cdots + \beta_P X_P$$

Correlation $\mathcal{R}$ is the population simple correlation between all such $Y$ and $Y'$ values. The numerical value of $\mathcal{R}$ cannot be negative. The maximum possible value for $\mathcal{R}$ is 1.0, indicating a perfect fit of the plane to the points in the population.

Another interpretation of the $\mathcal{R}$ coefficient for multivariate normal distributions involves the concept of the *conditional distribution*. This distribution describes all of the $Y$ values of individuals whose $X$ values are specified at certain levels. The variance of the conditional distribution is the variance of the $Y$ values about the regression plane in a vertical direction. For multivariate normal distributions this variance is the same at all combinations of levels of the $X$ variables and is denoted by $\sigma^2$. The following fundamental expression relates $\mathcal{R}$ to $\sigma^2$ and $\sigma_Y^2$:

$$\sigma^2 = \sigma_Y^2 (1 - \mathcal{R}^2)$$

Rearrangement of this expression shows that

$$\mathcal{R}^2 = \frac{\sigma_Y^2 - \sigma^2}{\sigma_Y^2} = 1 - \frac{\sigma^2}{\sigma_Y^2}$$

When the variance about the plane, or $\sigma^2$, is small relative to $\sigma_Y^2$, then the squared multiple correlation $\mathcal{R}^2$ is close to 1. When the variance $\sigma^2$

about the plane is almost as large as the variance $\sigma_Y^2$ of $Y$, then $\mathscr{R}^2$ is close to zero. In this case the regression plane does not fit the $Y$ values much better than $\mu_Y$. Thus the multiple correlation squared suggests the proportion of the variation accounted for by the regression plane. As in the case of simple linear regression, another interpretation of $\mathscr{R}$ is that $\sqrt{(1 - \mathscr{R}^2)} \times 100$ is the percentage of $\sigma_Y$ *not* explained by $X_1$ to $X_P$.

Note that $\sigma_Y^2$ and $\sigma^2$ can be estimated from a computer output as follows:

$$S_Y^2 = \frac{\sum (Y - \bar{Y})^2}{N - 1} = \frac{\text{total sum of squares}}{N - 1}$$

and

$$S^2 = \frac{\sum (Y - \hat{Y})^2}{N - P - 1} = \frac{\text{residual sum of squares}}{N - P - 1}$$

## Partial Correlation

Another correlation coefficient is useful in measuring the degree of dependence between two variables after adjusting for the linear effect of one or more of the other $X$ variables. For example, suppose an educator is interested in the correlation between the total scores of two tests, $T_1$ and $T_2$, given to twelfth graders. Since both scores are probably related to the student's IQ, it would be reasonable to first remove the linear effect of IQ from both $T_1$ and $T_2$ and then find the correlation between the adjusted scores. The resulting correlation coefficient is called a *partial correlation coefficient* between $T_1$ and $T_2$ given IQ.

For this example we first derive the regression equations of $T_1$ on IQ and $T_2$ on IQ. These equations are displayed in Figures 7.4a and 7.4b. Consider an individual whose IQ is IQ* and whose actual scores are $T_1^*$ and $T_2^*$. The test scores defined by the population regression lines are $\tilde{T}_1$ and $\tilde{T}_2$, respectively. The adjusted test scores are the residuals $T_1^* - \tilde{T}_1$ and $T_2^* - \tilde{T}_2$. These adjusted scores are calculated for each individual in the population, and the simple correlation between them is computed. The resulting value is defined to be the population partial correlation coefficient between $T_1$ and $T_2$ with the linear effects of IQ removed.

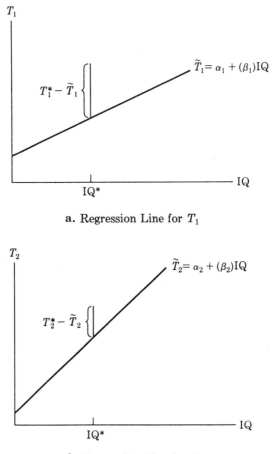

**a.** Regression Line for $T_1$

**b.** Regression Line for $T_2$

**FIGURE 7.4.** Hypothetical Population Regressions of $T_1$ and $T_2$ Scores, Illustrating the Computation of a Partial Correlation Coefficient

Formulas exist for computing partial correlations directly from the population simple correlations or other parameters without obtaining the residuals. In the above case suppose that the simple correlation for $T_1$ and $T_2$ is denoted by $\rho_{12}$, for $T_1$ and IQ by $\rho_{1q}$, and $T_2$ and IQ by $\rho_{2q}$. Then the partial correlation of $T_1$ and $T_2$ given IQ is derived as

$$\rho_{12.q} = \frac{\rho_{12} - \rho_{1q}\rho_{2q}}{\sqrt{(1 - \rho_{1q}^2)(1 - \rho_{2q}^2)}}$$

In general, for any three variables denoted by $i$, $j$, and $k$,

$$\rho_{ij.k} = \frac{\rho_{ij} - \rho_{ik}\rho_{jk}}{\sqrt{(1 - \rho_{ik}^2)(1 - \rho_{jk}^2)}}$$

It is also possible to compute partial correlations between any two variables after removing the linear effect of two or more other variables. In this case the residuals are deviations from the regression planes on the variables whose effects are removed. The partial correlation is the simple correlation between the residuals.

Packaged computer programs, discussed further in Section 7.10, calculate estimates of multiple and partial correlations. We note that the formula given above can be used with sample simple correlations to obtain sample partial correlations.

This section has covered the basic concepts in multiple regression and correlation. Additional items found in output will be presented in Section 7.8 and in Chapters 8 and 9.

## 7.6 HOW TO INTERPRET THE RESULTS: FIXED-X CASE

In this section we discuss interpretation of the estimated parameters and present some tests of hypotheses for the fixed-$X$ case.

As mentioned previously, the regression analysis might be performed for prediction: deriving an equation to predict $Y$ from the $X$ variables. This situation was discussed in Section 7.4. The analysis can also be performed for description: an understanding of the relationship between the $Y$ and $X$ variables.

In the fixed-$X$ case the number of $X$ variables is generally small. The values of the slopes describe how $Y$ changes with changes in the levels of $X$ variables. In most situations a plane does not describe the response surface for the whole possible range of $X$ values. Therefore it is important to define the region of concern for which the plane applies. The magnitudes of the $\beta$'s and their estimates depend on this choice. In the interpretation of the relative magnitudes of the estimated $B$'s, it is useful to compute $B_1\bar{X}_1$, $B_2\bar{X}_2$, ..., $B_P\bar{X}_P$ and compare the resulting values. A large (relative) value of the magnitude of $B_i\bar{X}_i$ indicates a relatively important contribution of the variable $X_i$. Here the $\bar{X}$'s represent typical values within the region of concern.

If we restrict the range of concern here, the *additive model,* $\alpha + \beta_1 X_1 + \beta_2 X_2 + \cdots + \beta_P X_P$, is often an appropriate description of the underlying relationship. It is also sometimes useful to incorporate interactions or nonlinearities. This step can partly be achieved by transformations and will be discussed in Section 7.8. Examination of residuals can help you assess the adequacy of the model, and this analysis is discussed in Section 7.8.

## Analysis of Variance

For a test of whether the regression plane is at all helpful in predicting the values of $Y$, the ANOVA table printed in most regression outputs can be used. The null hypothesis being tested is

$$H_0: \beta_1 = \beta_2 = \cdots = \beta_P = 0$$

That is, the mean of $Y$ is as accurate in predicting $Y$ as the regression plane. The ANOVA table for multiple regression has the form given in Table 7.1, which is similar to that of Table 6.1.

When the fitted plane differs from a horizontal plane in a significant fashion, then the term $\Sigma\,(\hat{Y} - \bar{Y})^2$ will be large relative to the residuals from the plane $\Sigma\,(Y - \hat{Y})^2$. This result is the case in the factory workers example, where $F = 144.6$ with 2 and 445 degrees of freedom, as indicated in Table 7.2. Note that only data on smokers are used in this table; regression analysis for the rest of the workers is given later in Section 7.9. The computed value of $F = 144.6$ with $P = 2$ degrees of freedom in the numerator and $N - P - 1 = 445$ degrees of freedom in the denominator is compared with the printed $F$ value in Appendix Table A.4 with $v_1$ (numerator) degrees of freedom across the top of the table and $v_2$

**TABLE 7.1.** ANOVA Table for Multiple Regression

| Source of Variation | Sums of Squares | df | Mean Square | F |
|---|---|---|---|---|
| Regression | $\Sigma\,(\hat{Y} - \bar{Y})^2$ | $P$ | $SS_{reg}/P$ | $MS_{reg}/MS_{res}$ |
| Residual | $\Sigma\,(Y - \hat{Y})^2$ | $N - P - 1$ | $SS_{res}/(N - P - 1)$ | |
| Total | $\Sigma\,(Y - \bar{Y})^2$ | $N - 1$ | | |

**TABLE 7.2.** ANOVA Example from Factory Workers Data (Smokers Only)

| Source of Variation | Sums of Squares | df | Mean Square | F |
|---|---|---|---|---|
| Regression | 117.492 | 2 | 58.746 | 144.6 |
| Residual | 180.746 | 445 | 0.406 | |
| Total | 298.238 | 447 | | |

(denominator) degrees of freedom in the first column of the table for an appropriate $\alpha$. For example, for $\alpha = 0.05$, $v_1 = 2$, and $v_2 = 120$ ($v_2 = 445$ is not tabled, so we take the next lower level), the tabled $F(1 - \alpha) = F(0.95) = 3.07$, from the body of the table. Since the computed $F$, 144.6, is much larger than the tabled $F$, 3.07, the null hypothesis is rejected.

The $P$ value can be determined more precisely. In fact, it is printed in the output as $P = 0.00000$. This value should be interpreted and reported as $P < 0.00001$. Thus the variables "age" and "height" together help significantly in predicting an individual's FEV1.

If this blanket hypothesis is rejected, then the degree to which the regression equation fits the data can be assessed by examining a quantity called the *coefficient of determination*, defined as

$$\text{coefficient of determination} = \frac{\text{sum of squares regression}}{\text{sum of squares total}}$$

If the regression sum of squares is not in the program output, it can be obtained by subtraction, as follows:

regression sum of squares = total sum of squares − residual sum of squares

For the factory workers example,

$$\text{coefficient of determination} = \frac{117.492}{298.238} = 0.394$$

This value is an indication of the reduction in the variance of $Y$ achieved by using $X_1$ and $X_2$ as predictors. In this case the variance around the regression plane is 39.4% less than the variance of the original $Y$ values. Numerically, the coefficient of determination is equal to the square of the multiple correlation coefficient, and therefore it is called RSQUARE in the output of some computer programs. Although the multiple correlation coefficient is not meaningful in the fixed-$X$ case, the interpretation of its square is valid.

## Other Tests of Hypotheses

Tests of hypotheses can be used to assess whether variables are contributing significantly to the regression equation. It is possible to test these variables either singularly or in groups. For any specific variable $X_i$ we can test the null hypothesis

$$H_0: \beta_i = 0$$

by computing

$$t = \frac{B_i - 0}{\text{SE}(B_i)}$$

and performing a one- or two-sided $t$ test with $N - P - 1$ degrees of freedom. These $t$ statistics are often printed in the output of packaged programs. Other programs print the corresponding $F$ statistics ($t^2 = F$), which can test the same null hypothesis against a two-sided alternative.

When this test is performed for each $X$ variable, the joint significance level cannot be determined. A method designed to overcome this uncertainty makes use of the so-called *Bonferroni inequality*. In this method, to compute the appropriate joint $P$ value for any test statistic, we multiply the single $P$ value obtained from the printout by the number of $X$ variables we wish to test. The joint $P$ value is then compared with the desired overall level.

For example, to test that age alone has an effect, we use

$$H_0: \beta_1 = 0$$

and from the computer output

$$t = \frac{-0.36 - 0}{0.0026} = -13.8$$

with 445 degrees of freedom. This result is also highly significant, indicating that the equation using age and height is significantly better than an equation using height alone.

Most computer outputs will display the $P$ level of the $t$ statistic for each regression coefficient. To get joint $P$ values according to the Bonferroni inequality, we multiply each individual $P$ value by the number of $X$ variables before we compare it with the overall significance level $\alpha$. For example, if $P = 0.015$ and there are two $X$'s, then $2(0.015) = 0.03$, which could be compared with an $\alpha$ level of 0.05.

## 7.7 HOW TO INTERPRET THE RESULTS: VARIABLE-X CASE

In the variable-$X$ case all of the $Y$ and $X$ variables are considered to be random variables. The means and standard deviations printed by packaged programs are used to estimate the corresponding population parameters. Frequently, the covariance and/or the correlation matrices are printed. In the remainder of this section we discuss the interpretation of the various correlation and regression coefficients found in the output.

### Correlation Matrix

Most people find the correlation matrix more useful than the covariance matrix since all of the correlations are limited to the range $-1$ to $+1$. Note that because of the symmetry of these matrices, some programs will print only the terms on one side of the diagonal. The diagonal terms of a correlation matrix will always be 1 and thus sometimes are not printed.

Initial screening of the correlation matrix is helpful in obtaining a preliminary impression of the interrelationships among the $X$ variables and between $Y$ and each of the $X$ variables. One way of accomplishing this screening is to first determine a cutoff point on the magnitude of the correlation, for example, 0.3 or 0.4. Then each correlation greater in magnitude than this number is underlined or highlighted. The resulting pattern can give a visual impression of the underlying interrelationships; i.e., highlighted correlations are indications of possible relationships. Correlations near $+1$ or $-1$ among the $X$ variables indicate that the two variables are nearly perfect functions of each other, and the investigator should consider dropping one of them since they convey nearly the same information. Also, if the physical situation leads the investigator to expect larger correlations than those found in the printed matrix, then the presence of outliers in the data or of nonlinearities among the variables may be causing the discrepancies.

If tests of significance are desired, the following test can be performed. To test the hypothesis

$$H_0: \rho = 0$$

use the statistic given by

$$t = \frac{r\sqrt{N-2}}{\sqrt{1-r^2}}$$

where $\rho$ is a particular population simple correlation and $r$ is the estimated simple correlation. This statistic can be compared with the $t$ table value with df $= N - 2$ to obtain one-sided or two-sided $P$ values. Again, $t^2 = F$ with 1 and $N - 2$ degrees of freedom can be used for two-sided alternatives.

Some programs, such as the SPSS–X procedure CORRELATIONS, will print the $P$ value for each correlation in the matrix. Since several tests are being performed in this case, it may be advisable to use the Bonferroni correction to the $P$ value; i.e., each $P$ is multiplied by the number of correlations (say $M$). This number of correlations can be determined as follows:

$$M = \frac{(\text{no. of rows})(\text{no. of rows} - 1)}{2}$$

After this adjustment to the $P$ values is made, they can be compared with the nominal significance level $\alpha$.

## Standardized Coefficients

The interpretations of the regression plane, the associated regression coefficients, and the standard error around the plane are the same as those in the fixed-$X$ case. All the tests presented in Section 7.6 apply here as well.

Another way to interpret the regression coefficients is to examine *standardized coefficients*. These are printed in the output of many regression programs and can be computed easily as

$$\text{standardized } B_i = B_i\left(\frac{\text{standard deviation of } X_i}{\text{standard deviation of } Y}\right)$$

These coefficients are the ones that would be obtained if the $Y$ and $X$ variables were standardized prior to performing the regression analysis. The standardized coefficients of the various $X$ variables can be directly compared in order to determine the relative contribution of each to the regression plane. The larger the magnitude of the standardized $B_i$, the more $X_i$ contributes to the prediction of $Y$. Comparing the unstandardized $B$ directly does not achieve this result because of the different units and degrees of variability of the $X$ variables. The regression equation itself should be reported for future use in terms of the unstandardized coefficients so that prediction can be made directly from the raw $X$ values.

## Multiple and Partial Correlations

As mentioned earlier, the multiple correlation coefficient is a measure of the strength of the linear relationship between $Y$ and the set of variables $X_1, X_2, \ldots, X_P$. The multiple correlation has another useful property: It is the highest possible simple correlation between $Y$ and any linear combination of $X_1$ to $X_P$. This property explains why $R$ (the computed correlation) is never negative. In this sense the least squares regression plane maximizes the correlation between the set of $X$ variables and the dependent variable $Y$. It therefore presents a numerical measure of how well the regression plane fits the $Y$ values. When the multiple correlation $R$ is close to zero, the plane barely predicts $Y$ better than simply using $\bar{Y}$ to predict $Y$. A value of $R$ close to 1 indicates a very good fit.

As in the case of simple linear regression, discussed in Chapter 6, $R^2$ is an estimate of the proportional reduction in the variance of $Y$ achieved by fitting the plane. Again, the proportion of the standard deviation of $Y$ around the plane is estimated by $\sqrt{1 - R^2}$. If a test of significance is desired, the hypothesis

$$H_0: \mathscr{R} = 0$$

can be tested by the statistic

$$F = \frac{R^2/P}{(1 - R^2)/(N - P - 1)}$$

which is compared with a tabled $F$ with $P$ and $N - P - 1$ degrees of freedom. Since this hypothesis is equivalent to

$$H_0: \beta_1 = \beta_2 = \cdots = \beta_P = 0$$

the $F$ statistic is equivalent to the one calculated in Table 7.1. (See Section 8.4 for an explanation of adjusted multiple correlations.)

As mentioned earlier, the partial correlation coefficient is a measure of the linear relationship between two variables after adjusting for the linear effect of a group of other variables. For example, suppose a regression of $Y$ on $X_1$ and $X_2$ is fitted. The square of the partial correlation of $Y$ on $X_3$ after adjusting $X_1$ and $X_2$ is the proportion of the variance of $Y$ reduced by using $X_3$ as an additional $X$ variable. The hypothesis

$$H_0: \text{partial } \rho = 0$$

can be tested by using a test statistic similar to the one for simple $r$, namely,

$$t = \frac{(\text{partial } r)\sqrt{N - Q - 2}}{\sqrt{1 - (\text{partial } r^2)}}$$

where $Q$ is the number of variables adjusted for. The value of $t$ is compared with the tabled $t$ value with $N - Q - 2$ degrees of freedom. The square of the $t$ statistic is an $F$ statistic with df $= 1$ and $N - Q - 2$; the $F$ statistic may be used to test the same hypothesis against a two-sided alternative.

In the special case where $Y$ is regressed on $X_1$ to $X_P$, a test that the partial correlation between $Y$ and any of the $X$ variables, say $X_i$, is zero after adjusting for the remaining $X$ variables is equivalent to testing that the regression coefficient $\beta_i$ is zero.

In Chapter 8, where variable selection is discussed, partial correlations will play an important role.

## 7.8 RESIDUAL ANALYSIS AND TRANSFORMATIONS

In this section we discuss residual analysis, transformations, polynomial regression, and the incorporation of interaction terms into the equation.

### Residuals

The use of residuals in the case of simple linear regression was discussed in Section 6.8. Residual analysis is even more important in multiple regression analysis because the scatter diagrams of $Y$ and all the $X$ variables simultaneously cannot be portrayed on two-dimensional paper. The three types of outliers discussed in Section 6.8 (outliers in $Y$, outliers in $X$, and influential points) are usually detected by analysis of residuals in multiple regression analysis. Scatter plots of the various types of residuals against $Y$ and against each $X$ variable can be obtained from regression programs (see Table 7.7). Initial scanning of these plots can help the investigator identify individual potential outliers as well as suspect observations. It is important either to scan the residuals against each $X$ or to obtain measures of leverage as well as residuals in $Y$ to assess possible outliers. A listing of the residuals together with an identifying variable allows the investigator to determine which cases contain the suspected outliers. It is recommended that the investigator request a list of at least one statistic for identifying each of the three types of outliers. This ensures that sufficient information will be available for identification of cases having a major influence on the regression equation. The information on types of residuals and the various statistics given in Section 6.8 all apply here. The references cited in that section are also useful in the case of multiple regression. The formulas for some statistics, for example $h$, become more complicated, but conceptually the same interpretations apply. The values of $h$ are obtained from the diagonal of a matrix called the "hat" matrix (hence the use of the symbol $h$), but large values of $h$ still indicate outliers in at least one $X$. The statistics that are useful for detecting influential cases (Cook's distance, modified Cook's distance, and DFFITS) are available in multiple regression programs. The Durbin–Watson statistic or the serial correlation of residuals can be used to assess independence of residuals (see Section 6.8). Normal probability plots of the residuals can be used to assess normality of the error variable.

The similarity of the spread of the residuals around the horizontal line through zero allows the investigator to assess the validity of the homogeneity-of-variance assumption. If, for example, a *fan-shaped pattern* of residuals (Figure 6.7b) emerges as a function of one of the $X$ variables and not of the others, transforming this variable might help. On the other hand, a fan-shaped appearance of the residuals with each of the $X_i$ may suggest taking the logarithm or square root of the $Y$ values and performing the regression on the transformed values. As in the case of simple linear regression, transformations of $Y$ should be avoided, when possible, because they tend to obscure the interpretation of the regression equation (see Section 6.9).

## Transformations

In examining the adequacy of the multiple linear regression model, the investigator may wonder whether the transformations of some or all of the $X$ variables might improve the fit. (See Section 6.9 for review.) For this purpose the residual plots against the $X$ variables may suggest certain transformations. For example, if the residuals against $X_i$ take the humped-shaped form shown in Figure 6.7c, it may be useful to transform $X_i$. Reference to Figure 6.9a suggests that log $X_i$ may be the appropriate transformation. In this situation log $X_i$ can be used in place of or together with $X_i$ in the same equation. When both $X_i$ and log $X_i$ are used as predictors, this method of analysis is an instance of attempting to improve the fit by including an additional variable. Other commonly used additional variables are the square $(X_i^2)$ and square root $(X_i^{1/2})$ of appropriate $X_i$ variables; other candidates may be new variables altogether. Plots against candidate variables should be obtained in order to check whether the residuals are related to them.

Caution should be taken at this stage not to be deluged by a multitude of additional variables. For example, it may be preferable in certain cases to use log $X_i$ instead of $X_i$ and $X_i^2$; the interpretation of the equation becomes more difficult as the number of predictors increases. The subject of variable selection will be discussed in Chapter 8.

## Polynomial Regression

A subject related to variable transformations is the so-called *polynomial regression*. For example, suppose that the investigator starts with a single variable $X$ and that the scatter diagram of $Y$ on $X$ indicates a curvilinear function, as shown in Figure 7.5a. An appropriate regression equation in this case has the form

$$\hat{Y} = A + B_1X + B_2X^2$$

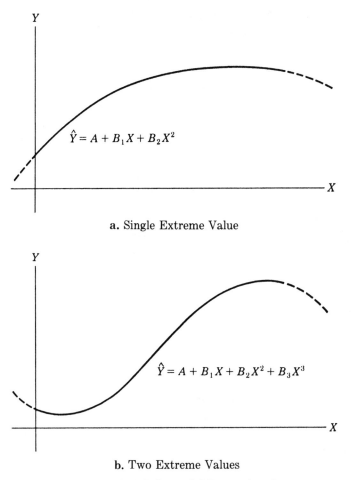

a. Single Extreme Value

b. Two Extreme Values

**FIGURE 7.5.** Hypothetical Polynomial Regression Curves

This equation is, in effect, a multiple regression equation with $X_1 = X$ and $X_2 = X^2$. Thus the multiple regression programs can be used to obtain the estimated curve. Similarly, for Figure 7.5b the regression curve has the form

$$\hat{Y} = A + B_1 X + B_2 X^2 + B_3 X^3$$

which is a multiple regression equation with $X_1 = X$, $X_2 = X^2$, and $X_3 = X^3$. Both of these equations are examples of polynomial regression equations with degrees two and three, respectively. Some computer packages have special programs that perform polynomial regression without the need for the user to make transformations (see Section 7.10).

## Interactions

Another issue in model fitting is to determine whether the $X$ variables interact. If the effects of two variables $X_i$ and $X_j$ are *not* interactive, then they appear as $B_i X_i + B_j X_j$ in the regression equation. In this case the effects of the two variables are said to be *additive*. Another way of expressing this concept is to say that there is *no interaction* between $X_i$ and $X_j$. If the additive terms for these variables do not completely specify their effects on the dependent variable $Y$, then interaction of $X_i$ and $X_j$ is said to be present. This phenomenon can be observed in many situations. For example, in chemical processes the additive effects of two agents are often not an accurate reflection of their combined effect since synergies or catalytic effects are often present. Similarly, in studies of economic growth the interactions between the two major factors labor and capital are important in predicting outcome. It is therefore often advisable to include additional variables in the regression equation to represent interactions. A commonly used practice is to add the product $X_i X_j$ to the set of $X$ variables to represent the interaction between $X_i$ and $X_j$. The use of so-called dummy variables is a method of incorporating interactions and will be discussed in Chapter 9.

As a check for the presence of interactions, tables can be constructed from a list of residuals, as follows: The ranges of $X_i$ and $X_j$ are divided into intervals, as shown in the examples in Tables 7.3 and 7.4. The residuals for each cell are found from the output and simply classified as positive or negative. The percentage of *positive residuals* in each cell is then recorded. If these percentages are around 50%, as shown in Table 7.3, then no

**TABLE 7.3.** Examination of Positive Residuals for Detecting Interactions: No Interaction

| | $X_j$ | | |
|---|---|---|---|
| $X_j$ | Low | Medium | High |
| Low | 50% | 52% | 48% |
| Medium | 52% | 48% | 50% |
| High | 49% | 50% | 51% |

**TABLE 7.4.** Examination of Positive Residuals for Detecting Interactions: Interactions Present

| | $X_j$ | | |
|---|---|---|---|
| $X_j$ | Low | Medium | High |
| Low | 20% | 40% | 45% |
| Medium | 40% | 50% | 60% |
| High | 55% | 60% | 80% |

interaction term is required. If the percentages vary greatly from 50% in the cells, as in Table 7.4, then an interaction term could most likely improve the fit. This search for interaction effects should be made after consideration of transformations and selection of additional variables.

## 7.9 OTHER OPTIONS IN COMPUTER PROGRAMS

Because regression analysis is a very commonly used technique, packaged programs offer a bewildering number of options to the user. But do not be tempted into using options that you do not understand well. As a guide in selecting options, we briefly discuss in this section some of the more popular ones. Other options are often, but not always, described in user guides.

The options for the use of *weights for the observations* and for *regression through the origin* were discussed in Section 6.10 for simple linear regression. These options are also available for multiple regression, and the discussion in Section 6.10 applies here as well. Be aware that regression through the origin means that the mean of $Y$ is assumed to be zero when *all* $X_i$ equal zero. This assumption is often unrealistic.

In Section 7.5 we described the matrix of correlations among the estimated slopes, and we indicated how this matrix can be used to find the covariances among the slopes. Covariances can be obtained directly as optional output in some programs.

## *Multicollinearity*

In practice, a problem called *multicollinearity* occurs when some of the $X$ variables are highly intercorrelated. A considerable amount of effort is devoted in the statistical literature to dealing with this problem (see Wonnacott and Wonnacott 1979; Chatterjee and Price 1977; Belsley, Kuh, and Welsch 1980). The main point is that when multicollinearity is present, the computed estimates of the regression coefficients are unstable and their interpretation becomes tenuous.

One simple way to check for multicollinearity is to examine the correlations among the $X$ variables. If, for example, $X_1$ and $X_2$ are highly correlated (say greater than 0.95), then it may be simplest to use only one of them since one variable conveys essentially all of the information contained in the other.

When more subtle patterns of correlations exist, the *tolerance option* available in many programs is useful. The tolerance associated with any variable $X_i$ is defined as 1 minus the squared multiple correlation between that $X_i$ and the remaining $X$ variables. When tolerance is small, say less than 0.01, then the investigator should discard the variable with the smallest tolerance. The inverse of the tolerance, called the *variance inflation factor* (VIF), is also printed for each variable by some programs. Some programs offer more complicated options for investigating multi-collinearity, but the use of tolerance as indicated is usually sufficient for handling the problem. If moderate multicollinearity is present, the investigator may consider using ridge regression, as discussed in Chapter 9, or principal components regression analysis discussed in Chapter 14.

## *Comparing Regression Planes*

Sometimes, it is desirable to examine the regression equations for *subgroups* of the population. For example, different regression equations for subgroups subjected to different treatments can be derived. In the program it is often convenient to designate the subgroups as various levels of a *grouping variable*. In the factory worker example, for instance, we presented the regression equation for 448 smokers:

$$FEV1 = -1.397 - 0.036(\text{age}) + 0.093(\text{height})$$

An equation was derived also for 231 nonsmokers, as

$$FEV1 = -0.906 - 0.030(\text{age}) + 0.084(\text{height})$$

In Table 7.5 the statistical results for both groups, as well as for smokers and nonsmokers combined, are presented. From the table we see that 448/679, or 66% of the males smoke. This result is much higher than the 42% male smokers found in the depression study described in Chapter

**TABLE 7.5.** Statistical Output from Factory Worker Example for Males Aged 20–64, Overall and by Smoking Status

| | Mean | Standard Deviation | Regression Coefficient | Standard Error | Standardized Regression Coefficient |
|---|---|---|---|---|---|
| **Overall ($N = 679$, df = 676)** | | | | | |
| Intercept | | | −1.163 | | 0 |
| Age | 38.45 | 12.09 | −0.034 | 0.002 | −0.525 |
| Height | 68.68 | 2.67 | 0.090 | 0.009 | 0.305 |
| FEV1 | 3.68 | 0.78 | | | |
| $R = 0.63$, $R^2 = 0.40$, $S = 0.61$ | | | | | |
| **Smokers ($N_S = 448$, df = 445)** | | | | | |
| Intercept | | | −1.397 | | 0 |
| Age | 39.43 | 11.87 | −0.036 | 0.003 | −0.517 |
| Height | 68.76 | 2.63 | 0.093 | 0.012 | 0.301 |
| FEV1 | 3.62 | 0.82 | | | |
| $R = 0.63$, $R^2 = 0.39$, $S = 0.64$ | | | | | |
| **Nonsmokers ($N_{ns} = 231$, df = 228)** | | | | | |
| Intercept | | | −0.906 | | 0 |
| Age | 36.57 | 12.33 | −0.030 | 0.003 | −0.527 |
| Height | 68.54 | 2.75 | 0.084 | 0.013 | 0.330 |
| FEV1 | 3.79 | 0.70 | | | |
| $R = 0.65$, $R^2 = 0.42$, $S = 0.54$ | | | | | |

3. The mean age for smokers among the factory workers is 3 years older than for nonsmokers, but the workers' heights are very similar.

In this type of analysis it is useful for an investigator to present the means and standard deviations along with the regression coefficients so that a reader of the report can try typical values of the independent variables in the regression equation in order to assess the numerical effects of the independent variables. Presentation of these statistics also enables readers to assess the characteristics of the sample under study. For example, suppose two 70-in.-tall males are 20 and 60 years old, respectively. From the two equations presented above we would predict FEV1 to be as follows:

| | Age (Years) | |
|---|---|---|
| Subgroup | 20 | 60 |
| Smoker | 4.39 | 2.95 |
| Nonsmoker | 4.37 | 3.17 |

Thus for the 20-year-old male the predicted FEV1 is almost the same whether or not he smokes, while for a 60-year-old male the FEV1 is predicted to be somewhat less if he smokes than it would be if he did not.

The standardized regression coefficient is also useful to the reader in assessing the relative effects of the two variables. Note that even though the unstandardized regression coefficients for age are smaller in magnitude than those for height, the magnitudes of the standardized regression coefficients for age are larger than those for height. Thus the relative contribution of age is greater than that of height.

If the regression coefficients are divided by their standard errors, then highly significant $t$ values are obtained. The levels of significance are sometimes included in the table, or asterisks are placed beside the regression coefficients to indicate their level of significance. The asterisks are then usually explained in footnotes to the table.

The standard errors could also be used to test equality of the individual coefficients for the two groups. For example, to compare the regression coefficients for height for the smokers and the nonsmokers and to test the null hypothesis of equal coefficients, we compute

$$Z = \frac{B_s - B_{ns}}{[SE^2(B_s) + SE^2(B_{ns})]^{1/2}}$$

or

$$Z = \frac{0.093 - 0.084}{(0.013^2 + 0.012^2)^{1/2}} = \frac{0.009}{(0.000313)^{1/2}} = 0.51$$

The computed value of $Z$ can be compared with the percentiles of the normal distribution in Appendix Table A.1 to obtain an approximate $P$ value for large samples. In this case the standard error for age was reported with too few significant digits, so its use would be questionable.

The standard deviations of the regression planes are useful to readers in making comparisons with their own results and in obtaining confidence intervals at the point where $X_i^* = \bar{X}$ (see Section 7.4).

The grouping option, as mentioned earlier, is useful in comparing relevant subgroups such as smokers and nonsmokers. An $F$ test is available as optional output from BMDP1R and the SAS GLM procedure to check whether the regression equations derived from the different subgroups are significantly different from each other. This test is an example of the general $F$ test that is often used in statistics and is given in its usual form in Section 8.5.

For our factory workers example the null hypothesis $H_0$ is that a single population plane for both smokers and nonsmokers combined is the true plane. The alternative hypothesis $H_1$ is that separate planes should be fitted for smokers and nonsmokers. The residual sum of squares obtained from $\Sigma(Y - \hat{Y})^2$ where the two separate planes are fitted will always be less than or equal to the residual sum of squares from the single plane for the overall sample. These residual sums of squares are printed in the analysis of variance table that accompanies the regression programs. Table 7.2 is an example of such a table for smokers.

The general $F$ test compares the residual sums of squares when a single plane is fitted to what is obtained when two planes are fitted. If these two residuals are not very different, then the null hypothesis cannot be rejected. If fitting two separate planes results in a much smaller residual sum of squares, then the null hypothesis is rejected. The smaller residual sum of squares for the two planes indicates that the regression coefficients differ beyond chance between the two groups. This difference could be a difference in either the intercepts or the slopes or both. In this test normality of errors is assumed.

The $F$ statistic for the test is

$$F = \frac{[SS_{res}(H_0) - SS_{res}(H_1)]/[df(H_0) - df(H_1)]}{SS_{res}(H_1)/df(H_1)}$$

In the factory workers example we have

$$F = \frac{[249.008 - (180.746 + 66.197)]/[676 - (445 + 228)]}{(180.746 + 66.197)/(445 + 228)}$$

where 249.008 is the residual sum of squares for the overall plane with 676 degrees of freedom, 180.746 is the residual sum of squares for the smokers with 445 degrees of freedom, and 66.197 is the residual sum of squares for the nonsmokers with 228 degrees of freedom. Thus

$$F = \frac{(249.008 - 246.943)/3}{246.943/673} = \frac{0.688}{0.367} = 1.88$$

with 3 and 673 degrees of freedom. The printed $P$ value from the program was 0.13, indicating that the two regression equations are not significantly different from each other.

Other topics of regression analysis and corresponding options will be discussed in the next three chapters. In particular, selecting the best predictors from a set of variables is given in Chapter 8.


## 7.10 DISCUSSION OF COMPUTER PROGRAMS

Almost every statistical package has a multiple regression program, although some packages have considerably fewer features than those found in BMDP, SAS, or SPSS–X. In this section, the 1R program from BMDP, the REG procedure from SAS, and the REGRESSION program from SPSS–X will be described. In Table 7.7, which summarizes the available options in these three programs, other programs are listed only if they contain useful additional features. Some available options are not listed in Table 7.7 since they will be discussed in Chapters 8 or 9. For example, the stepwise features in the REG procedure will be presented in Chapter 8, where methods of selecting independent variables are discussed. Also, the interactive features will be discussed in Chapter 8 as they are used mainly in variable selection. The three programs described in Table 7.7 are available for either a mainframe computer or a personal computer.

The regression analysis presented in Section 7.3 for the factory workers data was performed using the BMDP1R program run on a mainframe computer. Since this data set was used only for this analysis, no BMDP save file was made. The data are read from tape 9, which is one of the defined input tapes for the job control language on an IBM mainframe system. The data on the tape are represented as a series of codes for numeric digits or characters. The format statement is necessary to reassemble these individual digits or characters into real numbers, integers, or alphanumeric labels that represent the coded data variables. Note that we have skipped over some data on the tape using X indications in the format.

The BMDP statements are written in sentences that are grouped into paragraphs. Sentences always end with a period. Paragraphs are headed by a / (back slash) followed by the paragraph name. INPUT and VARIABLE paragraphs are the same for all BMDP programs (see Chapters 2, 3, and 4 in Volume 1 of the *BMDP Statistical Software Manual*).

```
/INPUT      UNIT=9.
            FORMAT IS '(18X, F2.0, 1X, F1.0, F2.0, 2F4.2, 15X, F1.0)'.
            VARIABLES ARE 6.
/VARIABLE   NAMES ARE AGE, SEX, HT, FVC, FEV1, SMOKER.
            GROUPING IS SMOKER.
```

In the INPUT paragraph, three essential pieces of information are given: where the data are located, the format of the data used, and the number of variables per case. The sentences used to describe where the data are located are quite specific to the data used for mainframe computers. In programs run on a personal computer, the data file name is specified directly in the input paragraph, such as FILE = 'FACTORY'. If the data file is in another directory, the file name must include the entire path. For example, FILE = '\USR\VAC\FACTORY'.

If there are missing values in the data set, it is important to limit the number of variables used by the program to those actually needed. The 1R program uses only complete cases, that is cases with no missing values. A variable that is read but not used in the regression equation could reduce the number of cases if it has missing values. One way to guard against this is to write a USE sentence in the VARIABLE paragraph, telling the program specifically what variables are to be used in the analysis. In the above example, a USE sentence was not needed, as all the variables in the FORMAT statement are used.

The format used for the factory workers data is a standard *F* format. The notation 18X tells the program to skip columns 1–18 and to start reading in column 19. The notation F2.0 specifies that the number is a whole number and occupies two columns (columns 19 and 20). The notation F4.2 signifies that the data have at most four digits and that the decimal point comes before the last two digits. The 2 in front of F4.2 (2F4.2) signifies that two numbers each with the F4.2 format occur.

The names used in the VARIABLE paragraph must be in the same order as the order used for the variables in the format sentence. The names must have eight or less characters and must be enclosed in apostrophes if they include blanks or symbols or do not begin with a letter. A grouping sentence is included to tell the computer to separate the cases into smokers and nonsmokers. Regression planes were obtained for all cases combined, for smokers, and for nonsmokers.

Five additional control statements are needed to complete the instructions for the computer. These statements for the factory workers example are given next and are explained in the paragraphs that follow.

```
/GROUP        CODES(6) ARE 1,2.
              NAMES(6) ARE YES,NO.
/REGRESSION   DEPENDENT = FEV1.
              INDEPENDENT = HT,AGE.
/PLOT         RESIDUAL.
              VARIABLES ARE HT,AGE.
              NORMAL.
/END
```

The paragraph GROUP is used to show how the program is to classify the cases into groups. The sentence CODES(6) ARE 1,2 says that the sixth variable (smoking status) takes on two values, 1 or 2. The names for the sixth variable are YES or NO, as signified in the next sentence.

The REGRESSION paragraph tells the computer which variables are the independent variables and which one is the dependent variable.

The PRINT paragraph lists which output, in addition to the standard or default output, the user wants printed. In this example just the standard output was needed, so this paragraph was omitted.

The PLOT paragraph specifies what scatter diagrams are to be printed. In this example, the residual and the residual squared versus the dependent variable are requested by using the RESIDUAL sentence. Then the VARIABLES ARE HT,AGE sentence is included to obtain the predicted and observed dependent variables plotted first against height. The residual is also plotted against height. To interpret this residual, we draw by hand a horizontal line through 0.0. If statistical tests or confidence limits are going to be reported, these residuals are assumed to be normally distributed. They should always be examined for outliers. Similar output follows for age. The normality of the residuals is examined by looking at the normal probability plot of the residuals obtained by including the NORMAL sentence.

The END paragraph tells the program that the investigator wishes to terminate the instructions for this problem. *Note*: Do not use a period after the word END.

We now have shown all the control statements for the factory workers regression analysis. The results shown in Table 7.5 were abstracted from the printout. In Table 7.6 we reproduce a part of the actual output for the total sample of smokers and nonsmokers. This output corresponds to the upper third of Table 7.5, where some rounding off was done. In addition to the items discussed in Table 7.5, Table 7.6 identifies the dependent variable, indicates the tolerance level used, and gives a printout of the ANOVA table. This ANOVA table presents a test of the hypothesis that $\beta_1$ and $\beta_2$ are both zero. The $P$ value is very low, indicating that the regression equation is useful in predicting FEV1.

Figure 7.6 shows a plot of the residuals from the regression plane plotted against height. The horizontal line at 0.0 has been drawn in. The average of all the residuals will always be zero. Note that there are no extreme

**TABLE 7.6.** BMDP1R Output for All Factory Workers

```
                    FEV1 ON HT,AGE 20-64

REGRESSION TITLE..........................FEV1 ON HT,AGE 20-64
DEPENDENT VARIABLE........................5 FEV1
TOLERANCE.................................0.0100
ALL DATA CONSIDERED AS A SINGLE GROUP

MULTIPLE R          0.6338
MULTIPLE R-SQUARE   0.4017          STD. ERROR OF EST.    0.6069
```

ANALYSIS OF VARIANCE

|  | SUM OF SQUARES | DF | MEAN SQUARE | F RATIO | P(TAIL) |
|---|---|---|---|---|---|
| REGRESSION | 167.150 | 2 | 83.575 | 226.887 | 0.00000 |
| RESIDUAL | 249.008 | 676 | 0.368 | | |

| VARIABLE | | COEFFICIENT | STD. ERROR | STD. REG COEFF | T | P(2 TAIL) | TOLERANCE |
|---|---|---|---|---|---|---|---|
| INTERCEPT | | -1.16281 | | | | | |
| AGE | 1 | -0.03403 | 0.002 | -0.525 | -17.562 | 0.0 | 0.989706 |
| HT | 3 | 0.08950 | 0.009 | 0.305 | 10.211 | 0.0 | 0.989705 |

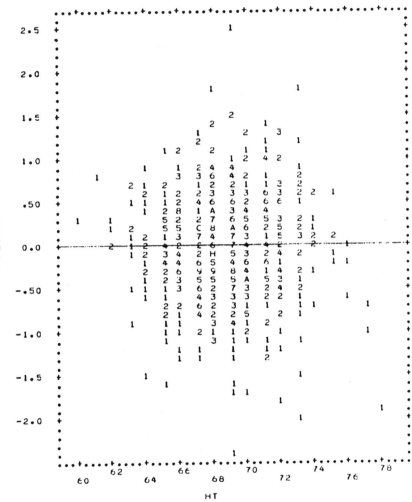

**FIGURE 7.6.** Residuals from Regression Plane Plotted against Height for Factory Workers; Data Set for Smokers and Nonsmokers

outliers, although several residuals may be suspect. Also, there is no clumping of residuals or fan-shaped pattern.

The normality of the residuals can be assessed by examining Figure 7.7, which gives the normal probability plot of the residuals from the regression plane of FEV1. The observed residuals are plotted on the horizontal axis. On the vertical axis the normal values of $Z$ are plotted, as described in

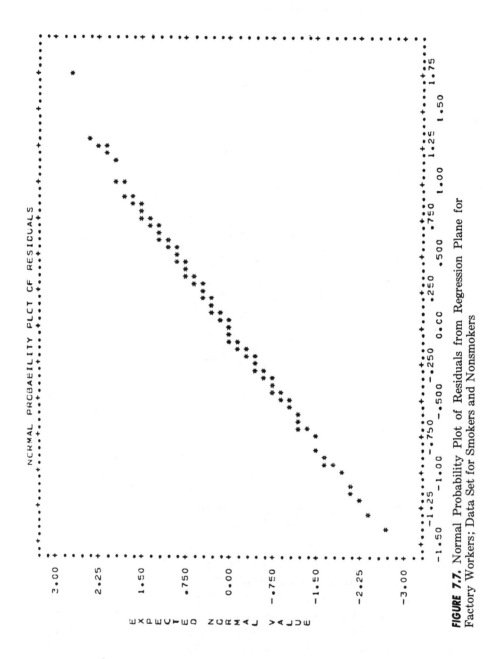

**FIGURE 7.7.** Normal Probability Plot of Residuals from Regression Plane for Factory Workers; Data Set for Smokers and Nonsmokers

Section 4.3. If the residuals are normally distributed, then the normal probability plot of the residuals should approximate a straight line. The extreme upper and lower points should be ignored, since they represent very few points. In this example the residuals appear to be either normally distributed or close enough to it so that the investigator can safely obtain confidence intervals or tests of hypotheses.

The use of weights mentioned in Section 6.10 is specific to each package. In BMDP, weights can be specified in the VARIABLE paragraph as well as in the INPUT paragraph of the regression program. In the VARIABLE paragraph, either "case frequency" or "case weights" can be used. If "case frequency" is used, the program assumes that the investigator has repeated cases with the same values and would enter a case three times if the case frequency was 3 for that case. The "case weights" option is useful, for example, when more than one determination may have been made by a laboratory. Suppose some observations are a single determination (weight $k_i = 1$), some an *average* of two determinations ($k_i = 2$), and so forth. Then observation number $i$ comes from a population with variance $\sigma^2/k_i$. If the "case weight" option is used, then the program would correctly estimate the population variance $\sigma^2$.

In addition to "case weights" and "case frequency" weights, any arbitrary weights can be used in the WEIGHT sentence of the REGRES-SION paragraph of BMDP1R.

When arbitrary weights are used in the "case weight" option, they should sum to the sample size $N$. If the weights (say $k_i$) do not sum to $N$, then we can use the transformation:

$$W_i = \frac{Nk_i}{\sum k_i}$$

These new weights, which add up to $N$, can be obtained by using the options available in the TRANSFORM paragraph (see Chapter 6 of the first manual).

The same regression analysis can also be done using the SAS procedure REG and data in the SAS data set LUNGFUN. The predicted values and residuals, along with the original variables, become another SAS data set and then PLOT is used to produce scatterplots of any pair of variables. An example of the statements needed to do this is given below and then described.

```
PROC REG DATA = LUNGFUN;
  MODEL FEV1 = HT AGE;
  BY SMOKER;
  OUTPUT OUT = LUNGPLT P = PRED R = RESID;
PROC PLOT DATA = LUNGPLT;
  PLOT RESID*FEV1 PRED*HT FEV1*HT;
```

Note that in both procedure statements the data set to be used was specified. If no data set is named, by default the data set most recently created or used in the program will be used again.

The MODEL statement in REG is required. It tells the computer what equation to fit by specifying the dependent variable to the left of the equality sign and the independent variables to the right. In the statements above, for example, we are asking for the regression plane of FEV1 on HT and AGE. A variety of options can be included in the MODEL statement by inserting a /, followed by the desired options, between the last independent variable and the semicolon. These options include a wide range of regression diagnostic statistics; consult the manual for more details.

All other statements are optional and can be listed in any order. We use the BY statement to perform regressions separately for smokers and nonsmokers. Note that the data set must be sorted by values of SMOKER. The OUTPUT statement creates a data set LUNGPLT that will automatically contain the original data, as well as any variables calculated during the regression analysis. We are requesting that the predicted values $P$ and the residuals $R$ be included in LUNGPLT with labels PRED and RESID, respectively. Finally, we ask for plots of the residuals against FEV1, and predicted and observed values of FEV1 against height.

Weighted regression is done in a straightforward manner by adding the statement WEIGHT = WVAR; in the regression procedure. The value of the variable WVAR is the relative weight of the corresponding observation.

The SPSS procedure REGRESSION performs multiple linear regression on both mainframe and personal computers, although there are differences between the two versions. These will be pointed out as we describe the procedure, repeating as an example the regression analysis of FEV1 on height and age.

REGRESSION consists of a series of subcommands, separated from each other by /. In SPSS/PC +, the final subcommand in a procedure must end in a period. The following code can be used to perform the regression analysis of FEV1 on HT and AGE, separately for smokers and non-smokers, in SPSS–X:

```
SORT CASES BY SMOKER
SPLIT FILE BY SMOKER
REGRESSION VARS = FEV1 HT AGE
/DEPENDENT = FEV1
/METHOD = ENTER HT AGE
/SCATTERPLOT (*RESID, FEV1) (FEV1,HT) (*PRED,HT)
```

The VARS subcommand names the variables to be used as independent or dependent variables in any of the regressions done during this call of the REGRESSION procedure (it is possible to fit more than one equation at a time). This subcommand is required in SPSS/PC + and optional in SPSS–X, where, if it is not used, all variables in the active data set are assumed available for analysis and a correlation matrix is calculated for all of them.

The DEPENDENT and METHOD subcommands are required in both versions of SPSS. The statement DEPENDENT = FEV1 indicates that the dependent variable is FEV1; all other variables, or all others named in the VARS subcommand, are now considered to be explanatory variables. The statement METHOD = ENTER HT AGE indicates that HT and AGE are to be used as independent variables. If only METHOD = ENTER is used, then all the independent variables (i.e., HT, AGE, and any others available from the VARS statement or by default) will be included in the regression. Other METHOD options will be listed in Chapter 8.

Scatterplots of the standardized residuals versus FEV1, as well as observed and standardized predicted values of FEV1 versus HT are produced with the SCATTERPLOT subcommand given above. Alternatively, the procedure PLOT, which gives the user more options and thus greater control over the final form of the graph, can be used. It *must* be used to get plots of unstandardized residuals and predicted values.

To perform regression (or any other analysis) separately for subsets of the data, it is necessary to sort the data with the SORT CASES command

**TABLE 7.7.** Summary of Computer Output from BMDP, SAS, and SPSS–X for Multiple Linear Regression

| Output | BMDP | SAS | SPSS-X |
|---|---|---|---|
| Matrix output | | | |
|   Covariance $X_i$ | 1R | CORR | REGRESSION |
|   Correlation $X_i$ | 1R | REG, CORR | REGRESSION |
|   Correlation $B_i$ | 1R | REG | REGRESSION |
| Regression equation | | | |
|   Slope coefficients, $B$ | 1R | REG | REGRESSION |
|   Intercept, $A$ | 1R | REG | REGRESSION |
|   Standardized $B$ | 1R | REG | REGRESSION |
|   Standard Error of $B$ | 1R | REG | REGRESSION |
|   Standard Error of $A$ | 6R,9R | REG | REGRESSION |
|   Test $\beta = 0$ | 1R | REG | REGRESSION |
|   Test $\alpha = 0$ | | REG | REGRESSION |
|   $P$ value for test $\beta = 0$ | 1R | REG | REGRESSION |
|   ANOVA | 1R | REG | REGRESSION |
|   Multiple correlation | 1R | REG | REGRESSION |
|   Test $\mathscr{R} = 0$ | 1R | REG | REGRESSION |
|   Partial correlations | 6R | REG | REGRESSION |
| Additional Options | | | |
|   Weighted regression | 1R | REG | REGRESSION |
|   Regression through zero | 1R | REG | REGRESSION |
|   Regression for subgroups | 1R | REG | REGRESSION |
|   Test equality of subgroups | 1R | GLM | |
|   Highlighting points in plots | | REG | |
| Outliers in $Y$ | | | |
|   Raw residual plots | 1R | REG | REGRESSION |
|   Raw residual lists | 1R | REG | REGRESSION |
|   Studentized residual plots | 2R* | REG | REGRESSION |
|   Studentized residual lists | 2R | REG | REGRESSION |
|   Deleted studentized residual plots | 2R | REG | REGRESSION |
|   Deleted studentized residual lists | 2R | REG | REGRESSION |
|   Other measures | 2R | REG | REGRESSION |
| Outliers in $X$ | | | |
|   $h$ "hat" plots | 2R | REG | |
|   $h$ "hat" lists | 2R | REG | |
|   Other leverage measures | 2R | | REGRESSION |
| Influential observations | | | |
|   Cook's distance $D$ | 2R | REG | REGRESSION |
|   Modified Cook's $D$ | 2R | | |
|   DFFITS | 2R | REG | |
|   Other influence measures | 2R | REG | |
| Checks for Other Assumptions | | | |
|   Tolerance | 1R | REG | REGRESSION |
|   Other multicollinearity measures | | REG | |
|   Serial correlation residuals | 1R | REG | |
|   Durbin–Watson test | 9R | REG | REGRESSION |
|   Normal probability plots | 1R | UNIVARIATE | REGRESSION |
|   Scatter plot $Y$ versus $X_i$ | 6D,1R | REG | REGRESSION |
|   Test homogeneity of variance | | REG | |

* Similar residual measures are available in 9R.

and then define the groups by the SPLIT FILE command. For weighted regression (not used in this example), either the WEIGHT command or REGWGT subcommand can be used. The WEIGHT command is entered before the REGRESSION procedure and, unless it is preceded by a TEMPORARY command, the weights are applied to the data for all subsequent analyses in the program. The REGWGT command is a subcommand in the REGRESSION procedure.

A useful feature of the REGRESSION procedure is that the CASE-WISE subcommand with the ALL option presents the standardized residuals in a form that makes it very easy to identify cases that are possible outliers. A vertical plot is given, with the listing of the values of the cases on the same line.

Some of the most used items of output are shown in Table 7.7 for the BMDP, SAS, and SPSS–X programs. These programs also include other output, which will be discussed in Chapters 8 and 9.

Regression analysis is one of the most widely used statistical techniques, so the programs have included a wide array of possible options, which can make it difficult to interpret the output. We recommend that you carefully review objectives in running a program before submitting program statements, to determine just what is needed. A second run can always be used to obtain additional output. The output should be read, and the results that are significant to you should be either highlighted, encircled, or written out in English on the printed output while it is still fresh. Many researchers find it useful to write a paragraph describing the output on the first page of the computer printout, which is equivalent to a paragraph in the results section of an article or a report. It is very easy to forget why a program was run and what the output signifies after a few months. Therefore such notations on the output can be useful reminders.


## 7.11 WHAT TO WATCH OUT FOR

The discussion and cautionary remarks given in Section 6.13 also apply to multiple regression analysis. In addition, because it is not possible to display all the results in a two-dimensional scatter diagram in the case of multiple regression, checks for some of the assumptions (such as independent normally distributed error terms) are somewhat more difficult to perform. But since regression programs are so widely used, they contain a large number of options to assist in checking the data. At a minimum, outliers and normality should routinely be checked. At least one measure of the three types (outliers in $Y$, outliers in $X$, and influential points) should be obtained along with a normal probability plot of the residuals against $\hat{Y}$ and against each $X$.

Often two investigators collecting similar data at two locations obtain slope coefficients that are not similar in magnitude in their multiple regression equation. One explanation of this inconsistency is that if the $X$ variables are not independent, the magnitudes of the coefficients depend on which other variables are included in the regression equation. If the two investigators do not have the same variables in their equations, then they should not expect the slope coefficients for a specific variable to be the same. This is one reason why they are called *partial regression coefficients*. For example, suppose $Y$ can be predicted very well from $X_1$ and $X_2$ but investigator A ignores the $X_2$. The expected value of the slope coefficient $B_1$ for investigator A will be different from $\beta_1$, the expected value for the second investigator. Hence, they will tend to get different results. The difference will be small if the correlation between the two $X$ variables is small, but if they are highly correlated even the sign of the slope coefficient can change. In order to get results that can be easily interpreted, it is important to include the proper variables in the regression model. In addition, the variables should be included in the model in the proper form such that the linearity assumption holds.

Some investigators discard all the variables that are not significant at a certain $P$ value, say $P < 0.05$, but this can sometimes lead to problems in interpreting other slope coefficients, as they can be altered by the omission of that variable. More on variable selection will be given in Chapter 8.

Another problem that can arise in multiple regression is multicollinearity, mentioned in Section 7.9. This is more apt to happen in economic data than in data on individuals, due to the high variability among persons. The programs will warn the user if this occurs. The problem is sometimes solved by discarding a variable or, if this does not make sense because of the problems discussed in the above paragraph, special techniques exist to obtain estimates of the slope coefficients when there is multicollinearity (see Chapter 9).

Another practical problem in performing multiple regression analyses is shrinking sample size. If many $X$ variables are used and each has some missing values, then many cases may be excluded because of missing values. The programs themselves can contribute to this problem. Suppose you specify a series of regression programs with slightly different $X$ variables. To save computer time the SAS REG procedure only computes the covariance or correlation matrix once using all the variables. Hence, any variable that has a missing value for a case causes that case to be excluded from *all* of the regression analyses, not just from the analyses that include that variable.

When a large portion of the sample is missing, then the problems of making inferences regarding the parent population can be difficult. In any case, the sample size should be "large" relative to the number of variables in order to obtain stable results. Some statisticians believe that if the sample size is not at least five to ten times the number of variables, then interpretation of the results is risky.

## SUMMARY

In this chapter we presented a subject in which the use of the computer is almost a must. The concepts underlying multiple regression were presented along with a fairly detailed discussion of packaged regression programs. Although much of the material in this chapter is found in standard textbooks, we have emphasized an understanding and presentation of the output of packaged programs. This philosophy is one we will follow in the rest of the book. In fact, in future chapters the emphasis on interpretation of output will be even more pronounced.

Certain topics included in this chapter but not covered in usual courses include how to handle interactions among the $X$ variables and how to compare regression equations from different subgroups. We also included discussion of recent developments, such as the examination of residuals and influential observations. A review of the literature on these and other topics in regression analysis is found in Hocking (1983).

Several texts contain excellent presentations of the subject of regression analysis. A partial list includes Neter, Wasserman, and Kutner (1985); Chatterjee and Price (1977); Wonnacott and Wonnacott (1979); Gunst and Mason (1980); Draper and Smith (1981); Achen (1982); and Myers (1986).

# BIBLIOGRAPHY

Achen, C. H. 1982. *Interpreting and using regression.* Beverly Hills: Sage.

Afifi, A. A., and Azen, S. P. 1979. *Statistical analysis: A computer oriented approach.* 2nd ed. New York: Academic Press.

Anscombe, F. J., and Tukey, J. W. 1963. The examination and analysis of residuals. *Technometrics* 5:141–160.

Atkinson, A. C. 1985. *Plots, transformations and regressions.* New York: Oxford University Press.

Belsley, D. A., Kuh, E., and Welsch, R. E. 1980. *Regression diagnostics: Identifying influential data and sources of collinearity.* New York: Wiley.

Brownlee, K. A. 1965. *Statistical theory and methodology in science and engineering.* 2nd ed. New York: Wiley.

Carroll, R. J., and Ruppert, D. 1988. *Transformation and weighting in regression.* London: Chapman and Hall.

Chatterjee, S., and Price, B. 1977. *Regression analysis by example.* New York: Wiley.

Cook, R. D. 1977. Detection of influential observations in linear regression. *Technometrics* 19:15–18.

*————. 1979. Influential observations in linear regression. *Journal of the American Statistical Association* 74:169–174.

*Cook, R. D., and Weisberg, S. 1980. Characterization of an empirical influence function for detecting influential cases in regression. *Technometrics* 22:495–508.

*Cook, R. D., and Weisberg, S. 1982. *Residuals and influence in regression.* London: Chapman and Hall.

Daniel, C., and Wood, F. S. 1980. *Fitting equations to data.* 2nd ed. New York: Wiley.

*Devlin, S. J., Gnanadesekan, R., and Kettenring, J. R. 1981. Robust estimation of dispersion matrices and principal components. *Journal of the American Statistical Association* 76:354–362.

Draper, N. R., and Smith, H. 1981. *Applied regression analysis.* 2nd ed. New York: Wiley.

Dunn, O. J., and Clark, V. A. 1987. *Applied statistics: Analysis of variance and regression.* 2nd ed. New York: Wiley.

*Graybill, F. A. 1976. *Theory and application of the linear model.* N. Scituate, Mass.: Duxbury Press.

Gunst, R. F., and Mason, R. L. 1980. *Regression analysis and its application.* New York: Dekker.

Hald, A. 1952. *Statistical theory with engineering applications.* New York: Wiley.

Hanushek, E. A., and Jackson, J. E. 1977. *Statistical methods for social scientists.* New York: Academic Press.

Hocking, R. R. 1983. Developments in linear regression methodology. *Technometrics* 25:219–229.

Myers, R. H. 1986. *Classical and modern regression with applications.* Boston: Duxbury Press.

Neter, J., Wasserman, W., and Kutner, M. H. 1985. *Applied linear statistical models.* Homewood, Ill.: Irwin.

Theil, H. 1971. *Principles of econometrics.* New York: Wiley.

Thorndike, R. M. 1978. *Correlational procedures for research.* New York: Gardner Press.

Velleman, P. F., and Welsch, R. E. 1981. Efficient computing of regression diagnostics. *The American Statistician.* 35:234–242.

Williams, E. J. 1959. *Regression analysis.* New York: Wiley.

Wonnacott, R. J., and Wonnacott, T. H. 1979. *Econometrics.* 2nd ed. New York: Wiley.

Younger, M. S. 1979. *A handbook for linear regression.* N. Scituate, Mass.: Duxbury Press.

## PROBLEMS

7.1 Using the chemical companies data given in Table 8.1, predict the price/earnings (P/E) ratio from the debt-to-equity (D/E) ratio, the annual dividends divided by the latest 12-months' earnings-per-share (PAYOUTR1), and the percentage net profit margin (NPM1). Obtain the correlation matrix, and check the intercorrelations of the variables. Summarize the results, including appropriate tests of hypotheses.

7.2 Fit regression planes for FEV1 on height and age for mothers alone, fathers alone, and overall from the lung function data given in Appendix B. Summarize the results in tabular form. Test whether the two regression planes for fathers and mothers are significantly different.

7.3 Compare the regression equation for fathers obtained in Problem 7.2 with the equation for nonsmokers given in Section 7.9.

7.4 In Problem 7.2, compute 95% confidence and prediction intervals for a 40-year-old father who is 68 in. tall. Do the same for a mother with the same age and height. Comment.

7.5 From the depression data set described in Table 3.2, predict the reported level of depression as given by CESD, using INCOME and AGE as independent variables, for females and males separately. Analyze the residuals and decide whether or not it is reasonable to assume that they follow a normal distribution.

7.6    The control cards for the BMDP6D program run on the factory workers data set used in Chapter 6 are as follows:

```
/PROBLEM      TITLE = 'FEV1 ON HT FOR MALE WORKERS AGE 20–24'.
/INPUT        UNIT = 9.
              FORMAT IS '(18X,F2.0,1X,F1.0,F2.0,2F4.2)'.
              CASES ARE 679.
              VARIABLES ARE 5.
/VARIABLE     NAMES ARE AGE,SEX,HT,FVC,FEV1.
              MAXIMUM IS (4)7.2,(5)7.2.
              MINIMUM IS (4)2,(5)1.
/TRANSFORM    IF (SEX NE 1 OR AGE GE 25 OR AGE LT 20) THEN USE = 0.
/PLOT         YVAR IS FEV1,FVC.
              XVAR IS HT.
              CROSS.
              STATISTICS.
/END
```

Only the first plot with FEV1 was given in Chapter 6. In this data set the variable SEX is coded 1 if male and 2 if female. Read the TRANSFORM and the VARIABLE paragraphs instructions in the manual to see how the cases are restricted to the desired ones. Also, check why the above instructions are used in the PLOT paragraph.

7.7    Create a hypothetical data set, which you will use for exercises in this chapter and some of the following chapters. Begin by generating 100 independent cases for each of 10 variables according to the standard normal distribution (means = 0, variances = one). Call these variables X1, X2, ..., X9, Y. This data set can be obtained by using BMDP1D, starting without input variables and generating them by means of transformation statements. The program specifications follow, interspersed with explanations. These specifications were prepared for a microcomputer and therefore are in lowercase letters.

```
/problem     title = 'generation of multivariate normal data'.
/input       format = free.
             cases = 100.
             variables = 0.
/variable    names are x1,x2,x3,x4,x5,x6,x7,x8,x9,y.
             add = 10.
```

```
/transform   x1 = rndg(36541).
             x2 = rndg(43893).
             x3 = rndg(45671).
             x4 = rndg(65431).
             x5 = rndg(98753).
             x6 = rndg(78965).
             x7 = rndg(67893).
             x8 = rndg(34521).
             x9 = rndg(98431).
             y = rndg(67895).
```

We now have 10 independent, random, normal numbers for each of 100 cases. The population mean of each variable is 0, and the population standard deviation is 1. Further transformations are done in order to make the variables intercorrelated. The transformations are accomplished by making some of the variables functions of other variables, as follows:

```
             x1 = 5*x1.
             x2 = 3*x2.
             x3 = x1 + x2 + 4*x3.
             x4 = x4.
             x5 = 4*x5.
             x6 = x5 − x4 + 6*x6.
             x7 = 2*x7.
             x8 = x7 + 2*x8.
             x9 = 4*x9.
             y = 5 + x1 + 2*x2 + x3 + 10*y.
/print       data.
/end
```

We now have created a random sample of 100 cases on 10 variables: X1, X2, X3, X4, X5, X6, X7, X8, X9, Y. The population distribution is multivariate normal. It can be shown that the population means and variances are as follows:

| Population | X1 | X2 | X3 | X4 | X5 | X6 | X7 | X8 | X9 | Y |
|---|---|---|---|---|---|---|---|---|---|---|
| Mean | 0 | 0 | 0 | 0 | 0 | 0 | 0 | 0 | 0 | 5 |
| Variance | 25 | 9 | 50 | 1 | 16 | 53 | 4 | 8 | 16 | 297 |

The population correlation matrix is as follows:

| | X1 | X2 | X3 | X4 | X5 | X6 | X7 | X8 | X9 | Y |
|----|------|------|------|------|------|-------|------|------|----|------|
| X1 | 1 | 0 | 0.71 | 0 | 0 | 0 | 0 | 0 | 0 | 0.58 |
| X2 | 0 | 1 | 0.42 | 0 | 0 | 0 | 0 | 0 | 0 | 0.52 |
| X3 | 0.71 | 0.42 | 1 | 0 | 0 | 0 | 0 | 0 | 0 | 0.76 |
| X4 | 0 | 0 | 0 | 1 | 0 | −0.14 | 0 | 0 | 0 | 0 |
| X5 | 0 | 0 | 0 | 0 | 1 | 0.55 | 0 | 0 | 0 | 0 |
| X6 | 0 | 0 | 0 | −0.14 | 0.55 | 1 | 0 | 0 | 0 | 0 |
| X7 | 0 | 0 | 0 | 0 | 0 | 0 | 1 | 0.71 | 0 | 0 |
| X8 | 0 | 0 | 0 | 0 | 0 | 0 | 0.71 | 1 | 0 | 0 |
| X9 | 0 | 0 | 0 | 0 | 0 | 0 | 0 | 0 | 1 | 0 |
| Y | 0.58 | 0.52 | 0.76 | 0 | 0 | 0 | 0 | 0 | 0 | 1 |

The population squared multiple correlation coefficient between Y and X1 to X9 is 0.6633, between Y and X1, X2, X3 is 0.6633, and between Y and X4 to X9 is zero. Also, the population regression line of Y on X1 to X9 has $\alpha = 5$, $\beta_1 = 1$, $\beta_2 = 2$, $\beta_3 = 1$, $\beta_4 = \beta_5 = \cdots = \beta_9 = 0$. The following table lists the first five cases from a sample obtained from the above program specification with its particular random number initializations:

| X1 | X2 | X3 | X4 | X5 |
|---------|--------|---------|---------|--------|
| −9.280 | 0.356 | −10.986 | 1.816 | −1.127 |
| −13.124 | 0.108 | −10.260 | 0.412 | −3.560 |
| −4.583 | 6.216 | 6.400 | −0.0383 | −5.559 |
| 10.387 | −0.442 | 10.709 | −1.146 | −0.381 |
| −4.261 | −1.341 | − 1.642 | −1.198 | 1.584 |

| X6 | X7 | X8 | X9 | Y |
|--------|--------|--------|--------|---------|
| −2.455 | 1.335 | 0.972 | −2.224 | −8.316 |
| −2.942 | 0.360 | −1.683 | −2.693 | −14.214 |
| 0.627 | 3.192 | 2.244 | 0.341 | 27.687 |
| 8.420 | 2.350 | 2.578 | 2.455 | 16.473 |
| −0.271 | −0.169 | −2.733 | −4.454 | 11.423 |

Now, using the data you generated, obtain the sample statistics from a computer packaged program, and compare them with the parameters given above. Using the Bonferroni inequality, test the simple correlations, and determine which are significantly different from zero. Comment.

7.8 Repeat Problem 7.7, using SAS.

7.9 Continuation of Problem 7.7: Calculate the population partial correlation coefficient between X2 and X3 after removing the linear effect of X1. Is it larger or smaller than $\rho_{23}$? Explain. Also, obtain the corresponding sample partial correlation. Test whether it is equal to zero.

7.10 Continuation of Problem 7.7: Using a multiple regression program, perform an analysis, with the dependent variable = Y and the independent variables = X1 to X9, on the 100 generated cases. Summarize the results and state whether they came out the way you expected them to, considering how the data were generated. Perform appropriate tests of hypotheses. Comment.

7.11 Continuation of Problem 7.5: Fit a regression plane for CESD on INCOME and AGE for males and females combined. Test whether the regression plane is helpful in predicting the values of CESD. Find a 95% prediction interval for a female with INCOME = 17 and AGE = 29 using this regression. Do the same using the regression calculated in Problem 7.5, and compare.

7.12 Continuation of Problem 7.11: For the regression of CESD on INCOME and AGE, choose 15 observations that appear to be influential or outlying. State your criteria, delete these points, and repeat the regression. Summarize the differences in the results of the regression analyses.

7.13 For the lung function data in Appendix B, find the regression of FEV1 on weight and height for the fathers. Divide each of the two explanatory variables into two intervals: greater than, and less than or equal to the respective median. Is there an interaction between the two explanatory variables?

7.14 Continuation of Problem 7.13: Find the partial correlation of FEV1 and age given height for the oldest child, and compare it to the simple correlation between FEV1 and age of the oldest child. Is either one significantly different from zero? Based on these results and without doing any further calculations, comment on the relationship between OCHEIGHT and OCAGE in these data.

7.15 Continuation of Problem 7.13: a. For the oldest child, find the regression of FEV1 on i) weight and age; ii) height and age; iii) height, weight, and age. Compare the three regression equations. In each regression, which coefficients are significantly different from zero? b. Find the correlation matrix for the four variables. Which pair of variables is most highly correlated? least correlated? Heuristically, how might this explain the results of part a?

7.16 Repeat Problem 7.15a for fathers' measurements instead of those of the oldest children. Are the regression coefficients more stable? Why?

## Chapter Eight

# VARIABLE SELECTION IN REGRESSION ANALYSIS

## 8.1 WHAT WILL YOU LEARN FROM THIS CHAPTER?

From this chapter you will learn how to select independent variables from the entire set of possible variables. This selection may be necessary for either descriptive or predictive purposes. In particular, you will learn:

- When variable selection methods are used (8.2, 8.3).
- How to use criteria for variable selection (8.4).
- About a general method for testing the significance of the contribution of a subgroup of independent variables (8.5).
- How to use forward and backward variable selection (8.6).
- How to use stepwise selection (8.6).
- About stopping rules for deciding how many and which variables to include (8.6).
- How to use all possible subsets and the best subsets of variables (8.7).
- How to force variables that you know are important into the equation (8.6, 8.7).
- Which options are available in the computer programs (8.8).

- About stagewise regression (8.9).
- What to watch out for in variable selection (8.10).

## 8.2 WHEN ARE VARIABLE SELECTION METHODS USED?

In Chapter 7 it was assumed that the investigators knew which variables they wished to include in the model. This is usually the case in the fixed-$X$ model, where often only two to five variables are being considered. It also is the case frequently when the investigator is working from a theoretical model and is choosing the variables that fit this model. But sometimes the investigator has only partial knowledge or is interested in finding out what variables can be used to best predict a given dependent variable.

Variable selection methods are used mainly in exploratory situations, where many independent variables have been measured and a final model explaining the dependent variable has not been reached. Variable selection methods are useful, for example, in a survey situation in which numerous characteristics and responses of each individual are recorded in order to determine some characteristic of the respondent, such as level of depression. The investigator may have prior justification for using certain variables but may be open to suggestions for the remaining variables. For example, age and gender have been shown to relate to levels of the CESD (the depression variable given in the depression study described in Chapter 3). An investigator might wish to enter age and gender as independent variables and try additional variables in an exploratory fashion. The set of independent variables can be broken down into logical subsets. First, the usual demographic variables that will be entered first, namely age and gender. Second, a set of variables that other investigators have shown to affect CESD, such as perceived health status. Finally, the investigator may wish to explore the relative value of including another set of variables after the effects of age, gender, and perceived health status have been taken into account. In this case, it is partly a model-driven regression analysis and partly an exploratory regression analysis. The programs described in this chapter allow analysis that is either partially or completely exploratory. The researcher may have one of two goals in mind:

1. To use the resulting regression equation to identify variables that best explain the level of the independent variable. The equation is then obtained for *descriptive* or *explanatory purposes.*

**2.** To obtain an equation that predicts the level of the dependent variable with as little error as possible. The purpose is then a *predictive purpose.*

The variable selection methods discussed in this chapter can sometimes serve one purpose better than the other, as will be discussed after the methods are presented.

## 8.3 DATA EXAMPLE

Table 8.1 presents various characteristics reported by the 30 largest chemical companies; the data are taken from a January 1981 issue of *Forbes.* This data set will henceforth be called the chemical companies data.

The variables listed in Table 8.1 are defined as follows (see Brigham 1980):

- $P/E$: Price-to-earnings ratio, which is the price of one share of common stock divided by the earnings per share for the past year. This ratio shows the dollar amount investors are willing to pay for the stock per dollar of current earnings of the company.

- $ROR5$: Percent rate of return on total capital (invested plus debt) averaged over the past five years.

- $D/E$: Debt-to-equity (invested capital) ratio for the past year. This ratio indicates the extent to which management is using borrowed funds to operate the company.

- $SALESGR5$: Percent annual compound growth rate of sales, computed from the most recent five years compared with the previous five years.

- $EPS5$: Percent annual compound growth in earnings per share, computed from the most recent five years compared with the preceding five years.

- $NPM1$: Percent net profit margin, which is the net profits divided by the net sales for the past year, expressed as a percentage.

- $PAYOUTR1$: Annual dividend divided by the latest 12-month earnings per share. This value represents the proportion of earnings paid out to shareholders rather than retained to operate and expand the company.

**TABLE 8.1.** Chemical Companies Financial Performance

| Company | P/E | ROR5 | D/E | SALESGR5 | EPS5 | NPM1 | PAYOUTR1 |
|---|---|---|---|---|---|---|---|
| Diamond Shamrock | 9 | 13.0% | 0.7 | 20.2% | 15.5% | 7.2% | 0.43 |
| Dow Chemical | 8 | 13.0% | 0.7 | 17.2% | 12.7% | 7.3% | 0.38 |
| Stauffer Chemical | 8 | 13.0% | 0.4 | 14.5% | 15.1% | 7.9% | 0.41 |
| E. I. du Pont | 9 | 12.2% | 0.2 | 12.9% | 11.1% | 5.4% | 0.57 |
| Union Carbide | 5 | 10.0% | 0.4 | 13.6% | 8.0% | 6.7% | 0.32 |
| Pennwalt | 6 | 9.8% | 0.5 | 12.1% | 14.5% | 3.8% | 0.51 |
| W. R. Grace | 10 | 9.9% | 0.5 | 10.2% | 7.0% | 4.8% | 0.38 |
| Hercules | 9 | 10.3% | 0.3 | 11.4% | 8.7% | 4.5% | 0.48 |
| Monsanto | 11 | 9.5% | 0.4 | 13.5% | 5.9% | 3.5% | 0.57 |
| American Cyanamid | 9 | 9.9% | 0.4 | 12.1% | 4.2% | 4.6% | 0.49 |
| Celanese | 7 | 7.9% | 0.4 | 10.8% | 16.0% | 3.4% | 0.49 |
| Allied Chemical | 7 | 7.3% | 0.5 | 15.4% | 4.9% | 5.1% | 0.27 |
| Rohm & Haas | 7 | 7.8% | 0.4 | 11.0% | 3.0% | 5.6% | 0.32 |
| Reichhold Chemicals | 10 | 6.5% | 0.4 | 18.7% | − 3.1% | 1.3% | 0.38 |
| Lubrizol | 13 | 24.9% | 0.0 | 16.2% | 16.9% | 12.5% | 0.32 |
| Nalco Chemical | 14 | 24.6% | 0.0 | 16.1% | 16.9% | 11.2% | 0.47 |
| Sun Chemical | 5 | 14.9% | 1.1 | 13.7% | 48.9% | 5.8% | 0.10 |
| Cabot | 6 | 13.8% | 0.6 | 20.9% | 36.0% | 10.9% | 0.16 |
| International Minerals & Chemical | 10 | 13.5% | 0.5 | 14.3% | 16.0% | 8.4% | 0.40 |
| Dexter | 12 | 14.9% | 0.3 | 29.1% | 22.8% | 4.9% | 0.36 |
| Freeport Minerals | 14 | 15.4% | 0.3 | 15.2% | 15.1% | 21.9% | 0.23 |
| Air Products & Chemicals | 13 | 11.6% | 0.4 | 18.7% | 22.1% | 8.1% | 0.20 |
| Mallinckrodt | 12 | 14.2% | 0.2 | 16.7% | 18.7% | 8.2% | 0.37 |
| Thiokol | 12 | 13.8% | 0.1 | 12.6% | 18.0% | 5.6% | 0.34 |
| Witco Chemical | 7 | 12.0% | 0.5 | 15.0% | 14.9% | 3.6% | 0.36 |
| Ethyl | 7 | 11.0% | 0.3 | 12.8% | 10.8% | 5.0% | 0.34 |
| Ferro | 6 | 13.8% | 0.2 | 14.9% | 9.6% | 4.4% | 0.31 |
| Liquid Air of North America | 12 | 11.5% | 0.4 | 15.4% | 11.7% | 7.2% | 0.51 |
| Williams Companies | 9 | 6.4% | 0.7 | 16.1% | − 2.8% | 6.8% | 0.22 |
| Akzona | 14 | 3.8% | 0.6 | 6.8% | −11.1% | 0.9% | 1.00 |
| | | | | | | | |
| Mean | 9.37 | 12.01% | 0.42 | 14.94% | 12.93% | 6.55% | 0.39 |
| Standard deviation | 2.80 | 4.50% | 0.23 | 4.05% | 11.15% | 3.92% | 0.16 |

Data abstracted from *Forbes*, 127, no. 1 (January 5, 1981).

The P/E ratio is usually high for growth stocks and low for mature or troubled firms. Company managers generally want high P/E ratios, because high ratios make it possible to raise substantial amounts of capital for a small number of shares and make acquisitions of other companies easier. Also, investors consider P/E ratios of companies, both over time and in relation to similar companies, as a factor in valuation of stocks for possible purchase and/or sale. Therefore it is of interest to investigate which of the other variables reported in Table 8.1 influence the level of the P/E ratio. In this chapter we use the regression of the P/E ratio on the remaining variables to illustrate various variable selection procedures. Note that all the firms of the data set are from the chemical industry; such a regression equation could vary appreciably from one industry to another.

To get a preliminary impression of the data, examine the means and standard deviations shown at the bottom of Table 8.1. Also examine Table 8.2, which shows the simple correlation matrix. Note that P/E, the dependent variable, is most highly correlated with D/E. The association, however, is negative. Thus investors tend to pay less per dollar earned for companies with relatively heavy indebtedness. The sample correlations of P/E with ROR5, NPM1, and PAYOUTR1 range from 0.32 to 0.35 and thus are quite similar. Investors tend to pay more per dollar earned for stocks of companies with higher rates of return on total capital, higher net profit margins, and higher proportion of earnings paid out to them as dividends. Similar interpretations can be made for the other correlations.

If a single independent variable were to be selected for "explaining" P/E, the variable of choice would be D/E since it is the most highly correlated. But clearly there are other variables that represent differences

**TABLE 8.2.** Correlation Matrix of Chemical Companies Data

|          | P/E   | ROR5  | D/E   | SALESGR5 | EPS5  | NPM1  | PAYOUTR1 |
|----------|-------|-------|-------|----------|-------|-------|----------|
| P/E      | 1     |       |       |          |       |       |          |
| ROR5     | 0.32  | 1     |       |          |       |       |          |
| D/E      | −0.47 | −0.46 | 1     |          |       |       |          |
| SALESGR5 | 0.13  | 0.36  | −0.02 | 1        |       |       |          |
| EPS5     | −0.20 | 0.56  | 0.19  | 0.39     | 1     |       |          |
| NPM1     | 0.35  | 0.62  | −0.22 | 0.25     | 0.15  | 1     |          |
| PAYOUTR1 | 0.33  | −0.30 | −0.16 | −0.45    | −0.57 | −0.43 | 1        |

in management style, dynamics of growth, and efficiency of operation that should also be considered.

It is possible to derive a regression equation using all the independent variables, as discussed in Chapter 7. However, some of the independent variables are strongly interrelated. For example, from Table 8.2 we see that growth of earnings (EPS5) and growth of sales (SALESGR5) are positively correlated, suggesting that both variables measure related aspects of dynamically growing companies. Further, growth of earnings (EPS5) and growth of sales (SALESGR5) are positively correlated with return on total capital (ROR5), suggesting that dynamically growing companies show higher returns than mature, stable companies. In contrast, growth of earnings (EPS5) and growth of sales (SALESGR45) are both negatively correlated with the proportion of earnings paid out to stockholders (PAYOUTR1), suggesting that earnings must be plowed back into operations to achieve growth. The profit margin (NPM1) shows the highest positive correlation with rate of return (ROR5), suggesting that efficient conversion of sales into earnings is consistent with high return on total capital employed.

Since the independent variables are interrelated, it may be better to use a subset of the independent variables to derive the regression equation. Most investigators prefer an equation with a small number of variables since such an equation will be easier to interpret. For future predictive purposes it is often possible to do at least as well with a subset as with the total set of independent variables. Methods and criteria for subset selection are given in the subsequent sections. It should be noted that these methods are fairly sensitive to gross outliers. The data sets should therefore be carefully edited prior to analysis.

## 8.4 CRITERIA FOR VARIABLE SELECTION

In many situations where regression analysis is useful, the investigator has strong justification for including certain variables in the equation. The justification may be to produce results comparable to previous studies or to conform to accepted theory. But often the investigator has no preconceived assessment of the importance of some or all of the independent variables. It is in the latter situation that variable selection procedures can be useful.

Any variable selection procedure requires a criterion for deciding how many and which variables to select. As discussed in Chapter 7, the least squares method of estimation minimizes the *residual sum of squares* (RSS) about the regression plane [$RSS = \Sigma (Y - \hat{Y})^2$]. Therefore an implicit criterion is the value of RSS. In deciding between alternative subsets of variables, the investigator would select the one producing the smaller RSS if this criterion were used in a mechanical fashion. Note, however, that

$$RSS = \sum (Y - \bar{Y})^2 (1 - R^2)$$

where $R$ is the multiple correlation coefficient. Therefore minimizing RSS is equivalent to maximizing the multiple correlation coefficient. If the criterion of maximizing $R$ were used, the investigator would always select all of the independent variables, because the value of $R$ will never decrease by including additional variables.

Since the multiple correlation coefficient, on the average, overestimates the population correlation coefficient, the investigator may be misled into including too many variables. For example, if the population multiple correlation coefficient is, in fact, equal to zero, the average of all possible values of $R^2$ from samples of size $N$ from a multivariate normal population is $P/(N - 1)$, where $P$ is the number of independent variables (see Kendall and Stuart 1979). An estimated multiple correlation coefficient that reduces the bias is the *adjusted multiple correlation coefficient*, denoted by $\bar{R}$. It is related to $R$ by the following equation:

$$\bar{R}^2 = R^2 - \frac{P(1 - R^2)}{N - P - 1}$$

where $P$ is the number of independent variables *in the equation* (see Theil 1971). Note that in this chapter the notation $P$ will sometimes signify less than the total number of available independent variables.

The investigator may proceed now to select the independent variables that maximize $\bar{R}^2$. Note that excluding some independent variables may, in fact, result in a higher value of $\bar{R}^2$. As will be seen, this result occurs for the chemical companies data. Note also that if $N$ is very large relative to $P$, then $\bar{R}^2$ will be approximately equal to $R^2$. Conversely, maximizing $\bar{R}^2$ can give different results from those obtained in maximizing $R^2$ when $N$ is small.

Another method suggested by statisticians is to minimize the *residual mean square*, which is defined as

$$\text{RMS or RES. MS.} = \frac{\text{RSS}}{N - P - 1}$$

This quantity is related to the adjusted $\bar{R}^2$ as follows:

$$\bar{R}^2 = 1 - \frac{(\text{RMS})}{S_Y^2}$$

where

$$S_Y^2 = \frac{\sum (Y - \bar{Y})^2}{N - 1}$$

Since $S_Y^2$ does not involve any independent variables, minimizing RMS is equivalent to maximizing $\bar{R}^2$.

Another quantity used in variable selection and found in standard computer packages is the so-called $C_p$ *criterion*. The theoretical derivation of this quantity is beyond the scope of this book (see Mallows 1973). However, $C_p$ can be expressed as follows:

$$C_p = (N - P - 1)\left(\frac{\text{RMS}}{\hat{\sigma}^2} - 1\right) + (P + 1)$$

where RMS is the residual mean square based on the $P$ selected variables and $\hat{\sigma}^2$ is the RMS derived from the total set of independent variables. The quantity $C_p$ is the sum of two components, $(P + 1)$ and the remainder of the expression. While $(P + 1)$ increases as we choose more independent variables, the other part of $C_p$ will tend to decrease. When all variables are chosen, $P + 1$ is at its maximum but the other part of $C_p$ is zero since RMS $= \hat{\sigma}^2$. Many investigators recommend selecting those independent variables that minimize the value of $C_p$.

Another criterion, which is appropriate for the variable-$X$ case, is the theoretical *average of the mean squared error* averaged over the values of the $X$ variables. An estimate of this quantity is found in Bendel and Afifi

(1977). The variables can be selected to minimize this average mean squared error, which is equivalent to minimizing the quantity

$$U_p = \frac{1 - R^2}{(N - P - 1)(N - P - 2)}$$

The above criteria are highly interrelated. No criterion will serve under all circumstances, and the criteria should not be used mechanically, particularly when the sample size is small. Practical methods for using these criteria for variable selection will be discussed in Sections 8.6 and 8.7. Hocking (1976) compared these and other criteria and recommends the following uses:

1. In a given sample the value of the unadjusted $R^2$ can be used as a measure of data description. Many investigators exclude variables that add only a very small amount to the $R^2$ value obtained from the remaining variables.

2. If the object of the regression is extrapolation or estimation of the regression parameters, the investigator is advised to select those variables that maximize the adjusted $\bar{R}^2$ (or, equivalently, minimize RMS).

3. For prediction, finding the variables that make the value of $C_p$ approximately equal to $P + 1$ is a reasonable strategy.

Bendel and Afifi (1977) also examined these and other criteria in the context of stepwise regression (see Section 8.6). They found that minimizing $C_p$ or $U_p$ often selects the best variables when $N$ is large relative to $P$.

The discussion above was concerned with criteria for judging among alternative subsets of independent variables. How to select the candidate subsets is our next subject of concern. Before describing methods to implement the selection, though, we first discuss the use of the $F$ test for determining the effectiveness of a subset of independent variables relative to the total set.

## 8.5 A GENERAL F TEST

Suppose we are convinced that the variables $X_1, X_2, \ldots, X_P$ should be used in the regression equation. Suppose also that measurements on $Q$ additional variables, $X_{P+1}, X_{P+2}, \ldots, X_{P+Q}$, are available. Before deciding whether any of the additional variables should be included, we can test the hypothesis that, as a group, the $Q$ variables do not improve the regression equation.

If the regression equation in the population has the form

$$Y = \alpha + \beta_1 X_1 + \beta_2 X_2 + \cdots + \beta_P X_P + \beta_{P+1} X_{P+1} + \cdots + \beta_{P+Q} X_{P+Q} + e$$

we test the hypothesis $H_0: \beta_{P+1} = \beta_{P+2} = \cdots = \beta_{P+Q} = 0$. To perform the test, we first obtain an equation that includes all the $P + Q$ variables, and we obtain the residual sum of squares $(\text{RSS}_{P+Q})$. Similarly, we obtain an equation that includes only the first $P$ variables and the corresponding residual sum of squares $(\text{RSS}_P)$. Then the test statistic is computed as

$$F = \frac{(\text{RSS}_P - \text{RSS}_{P+Q})/Q}{\text{RSS}_{P+Q}/(N - P - Q - 1)}$$

The numerator measures the improvement in the equation from using the additional $Q$ variables. This quantity is never negative. The hypothesis is rejected if the computed $F$ exceeds the tabled $F(1 - \alpha)$ with $Q$ and $N - P - Q - 1$ degrees of freedom.

This very general test is sometimes referred to as the *generalized linear hypothesis test*. Essentially, this same test was used in Section 7.9 to test whether or not it is necessary to report the regression analyses by subgroups. The quantities $P$ and $Q$ can take on any integer values greater than or equal to one. For example, suppose that six variables are available. If we take $P$ equal to 5 and $Q$ equal to 1, then we are testing $H_0: \beta_6 = 0$ in the equation $Y = \alpha + \beta_1 X_1 + \beta_2 X_2 + \beta_3 X_3 + \beta_4 X_4 + \beta_5 X_5 + \beta_6 X_6 + e$. This test is the same as the test that was discussed in Section 7.6 for the significance of a single regression coefficient.

As another example, in the chemical companies data it was already observed that D/E is the best single predictor of the P/E ratio. A relevant hypothesis is whether the remaining five variables improve the prediction obtained by D/E alone. Two regressions were run (one with all six variables and one with just D/E), and the results were as follows:

$$\text{RSS}_6 = 103.86$$

$$\text{RSS}_1 = 176.08$$

$$P = 1, Q = 5$$

Therefore the test statistic is

$$F = \frac{(176.08 - 103.86)/5}{103.86/(30 - 1 - 5 - 1)} = \frac{14.44}{4.52} = 3.20$$

with 5 and 23 degrees of freedom. Comparing this value with the value found in Appendix Table A.4, we find the $P$ value to be less than 0.025. Thus at the 5% significance level we conclude that one or more of the other five variables significantly improves the prediction of the P/E ratio. Note, however, that the D/E ratio is the most highly correlated variable with the P/E ratio.

This selection process affects the inference so that the true $P$ value of the test is unknown, but it is perhaps greater than 0.025. Strictly speaking, the general linear hypothesis test is valid only when the hypothesis is determined prior to examining the data.

The general linear hypothesis test is the basis for several selection procedures, as will be discussed in the next two sections.

## 8.6 STEPWISE REGRESSION

The variable selection problem can be described as considering certain subsets of independent variables and selecting that subset that either maximizes or minimizes an appropriate criterion. Two obvious subsets are the *best single variable* and the *complete set of independent variables*, as considered in Section 8.5. The problem lies in selecting an *intermediate subset* that may be better than both these extremes. In this section we discuss some methods for making this choice.

### Forward Selection Method

Selecting the best single variable is a simple matter: We choose the variable with the highest absolute value of the simple correlation with $Y$. In the chemical companies data example D/E is the best single predictor of P/E because the correlation between D/E and P/E is $-0.47$, which has a larger magnitude than any other correlation with P/E (see Table 8.2).

To choose a second variable to combine with D/E, we could naively select NPM1 since it has the second highest absolute correlation with P/E

(0.35). This choice may not be wise, however, since another variable together with D/E may give a higher multiple correlation than D/E combined with NPM1. One strategy therefore is to search for that variable that maximizes multiple $R^2$ when combined with D/E. Note that this procedure is equivalent to choosing the second variable to minimize the residual sum of squares given that D/E is kept in the equation. This procedure is also equivalent to choosing the second variable to maximize the magnitude of the partial correlation with P/E after removing the linear effect of D/E. The variable thus selected will also maximize the $F$ statistic for testing that the above-mentioned partial correlation is zero (see Section 7.7). This $F$ test is a special application of the general linear hypothesis test described in Section 8.5.

In the chemical companies data example the partial correlations between P/E and each of the other variables after removing the linear effect of D/E are as follows:

| | *Partial Correlations* |
| --- | --- |
| ROR5 | 0.126 |
| SALESGR5 | 0.138 |
| EPS5 | −0.114 |
| NPM1 | 0.285 |
| PAYOUTR1 | 0.286 |

Therefore, the method described above, called the *forward selection method*, would choose D/E as the first variable and PAYOUTR1 as the second variable. Note that it is almost a toss-up between PAYOUTR1 and NPM1 as the second variable. The computer programs implementing this method will ignore such reasoning and select PAYOUTR1 as the second variable because 0.286 is greater than 0.285. But to the investigator the choice may be more difficult since NPM1 and PAYOUTR1 measure quite different aspects of the companies.

Similarly, the third variable chosen by the forward selection procedure is the variable with the highest absolute partial correlation with P/E after removing the linear effects of D/E and PAYOUTR1. Again, this procedure is equivalent to maximizing multiple $R$ and $F$ and minimizing the residual mean square, given that D/E and PAYOUTR1 are kept in the equation.

The forward selection method proceeds in this manner, each time adding one variable to the variables previously selected, until a specified *stopping rule* is satisfied. The most commonly used stopping rule in packaged programs is based on the $F$ test of the hypothesis that the partial correlation of the variable entered is equal to zero. One version of the stopping rule terminates entering variables when the computed value of $F$ is less than a specified value. This cutoff value is often called the *minimum F-to-enter*.

Equivalently, the $P$ value corresponding to the computed $F$ statistic could be calculated and the forward selection stopped when this $P$ value is greater than a specified level. Note that here also the $P$ value is affected by the fact that the variables are selected from the data and therefore should not be used in the hypothesis-testing context.

Bendel and Afifi (1977) compared various levels of the minimum $F$-to-enter used in forward selection. A recommended value is the $F$ percentile corresponding to a $P$ value equal to 0.15. For example, if the sample size is large, the recommended minimum $F$-to-enter is the 85th percentile of the $F$ distribution with 1 and $\infty$ degrees of freedom, or 2.07.

Any of the criteria discussed in Section 8.4 could also be used as the basis for a stopping rule. For instance, the multiple $R^2$ is always presented in the output of the commonly used programs whenever an additional variable is selected. The user can examine this series of values of $R^2$ and stop the process when the increase in $R^2$ is a very small amount. Alternatively, the series of adjusted $\bar{R}^2$ values can be examined. The process stops when the adjusted $\bar{R}^2$ is maximized.

As an example, data from the chemical companies analysis are given in Table 8.3. Using the $P$ value of 0.15 (or the minimum $F$ of 2.07) as a cutoff, we would enter only the first four variables since the computed $F$-to-enter for the fifth variable is only 0.37. In examining the multiple $R^2$, we note

**TABLE 8.3.** Forward Selection of Variables in Chemical Companies Data Example

| Variables Added | Computed F-to-enter | Multiple $R^2$ | Multiple $\bar{R}^2$ |
|---|---|---|---|
| 1. D/E | 8.09 | 0.224 | 0.197 |
| 2. PAYOUTR1 | 2.41 | 0.288 | 0.235 |
| 3. NPM1 | 9.05 | 0.472 | 0.411 |
| 4. SALESGR5 | 3.33 | 0.534 | 0.459 |
| 5. EPS5 | 0.37 | 0.541 | 0.445 |
| 6. ROR5 | 0.08 | 0.542 | 0.423 |

that going from five to six variables increased $R^2$ only by 0.001. It is therefore obvious that the sixth variable should not be entered. However, other readers may disagree about the practical value of a 0.007 increase in $R^2$ obtained by including the fifth variable. We would be inclined not to include it. Finally, in examination of the adjusted multiple $\bar{R}^2$ we note that it is also maximized by including only the first four variables.

This selection of variables by the forward selection method for the chemical companies data agrees well with our understanding gained from study of the correlations between variables. As we noted in Section 8.3, return on total capital (ROR5) is correlated with NPM1, SALESGR5, and EPS5 and so adds little to the regression when they are already included. Similarly, the correlation between SALESGR5 and EPS5, both measuring aspects of company growth, corroborates the result shown in Table 8.3 that EPS5 adds little to the regression after SALESGR5 is already included.

It is interesting to note that the computed $F$-to-enter follows no particular pattern as we include additional variables. Note also that even though $R^2$ is always increasing, the amount by which it increases varies as new variables are added. Finally, this example verifies that the adjusted multiple $\bar{R}^2$ increases to a maximum and then decreases.

The other criteria discussed in Section 8.4 are $C_p$ and $U_p$. These criteria can also be used for stopping rules. Specifically, the combination of variables minimizing their values can be chosen. A numerical example for this procedure will be given in Section 8.7.

Computer programs may also terminate the process of forward selection when the tolerance level (see Section 7.9) is smaller than a specified minimum tolerance.

## Backward Elimination Method

An alternative strategy for variable selection is the *backward elimination method*. This technique begins with all of the variables in the equation and proceeds by eliminating the least useful variables one at a time.

As an example, Table 8.4 lists the regression coefficient (both standardized and unstandardized), the $P$ value, and the corresponding computed $F$ for testing that each coefficient is zero for the chemical company data. The $F$ statistic here is called the *computed F-to-remove*.

Since ROR5 has the smallest computed $F$-to-remove, it is a candidate for removal. The user must specify the *maximum F-to-remove*; if the

**TABLE 8.4.** Computed Coefficients and $F$ and $P$ Values for Chemical Companies Data

| Variable | Coefficients | | Computed F-to-Remove | P |
| | Unstandardized | Standardized | | |
|---|---|---|---|---|
| Intercept | 1.26 | — | — | — |
| D/E | −2.51 | −0.20 | 1.03 | 0.32 |
| PAYOUTR1 | 9.76 | 0.57 | 8.48 | 0.01 |
| NPM1 | 0.35 | 0.49 | 6.25 | 0.02 |
| SALESGR5 | 0.20 | 0.29 | 3.13 | 0.09 |
| EPS5 | −0.04 | −0.15 | 0.40 | 0.53 |
| ROR5 | 0.05 | 0.08 | 0.08 | 0.78 |

computed $F$-to-remove is less than that maximum, ROR5 is removed. No recommended value can be given for the maximum $F$-to-remove. We suggest, however, that a reasonable choice is the 70th percentile of the $F$ distribution (or, equivalently, $P = 0.30$). For a large value of the residual degrees of freedom, this procedure results in a maximum $F$-to-remove of 1.07. In Table 8.4, since the computed $F$-to-remove for ROR5 is 0.08, this variable is removed first. Note that ROR5 also has the smallest standardized regression coefficient and the largest $P$ value.

The backward elimination procedure proceeds by computing a new equation with the remaining five variables and examining the computed $F$-to-remove for another likely candidate. The process continues until no variable can be removed according to the stopping rule.

In comparing the forward selection and the backward elimination methods, we note that one advantage of the former is that it involves a smaller amount of computation than the latter. However, it may happen that two or more variables can together be a good predictive set while each variable taken alone is not very effective. In this case backward elimination would produce a better equation than forward selection. Neither method is expected to produce the best possible equation for a given number of variables to be included (other than one or the total set).

## Stepwise Procedure

One very commonly used technique that combines both of the above methods is called the *stepwise procedure*. In fact, the forward selection method is often called the *forward stepwise method*. In this case a step consists of adding a variable to the predictive equation. At step 0 the only "variable" used is the mean of the $Y$ variable. At that step the program normally prints the computed $F$-to-enter for each variable. At step 1 the variable with the highest computed $F$-to-enter is entered; and so forth.

Similarly, the backward elimination method is often called the *backward stepwise method*. At step 0 the computed $F$-to-remove is calculated for each variable. In successive steps the variables are removed one at a time, as described above.

The standard stepwise regression programs do forward selection with the option of removing some variables already selected. Thus at step 0 only the $Y$ mean is included. At step 1 the variable with the highest computed $F$-to-enter is selected. At step 2 a second variable is entered if any variable qualifies (i.e., if at least one computed $F$-to-enter is greater than the minimum $F$-to-enter). After the second variable is entered, the $F$-to-remove is computed for both variables. If either of them is lower than the maximum $F$-to-remove, that variable is removed. If not, a third variable is included if its computed $F$-to-enter is large enough. In successive steps this process is repeated. For a given equation, variables with small enough computed $F$-to-remove values are removed, and the variables with large enough computed $F$-to-enter values are included. The process stops when no variables can be deleted or added.

The choice of the *minimum $F$-to-enter* and the *maximum $F$-to-remove* affects both the nature of the selection process and the number of variables selected. For instance, if the maximum $F$-to-remove is much smaller than the minimum $F$-to-enter, then the process is essentially forward selection. In any case the minimum $F$-to-enter must be larger than the maximum $F$-to-remove; otherwise, the same variable will be entered and removed continuously. In many situations it is useful for the investigator to examine the full sequence until all variables are entered. This step can be accomplished by setting the minimum $F$-to-enter equal to a small value, such as 0.1 (or a corresponding $P$ value of 0.99). In that case the maximum $F$-to-remove must then be smaller than 0.1. After examining this sequence, the investigator can make a second run, using other $F$ values. An alternative using BMDP2R is to choose a double set of $F$-to-enter and $F$-to-remove values. The first set is chosen with small $F$ levels and allows numerous variables, but not necessarily all, to enter in a forward stepwise mode. After all the variables are entered that meet this liberal set of $F$

levels, the program operates in a backward stepwise manner using the second set of $F$ levels, which are the levels the investigator wishes to actually use (see the manual for example). In this way, stepwise is used on not all of the candidate variables but rather on those that have a better chance of being useful.

Values that have been found to be useful in practice are a minimum $F$-to-enter equal to 2.07 and a maximum $F$-to-remove of 1.07. With these recommended values, for example, the results for the chemical companies data are identical to those of the forward selection method presented in Table 8.3 up to step 4. At step 4 the variables in the equation and their computed $F$-to-remove are as shown in Table 8.5. Also shown in the table are the variables not in the equation and their computed $F$-to-enter. Since all the computed $F$-to-remove values are larger than the minimum $F$-to-remove of 1.07, none of the variables are removed. Also, since both of the computed $F$-to-enter values are smaller than the minimum $F$-to-enter of 2.07, no new variables are entered. The process terminates with four variables in the equation.

There are many situations in which the investigator may wish to include certain variables in the equation. For instance, the theory underlying the subject of the study may dictate that a certain variable or variables be used. The investigator may also wish to include certain variables in order for the results to be comparable with previously published studies. Similarly, the investigator may wish to consider two variables when both provide nearly the same computed $F$-to-enter. If the variable with the slightly lower $F$-to-enter is the preferred variable from

**TABLE 8.5.** Step 4 of the Stepwise Procedure for Chemical Companies Data Example

| Variables in Equation | Computed F-to-remove | Variables Not in Equation | Computed F-to-enter |
|---|---|---|---|
| D/E | 3.03 | | |
| PAYOUTR1 | 13.03 | | |
| NPM1 | 9.37 | | |
| SALESGR5 | 3.33 | | |
| | | EPS5 | 0.37 |
| | | ROR5 | 0.03 |

other viewpoints, the investigator may force it to be included. This preference may arise from cost or simplicity considerations. Most stepwise computer programs offer the user the option of *forcing variables* at the beginning of or later in the stepwise sequence, as desired.

We emphasize that none of the procedures described here are guaranteed or even expected to produce the best possible regression equation. This comment applies particularly to the intermediate steps where the number of variables selected is more than one but less than $P - 1$. In the majority of applications of the stepwise method, investigators attempt to select two to five variables from the available set of independent variables. Programs that examine the best possible subsets of a given size are also available. These programs are discussed next.

## 8.7 SUBSET REGRESSION

The availability of the computer makes it possible to compute the multiple $R^2$ and the regression equation for all possible subsets of variables. For a specified subset size the "best" subset of variables is the one that has the largest multiple $R^2$. The investigator can thus compare the best subsets of different sizes and choose the preferred subset. The preference is based on a compromise between the value of the multiple $R^2$ and the subset size. In Section 8.4 three criteria were discussed for making this comparison, namely, adjusted $\bar{R}^2$, $C_p$, and $U_p$.

For a small number of independent variables, say three to five, the investigator may indeed be well advised to obtain all possible subsets. This technique allows detailed examination of all the regression models so that an appropriate one can be selected. Programs are available for computing the necessary statistics for all possible subsets.

If the number of independent variables is larger than five, the number of possible subsets becomes too large for practical purposes. Furnival and Wilson (1974) have devised a technique for finding the subset of a given size (number of independent variables used) having a maximum $R^2$ without examining all possible subsets. Furthermore, this technique will identify the nearly best subsets of a given size. This technique has been implemented in BMDP9R. The program will further select the subsets that either (1) maximize $R^2$, (2) maximize adjusted $\bar{R}^2$, or (3) minimize $C_p$

for *different* subset sizes. For a given size these three criteria are equivalent. But if size is not specified, they may select subsets of different sizes.

Table 8.6 includes part of the output produced by running BMDP9R for the chemical companies data. As before, the best single variable is D/E, with a multiple $R^2$ of 0.224. Included in Table 8.6 are the next two best candidates if only one variable is to be used. Note that if either NPM1 or PAYOUTR1 is chosen instead of D/E, then the value of $R^2$ drops by about one-half. The relatively low value of $R^2$ for any of these variables reinforces our understanding that any one independent variable alone does not do well in "explaining" the variation in the dependent variable P/E.

The best combination of two variables is NPM1 and PAYOUTR1, with a multiple $R^2$ of 0.404. This combination does not include D/E, as would be the case in stepwise regression (Table 8.3). In stepwise regression the two variables selected were D/E and PAYOUTRI with a multiple $R^2$ of 0.228. So in this case stepwise regression does not come close to selecting the best combination of two variables. Here stepwise regression resulted in the second-best choice. Also, the third-best combination of two variables is essentially as good as the second best. The best combination of two variables, NPM1 and PAYOUTR1, as chosen by BMDP9R, is interesting in light of our earlier interpretation of the variables in Section 8.3. Variable NPM1 measures the efficiency of the operation in converting sales to earnings, while PAYOUTR1 measures the intention to plow earnings back into the company or distribute them to stockholders. These are quite different aspects of *current* company behavior. In contrast, the debt-to-equity ratio (D/E) may be, in large part, a *historical* carry-over from past operations or a reflection of management style.

For the best combination of three variables the value of the multiple $R^2$ is 0.477, only slightly better than the stepwise choice. If D/E is a lot simpler to obtain than SALESGR5, the stepwise selection might be preferred since the loss in the multiple $R^2$ is negligible. Here the investigator, when given the option of different subsets, might prefer the first (NPM1, PAYOUTR1, SALESGR5) on theoretical grounds, since it is the only option that explicitly includes a measure of growth (SALESGR5). (You should also examine the four-variable combinations in light of the above discussion.)

**TABLE 8.6.** Best Three Subsets for One, Two, Three, or Four Variables for Chemical Companies Data

| Number of Variables | Names of Variables | $R^2$ | Adjusted $\overline{R}^2$ | $C_p$ | $U_p \times 1000$ |
|---|---|---|---|---|---|
| 1 | D/E* | 0.224 | 0.197 | 13.0 | 1.03 |
| 1 | NPM1 | 0.123 | 0.092 | 18.1 | |
| 1 | PAYOUTR1 | 0.106 | 0.074 | 18.9 | |
| 2 | NPM1, PAYOUTR1 | 0.404 | 0.360 | 6.0 | 0.85 |
| 2 | D/E, PAYOUTR1* | 0.288 | 0.235 | 11.8 | |
| 2 | D/E, NPM1 | 0.287 | 0.234 | 11.8 | |
| 3 | NPM1, PAYOUTR1, SALESGR5 | 0.477 | 0.417 | 4.3 | 0.80 |
| 3 | D/E, PAYOUTR1, NPM1* | 0.472 | 0.410 | 4.6 | |
| 3 | NPM1, PAYOUTR1, ROR5 | 0.428 | 0.362 | 6.8 | |
| 4 | D/E, NPM1, PAYOUTR1, SALESGR5 | 0.534 | 0.459 | 3.4 | 0.78 |
| 4 | NPM1, PAYOUTR1, SALESGR5, EPS5* | 0.494 | 0.413 | 5.4 | |
| 4 | NPM1, PAYOUTR1, SALESGR5, ROR5 | 0.484 | 0.402 | 5.9 | |
| Best 5 | D/E, NPM1, PAYOUTR1, SALESGR5, EPS5* | 0.541 | 0.445 | 5.1 | 0.83 |
| All 6 | All variables | 0.542 | 0.423 | 7.0 | 0.90 |

*Combinations selected by the stepwise procedure.

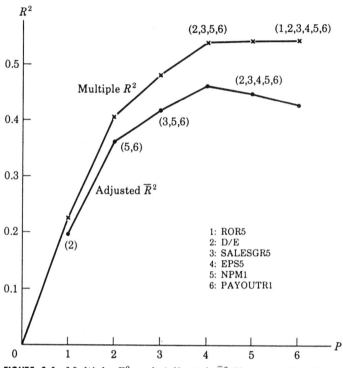

**FIGURE 8.1.** Multiple $R^2$ and Adjusted $\bar{R}^2$ Versus $P$ for Best Subset with $P$ Variables for Chemical Companies Data

Summarizing the results in the form of Table 8.6 is advisable. Then graphing the best combination of one, two, three, four, five, or six variables helps the investigator decide how many variables to use. For example, Figure 8.1 shows a plot of the multiple $R^2$ and the adjusted $\bar{R}^2$ versus the number of variables included in the best combination for the chemical companies data. Note that $R^2$ is a nondecreasing function. However, it levels off after four variables. The adjusted $\bar{R}^2$ reaches its maximum with four variables (D/E, NPM1, PAYOUTR1, and SALESGR5) and decreases with five and six variables.

Figure 8.2 shows $C_p$ versus $P$ (the number of variables) for the best combinations for the chemical companies data. The same combination of four variables selected by the $\bar{R}^2$ criterion minimizes $C_p$. A similar graph in Figure 8.3 shows that $U_p$ is also minimized by the same choice of four variables.

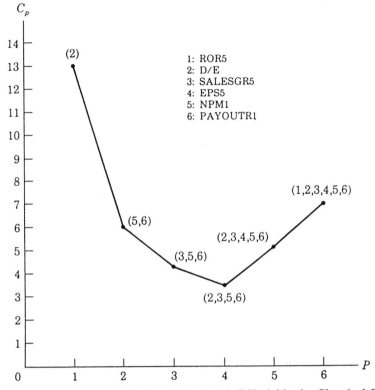

**FIGURE 8.2.** $C_p$ Versus $P$ for Best Subset with $P$ Variables for Chemical Companies Data

In this particular example all three criteria agree. However, in other situations the criteria may select different numbers of variables. In such cases the investigator's judgment must be used in making the final choice. Even in this case an investigator may prefer to use only three variables, such as NPM1, PAYOUTR1, and SALESGR5. The value of having these tables and figures is that they inform the investigator of how much is being given up, as estimated by the sample on hand.

You should be aware that variable selection procedures are highly dependent on the particular sample being used. For example, data from two different years could give very different results. Also, any tests of hypotheses should be viewed with extreme caution since the significance levels are not valid when variable selection is performed.

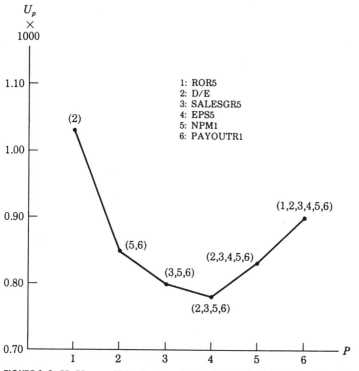

**FIGURE 8.3.** $U_p$ Versus $P$ for Best Subset with $P$ Variables for Chemical Companies Data

## 8.8 DISCUSSION OF COMPUTER PROGRAMS

The techniques described in this chapter require extensive computation. It is difficult to imagine situations where variable selection using these techniques is done without the aid of a computer. In Table 8.7 a summary of options suitable for variable selection is presented for the three major packages. In addition to the variable selection techniques discussed in this chapter, other less-used methods are given in the SAS REG procedure and the BMDP2R program. These methods (MAXR and MINR in REG, and FSWAP and RSWAP in 2R) require more computing time than the stepwise methods. Other multiple linear regression options were discussed in Section 7.10 and in Table 7.7.

In Section 8.6 a description of the stepwise regression procedures was presented with results from the analysis of the chemical companies data.

**TABLE 8.7.** Summary of Computer Output from BMDP, SAS, and SPSS–X for Various Variable Selection Methods

| OUTPUT | BMDP | SAS | SPSS–X |
|---|---|---|---|
| **Variable selection criteria** | | | |
| Maximize $R^2$ | 2R,9R | REG | |
| Maximize adjusted $\bar{R}^2$ | 9R | REG | |
| Minimize $C_p$ | 9R | REG | |
| $F$-to-enter | 2R | REG | REGRESSION |
| $F$-to-remove | 2R | REG | REGRESSION |
| Forcing variable entry | 2R | REG | REGRESSION |
| **Selection method** | | | |
| Forward | 2R | REG | REGRESSION |
| Backward | 2R | REG | REGRESSION |
| Stepwise | 2R | REG | REGRESSION |
| Best subsets | 9R | REG | |
| All possible subsets | 9R | REG | |
| Sets of variables entered | 2R | REG | REGRESSION |
| Interactive mode of entry | 2R | REG | REGRESSION |
| Swapping of variables | 2R | REG | REGRESSION |

*Note:* See Table 7.7 for standard regression output.

These results were obtained from running the stepwise regression program BMDP2R. Some further details of specification of the computer run and explanation of some of the output are presented here to help you master the mechanics of the process.

The program statements for the BMDP2R example are as follows:

```
/input     file is 'chem80'.
           format is free.
           variables = 8.
/variable  names are symb, 'p/e',
           ror5,'d/e',salesgr5,eps5,npm1,payoutr1.
           label = 1.
           use = 2 to 8.
/regress   dependent is 'p/e'.
           independent are
           ror5, 'd/e',salesgr5,eps5,npm1,payoutr1.
           enter = 2.1.
           remove = 1.0.
```

```
/print      data.
            fratio.
/plot       var = ror5,'d/e',salesgr5,eps5,npm1,payoutr1.
            resid.
/end
```

The data file can be created before the run. The data are entered in free format with the values for each case on one line, separated by blanks. The data file is named as "chem80". The named file is retrieved by the BMDP program as a consequence of the input paragraph statement "file is 'chem80'." The data option is included in the print paragraph in order to examine the original input data for entry errors.

The stock symbol (a symbol used in the stock market) for each of the companies (cases) is entered as the first variable—as a label for that case—and the "label = 1" sentence is included to tell the computer to read an alphanumeric label of up to four characters for that variable. The variable names 'p/e' and 'd/e' are enclosed in apostrophes, or single quotation marks, as required by BMDP for names including characters other than letters or numbers. The same technique would be necessary for names beginning with a number digit. Note that BMDP also restricts names to a maximum of eight characters in length. (It should be obvious from the examples we include in these chapters that each of the major computer program packages has its particular rules of syntax, formats, conventions, and restrictions for problem specification.)

The manuals published for the packages provide extensive guides to organization and form of the statements. After working with a package, you will develop some semblance of personal style that helps avoid omissions or errors. For example, we find that in writing the input statements in BMDP, we make fewer mistakes by proceeding from the general to the specific. Thus we begin with "Where are the data found?" which corresponds to "file is chem80." Next, "How are the data organized?" corresponds to "format is free." Finally, "How many variables per case?" becomes "variables = 8."

The "enter = 2.1" and "remove = 1.0" statements in the regress paragraph specify an $F$-to-enter of 2.1 and an $F$-to-remove of 1.0. Note that the $F$-to-remove is smaller than the $F$-to-enter and that they are set at the levels suggested in Section 8.6.

The "fratio" option is included in the print paragraph to obtain a summary table of the computed $F$-to-enter and $F$-to-remove values from the analysis. The results of the summary table are given in Table 8.8. The variables ROR5 and EPS5 were not entered in the multiple regression because their $F$-to-enter values were 0.03 and 0.37, respectively, well below the 2.10 level we specified.

The results of Table 8.8 show that no variables were removed once they were entered, so the results are the same as those we would have obtained with forward selection.

The increase in the squared multiple correlation (RSQ) did not steadily decrease in successive steps. In step 0 the mean of $Y$ is entered, and the printed mean square in the analysis of variance table is simply the variance of the dependent variable $Y$. When the first independent variable chosen by the program (D/E) was entered in step 1, RSQ was computed as 0.2242. Since it was the first variable entered, the increase in RSQ was also 0.2242. At step 2, RSQ went to 0.2879, for an increase of 0.0637. At step 3, however, RSQ jumped to 0.4717, for an increase of 0.1838. This pattern is not uncommon. The combination of D/E, PAYOUTR1, and NPM1 gave much better prediction than just D/E and PAYOUTR1. Note that in the *best-subsets regression* given earlier in Table 8.6, NPM1 and PAYOUTR1 were the best choice when the number of independent variables was two.

Another interesting result is that the order of entry does not correspond to the size of the standardized regression coefficients obtained when the four variables have been entered. The largest standard regression coefficient is 0.623 for PAYOUTR1, followed by 0.492 for NPM1, then 0.280 for SALESGR5, and finally 0.256 for D/E. The order of magnitude

**TABLE 8.8.** Summary Table for Stepwise Regression of Chemical Companies Data

| | Variable | | Multiple | | Increase | F-to- |
|------|----------|---------|--------|--------|---------|-------|
| Step | Entered | Removed | R | RSQ | in RSQ | Enter |
| 1 | D/E | — | 0.4735 | 0.2242 | 0.2242 | 8.0924 |
| 2 | PAYOUTR1 | — | 0.5366 | 0.2879 | 0.0637 | 2.4161 |
| 3 | NPM1 | — | 0.6868 | 0.4717 | 0.1838 | 9.0463 |
| 4 | SALESGR5 | — | 0.7306 | 0.5338 | 0.0620 | 3.3267 |

of the standardized regression coefficients does, however, relate directly to the size of the computed $F$-to-remove, with larger $F$-to-remove values corresponding to larger coefficients.

The various plots can be examined for outliers and for nonnormality. In this example no striking outliers were found. The case with the largest positive residual was for the company Air Products and Chemicals.

## 8.9 DISCUSSION AND EXTENSIONS

The process leading to a final regression equation involves several steps and tends to be iterative. The investigator should begin with data screening, eliminating blunders and possibly some outliers and deleting variables with an excessive proportion of missing values. Initial regression equations may be forced to include certain variables as dictated by the underlying situation. Other variables are then added on the basis of selection procedures outlined in this chapter. The choice of the procedure depends on which computers and programs are accessible to the investigator.

The results of this phase of the computation will include various alternative equations. For example, if the forward selection procedure is used, a regression equation at each step is printed. Thus in the chemical companies data the equations resulting from the first four steps are as given in Table 8.9. Note that the coefficients for given variables could vary appreciably from one step to another. In general, coefficients for variables that are independent of the remaining predictor variables tend to have stable coefficients. The effect of a variable whose coefficient is

**TABLE 8.9.** Regression Coefficients for Chemical Companies Data: First Four Steps

| Step | Intercept | D/E | PAYOUTR1 | NPM1 | SALESGR5 |
|------|-----------|------|----------|------|----------|
| | | | **Regression Coefficients** | | |
| 1 | 11.8 | −5.86 | | | |
| 2 | 9.9 | −5.35 | 4.38 | | |
| 3 | 5.1 | −3.46 | 8.54 | 0.36 | |
| 4 | 1.3 | −3.17 | 10.67 | 0.35 | 0.19 |

highly dependent on the other variables in the equation is more difficult to interpret than the effect of a variable with stable coefficients.

A particularly annoying situation occurs when the sign of a coefficient changes from one step to the next. A method for avoiding this difficulty is to impose restrictions on the coefficients (see Chapter 9).

For some of the promising equations in the sequence given in Table 8.9, another program could be run to obtain extensive residual analysis and to determine influential cases. Examination of this output may suggest further data screening. The whole process could then be repeated. In 2R it is possible to obtain the diagnostic plots after each step by using the STEP option in the PLOT paragraph. The case plots may be a good option to choose in this case, as it provides one measure of outliers in $Y$, one in $X$, and one influential measure. Note that if you decide not to report the equation as given in the final step the diagnostics associated with the final step would not apply to your regression equation, and the program should be re-run to obtain the appropriate diagnostics.

Programs such as 2R can also be run in an interactive mode. In this case the $F$-to-enter values and partial correlations for the next step appear on the console screen and the investigators can either choose the variable which has the largest $F$-to-enter by default or substitute a different variable of their choice at each step. The use of the interactive mode makes it very convenient to perform stepwise regression when the investigator wishes to perform a partial exploratory regression analysis. SAS REG, when run in an interactive mode, allows for addition and deletion of variables.

As discussed in Chapters 6 and 7, some regression programs allow the user to weight each case differently. This option is available in some variable selection programs as well. Deleting an observation is equivalent to giving it a weight of zero. This device can be used to adjust the influence of certain observations without necessarily totally deleting them, namely by giving them small weight relative to the rest of the observations. Such adjustments are cumbersome and require thorough knowledge of the underlying situation. Furthermore, for a large data set the influence of a single observation is less than it is for a small data set, other things being equal.

An alternative approach to variable selection is the so-called *stagewise regression procedure*. In this method the first stage (step) is the same as in forward stepwise regression; i.e., the ordinary regression equation of $Y$ on the most highly correlated independent $X$ variable is computed. Residuals from this equation are also computed. In the second step the $X$ variable that is most highly correlated with these residuals is selected. Here the residuals are considered the "new" $Y$ variables. The regression coefficient of these residuals on the $X$ variable selected in step 2 is computed. The constant and the regression coefficient from the first step are combined with the regression coefficient from the second step to produce the equation with two $X$ variables. This equation is the two-stage regression equation. The process can be continued to any number of desired stages.

The resulting *stagewise regression equation* does not fit the data as well as the least squares equation using the same variables (unless the $X$ variables are independent in the sample). However, stagewise regression has some desirable advantages in certain applications. In particular, econometric models often use stagewise regression to adjust the data source factor, such as a trend or seasonality. Another feature is that the coefficients of the variables already entered are preserved from one stage to the next. For example, for the chemical companies data the coefficient for D/E would be preserved over the successive stages of a stagewise regression. This result can be contrasted with the changing coefficient in the stepwise process summarized in Table 8.9. In behavioral science applications the investigator can determine whether the addition of a variable reflecting a score or an attitude scale improves prediction of an outcome over using only demographic variables.

## 8.10 WHAT TO WATCH OUT FOR

The variable selection techniques described in this chapter are used for the variable-$X$ regression model. Therefore, the problem areas discussed at the ends of Chapters 6 and 7 also apply to this chapter. Only possible problems connected with variable selection will be mentioned here. Areas of special concern are:

   *1.* Selection of the best possible variables to be checked for inclusion in the model is extremely important. As discussed in Section 7.11, if an appropriate set of variables is not chosen, then the coefficients obtained for the "good" variables will be biased by the lack of inclusion of other needed variables (if the "good" variables and the needed variables are correlated). Lack of inclusion of needed or missing variables also results in an overestimate of the standard errors of the slope of the variables used. Also, the predicted $Y$ is biased unless the missing variables are uncorrelated with the variables included or have zero slope coefficients (see Chatterjee and Price 1977). If the best candidate variables are not measured in the first place, then no method of variable selection can make up for it later.

**2.** Since the programs select the "best" variable or set of variables for testing in the various variable-selection methods, the level of significance of the resulting $F$ test cannot be read from a table of the $F$ distribution. Any levels of significance should be viewed with caution. The test should be viewed as a screening device, not as a test of significance.

**3.** Minor differences can affect the order of entry of the variables. For example, in forward stepwise selection the next variable chosen is the one with the largest $F$-to-enter. If the $F$-to-enter is 4.444444 for variable one and 4.444443 for variable two, then variable one enters first unless some method of forcing is used or the investigator runs the program in an interactive mode and intervenes. The rest of the analysis can be affected greatly by the result of this toss-up situation.

**4.** Spurious variables can enter especially for small samples. These variables may enter due to outliers or other problems in the data set. Whatever the cause, investigators will occasionally obtain a variable from mechanical use of variable-selection methods that no other investigator can duplicate and which would not be expected for any theoretical reason. Flack and Chang (1987) discuss situations in which the frequency of selecting such "noise" variables is high.

**5.** Methods using $C_p$, which depends on a good estimate of $\sigma^2$, may not actually be better in practice than other methods if a proper estimate for $\sigma^2$ does not exist.

**6.** Especially in forward stepwise selection, it may be useful to perform regression diagnostics on the entering variable given the variables already entered (see Chatterjee and Hadi 1988 or Cook and Weisberg 1982). Several diagnostic plots are available in the computer programs. Most of the plots involve the use of residuals and are useful in checking the assumptions of the model in connection with the additional variable. Suppose $P$ variables have been entered and the $P + 1$st variable is being considered. SAS includes the option to obtain partial regression plots. Here, the residuals from the equation fitted predicting $Y$ using the first $P$ variables are plotted on the vertical axis, versus the residuals obtained from predicting $X_{P+1}$ from the first $P$ independent variables on the horizontal axis. If a straight line is fitted to the resulting points, it would go through the origin and have a slope of $B_{P+1}$. In addition, the residuals from the plane fitted with all the $P + 1$ variables equal the residuals obtained in the partial regression plot. The position of the various points is useful in determining which observations influence the estimate of $B_{P+1}$ the most. Points which are outliers on the vertical or horizontal axes have the most influence.

BMDP includes the option to obtain partial residual plots (also called component plus residual plots). A plot of the residuals from the plane when all $P + 1$ are used to predict $Y$ is plotted on the vertical axis versus $X_{P+1}$ on the horizontal axis. The line fitted to this set of points also has a slope of $B_{P+1}$, and significant deviations of the points from the line fitted to them can indicate a need for a transformation of $X_{P+1}$. In SPSS–X and SPSS/PC+ there is a PARTIALPLOT subcommand in the REGRESSION procedure that will produce partial regression plots. These are similar to the partial regression plots produced by SAS and described above, with the difference being that all residuals have been standardized before plotting.

**7.** For personal computers with slow central processing units and small memory, some of the options described in this chapter can be quite time-consuming. Subset regression is recommended for these machines only if one limits the number of independent variables to, say, less than 10 and does not print out very many best subsets. So many options are available in the regression packages that it is possible to overload a slower machine.

## SUMMARY

Variable selection techniques are helpful in situations in which many independent variables exist and no standard model is available. The methods we described in this chapter constitute one part of the process of making the final selection of one or more equations.

Although these methods are useful in exploratory situations, they should not be used as a substitute for modeling based on the underlying scientific problem. In particular, the variables selected by the methods described here depend on the particular sample on hand, especially when the sample size is small. Also, significance levels are not valid in variable selection situations.

These techniques can, however, be an important component in the modeling process. They frequently suggest new variables to be considered. They also help cut down the number of variables so that the problem becomes less cumbersome both conceptually and computationally. For these reasons variable selection procedures are among the most popular of the packaged options. The ideas underlying variable selection have also been incorporated into other multivariate programs (see, e.g., Chapters 11, 12, and 13).

In the intermediate stages of analysis the investigator can consider several equations as suggested by one or more of the variable selection procedures. The equations selected should be considered tentative until the same variables are chosen repeatedly. Consensus will also occur when the same variables are selected by different subsamples within the same data set or by different investigators using separate data sets.

## BIBLIOGRAPHY

*Beale, E. M. L., Kendall, M. G., and Mann, D. W. 1967. The discarding of variables in multivariate analysis. *Biometrika* 54:357–366.

Bendel, R. B., and Afifi, A. A. 1977. Comparison of stopping rules in forward stepwise regression. *Journal of the American Statistical Association* 72:46–53.

Brigham, E. F. 1980. *Fundamentals of financial management.* 2nd ed. Hinsdale, Ill.: Dryden Press.

Chatterjee, S., and Hadi, A. S. 1988. *Sensitivity analysis in linear regression.* New York: Wiley.

Chatterjee, S., and Price, B. 1977. *Regression analysis by example*. New York: Wiley.

Cook, R. D., and Weisberg, S. 1982. *Residuals and influence in regression*. London: Chapman and Hall.

*Efroymson, M. A. 1960. Multiple regression analysis. In *Mathematical methods for digital computers*, ed. Ralston and Wilf. New York: Wiley.

Flack, V. F., and Chang, P. C. 1987. Frequency of selecting noise variables in a subset regression analysis: A simulation study. *The American Statistician* 41:84–86.

Forsythe, A. B., Engleman, L., Jennrich, R. and May, P. R. A. 1973. Stopping rules for variable selection in multiple regression. *Journal of the American Statistical Association* 68:75–77.

*Furnival, G. M., and Wilson, R. W. 1974. Regressions by leaps and bounds. *Technometrics* 16:499–512.

Gorman, J. W., and Toman, R. J. 1966. Selection of variables for fitting equations to data. *Technometrics* 8:27–51.

Hocking, R. R. 1972. Criteria for selection of a subset regression: Which one should be used. *Technometrics* 14:967–970.

————. 1976. The analysis and selection of variables in linear regression. *Biometrics* 32:1–50.

*Kendall, M. G., and Stuart, A. 1979. *The advanced theory of statistics*, Vol. 2. 4th ed. New York: Hafner.

Mallows, C. L. 1973. Some comments on Cp. *Technometrics* 15:661–676.

Mantel, N. 1970. Why stepdown procedures in variable selection. *Technometrics* 12:621–626.

Myers, R. H. 1986. *Classical and modern regression with applications*. Boston: Duxbury Press.

Pope, P. T., and Webster, J. T. 1972. The use of an *F*-statistic in stepwise regression procedures. *Technometrics* 14:327–340.

Pyne, D. A., and Lawing, W. D. 1974. *A note on the use of the Cp statistic and its relation to stepwise variable selection procedures*. Technical Report no. 210, Johns Hopkins University.

Theil, H. 1971. *Principles of econometrics*. New York: Wiley.

Wilkinson, L., and Dallal, G. E. 1981. Tests of significance in forward selection regression with an *F*-to-enter stopping rule. *Technometrics* 23:377–380.

## PROBLEMS

8.1 Use the depression data set given in Table 3.3. Using CESD as the dependent variable and age, income, and level of education as the independent variables, run a forward stepwise regression program to determine which of the independent variables predict level of depression for women.

8.2 Repeat Problem 8.1, using subset regression, and compare the results.

8.3 *Forbes* gives, each year, the same variables listed in Table 8.1 for the chemical industry. The changes in lines of business and company mergers resulted in a somewhat different list of chemical companies in 1982. We have selected a subset of 13 companies that are listed in both years and whose main product is chemicals. Table 8.10 includes data for both years. Do a forward stepwise regression analysis, using P/E as the dependent variable and ROR5, D/E, SALESGR5, EPS5, NPM1, and PAYOUTR1 as independent variables, on both years' data and compare the results. Note that this period showed little growth for this subset of companies, and the variable(s) entered should be evaluated with that idea in mind.

8.4 For adult males it has been demonstrated that age and height are useful in predicting FEV1. Using the data given in Appendix B, determine whether the regression plane can be improved by also including weight.

8.5 Using the data given in Table 8.1, repeat the analyses described in this chapter with $\sqrt{P/E}$ as the dependent variable instead of P/E. Do the results change much? Does it make sense to use the square root transformation?

8.6 Use the data you generated from Problem 7.7, where X1, X2, ..., X9 are the independent variables and Y is the dependent variable. Use the generalized linear hypothesis test to test the hypothesis that $\beta_4 = \beta_5 = \cdots = \beta_9 = 0$. Comment in light of what you know about the population parameters.

8.7 For the data from Problem 7.7, perform a variable selection analysis, using the methods described in this chapter. Comment on the results in view of the population parameters.

8.8 In Problem 7.7 the population multiple $\mathscr{R}^2$ of Y on X4, X5, ..., X9 is zero. However, from the sample alone we don't know this result. Perform a variable selection analysis on X4 to X9, using your sample, and comment on the results.

8.9 a. For the lung function data set in Appendix B, with age, height, weight, and FVC as the candidate independent variables, use subset regression to find which variables best predict FEV1 in the oldest child. State the criteria you use to decide.

b. Repeat, using forward selection and backward elimination. Compare with part a.

8.10 Force the variables you selected in Problem 8.9a into the regression equation with OCFEV1 as the dependent variable, and test whether including the FEV1 of the parents (i.e., the variables MFEV1 and FFEV1 taken as a pair) in the equation significantly improves the regression.

8.11 Using the methods described in this chapter and the family lung function data in Appendix B, and choosing from among the variables OCAGE, OCWEIGHT, MHEIGHT, MWEIGHT, FHEIGHT, and FWEIGHT, select the variables that best predict height in the oldest child. Show your analysis.

**TABLE 8.10.** Chemical Companies Financial Performance as of 1980 and 1982 (Subset of 13 of the Chemical Companies)

| Company | Symbol | P/E | ROR5 | D/E | SALESGR5 | EPS5 | NPM1 | PAYOUTR1 |
|---|---|---|---|---|---|---|---|---|
| **1980** | | | | | | | | |
| Diamond Shamrock | dia | 9 | 13.0 | 0.7 | 20.2 | 15.5 | 7.2 | 0.43 |
| Dow Chemical | dow | 8 | 13.0 | 0.7 | 17.2 | 12.7 | 7.3 | 0.38 |
| Stauffer Chemical | stf | 8 | 13.0 | 0.4 | 14.5 | 15.1 | 7.9 | 0.41 |
| E. I. du Pont | dd | 9 | 12.2 | 0.2 | 12.9 | 11.1 | 5.4 | 0.57 |
| Union Carbide | uk | 5 | 10.0 | 0.4 | 13.6 | 8.0 | 6.7 | 0.32 |
| Pennwalt | psm | 6 | 9.8 | 0.5 | 12.1 | 14.5 | 3.8 | 0.51 |
| W. R. Grace | gra | 10 | 9.9 | 0.5 | 10.2 | 7.0 | 4.8 | 0.38 |
| Hercules | hpc | 9 | 10.3 | 0.3 | 11.4 | 8.7 | 4.5 | 0.48 |
| Monsanto | mtc | 11 | 9.5 | 0.4 | 13.5 | 5.9 | 3.5 | 0.57 |
| American Cyanamid | acy | 9 | 9.9 | 0.4 | 12.1 | 4.2 | 4.6 | 0.49 |
| Celanese | cz | 7 | 7.9 | 0.4 | 10.8 | 16.0 | 3.4 | 0.49 |
| Allied Chemical | acd | 7 | 7.3 | 0.6 | 15.4 | 4.9 | 5.1 | 0.27 |
| Rohm & Haas | roh | 7 | 7.8 | 0.4 | 11.0 | 3.0 | 5.6 | 0.32 |
| **1982** | | | | | | | | |
| Diamond Shamrock | dia | 8 | 9.8 | 0.5 | 19.7 | -1.4 | 5.0 | 0.68 |
| Dow Chemical | dow | 13 | 10.3 | 0.7 | 14.9 | 3.5 | 3.6 | 0.88 |
| Stauffer Chemical | stf | 9 | 11.5 | 0.3 | 10.6 | 7.6 | 8.7 | 0.46 |
| E. I. du Pont | dd | 9 | 12.7 | 0.6 | 19.7 | 11.7 | 3.1 | 0.65 |
| Union Carbide | uk | 9 | 9.2 | 0.3 | 10.4 | 4.1 | 4.6 | 0.56 |
| Pennwalt | psm | 10 | 8.3 | 0.4 | 8.5 | 3.7 | 3.1 | 0.74 |
| W. R. Grace | gra | 6 | 11.4 | 0.6 | 10.4 | 9.0 | 5.5 | 0.39 |
| Hercules | hpc | 14 | 10.0 | 0.4 | 10.3 | 9.0 | 3.2 | 0.71 |
| Monsanto | mtc | 10 | 8.2 | 0.3 | 10.8 | -0.2 | 5.6 | 0.45 |
| American Cyanamid | acy | 12 | 9.5 | 0.3 | 11.3 | 3.5 | 4.0 | 0.60 |
| Celanese | cz | 15 | 8.3 | 0.6 | 10.3 | 8.9 | 1.7 | 1.23 |
| Allied Corp* | ald | 6 | 8.2 | 0.3 | 17.1 | 4.8 | 4.4 | 0.36 |
| Rohm & Haas | roh | 12 | 10.2 | 0.3 | 10.6 | 16.3 | 4.3 | 0.45 |

Data abstracted from *Forbes*, 127, no. 1 (January 5, 1981) and *Forbes*, 131, no. 1 (January 3, 1983).
* Name and symbol changed.

8.12 From among the candidate variables given in Problem 8.11, find the subset of three variables that best predicts height in the oldest child, separately for boys and girls. Are the two sets the same? Find the best subset of three variables for the group as a whole. Does adding OCSEX into the regression equation improve the fit?

# Chapter Nine

# SPECIAL REGRESSION TOPICS

## 9.1 WHAT WILL YOU LEARN FROM THIS CHAPTER?

This chapter presents brief descriptions of several topics and specialized uses related to regression. From this chapter you will learn:

- How to handle missing values in regression analysis (9.2).
- How to use dummy variables as independent variables in regression analysis (9.3).
- How to interpret the results when dummy variables are used (9.3).
- How to obtain regression equations when constraints are placed on the parameter estimates (9.4).
- How to obtain segmented regression equations (9.4).
- How to look for indications of multicollinearity (9.5).
- How to determine when to use ridge regression (9.6).
- How to perform ridge regression by using standard computer packages (9.6).

## 9.2 MISSING VALUES

Missing values are often classified into three types (see Little and Rubin 1987). The first of these types is called missing completely at random (the other two types will be discussed later in this section). For example, if a chemist knocks over a test tube or an interviewer erroneously skips a page in an interview when assembling the questionaire, then these observations would be missing completely at random. Here, we assume that being missing is independent of both $X$ and $Y$. In other words, the missing cases are a random subsample of the original sample. In such situations there should be no significant differences between those cases with missing values and those without, on either the $X$ variables or the $Y$ variable. This can be checked by a series of $t$ tests or by more complicated multivariate analyses (for example, the analysis given in Chapter 12 of this book). But it should be noted that lack of a significant difference does not necessarily prove that the data are missing completely at random. Sometimes the chance of making a beta error is quite large. It is always recommended that the investigator examine the pattern of missing values. If they occur in clusters, this may be an indication of lack of missing completely at random. Programs such as BMDPAM can be used to determine if patterns exist among the missing values.

The subject of randomly missing observations in regression analysis has received wide attention from statisticians. Afifi and Elashoff (1966) present a review of the literature up to 1966, and Hartley and Hocking (1971) present a simple taxonomy of incomplete data problems (see also Little 1982). When some observations are missing completely at random, the simplest option is again to use only the cases with complete data. Since this procedure may result in a reduced sample size, various methods have been developed to fill in, or *impute*, the missing data and then perform the regression analysis on the completed data set.

One method for filling in data, called *mean substitution*, replaces the missing value for a given variable by the mean of the observed values of that variable. A more complicated method involves first computing a *regression equation* of a given variable on one or more of the remaining variables and then using that equation to "predict" the missing values. For example, suppose that for a sample of schoolchildren some heights are missing. The investigator could first derive the regressions of height on age for boys and girls and then use these regressions to predict the missing heights. If the children have a wide range of ages, this method presents an improvement over mean substitution (for simple linear regression, see Afifi and Elashoff 1969*a* and 1969*b*). A third method, which applies regression with normally distributed $X$ variables, was studied by Orchard and Woodbury (1972) and further developed by Beale and Little (1975). This method is based on the theoretical principle of *maximum likelihood* used in statistical estimation.

If the investigator wishes to make some use of the incomplete cases, mean substitution can be useful when the intercorrelations among the variables are small. This method, however, may result in biases of unknown magnitude in the resulting estimates. The regression method of filling in the data is preferable when higher correlations exist. Some biases are incurred here as well. The maximum likelihood method is known to have theoretically appealing features if the data are, indeed, normally distributed. Little is known about the properties of this method of dealing with missing values when the underlying distribution is not normal.

All three methods are available in the BMDPAM program, in addition to estimation based on complete data only. Output from this program can then be used for a regression analysis or other multivariate analyses. We believe that the most understandable analysis is the one based on complete data only. To obtain a complete data set that is as large as possible, though, the investigator may have to delete some variables from the analysis.

In the SPSS–X REGRESSION program the investigator has the option of using only the cases with no missing values on all variables, mean substitution, or two other option. One option is to include cases with missing values, with the missing value code considered a valid observation. If the data are continuous, missing values are often coded as extreme values, such as 99 or blanks. If so, then this option would not appear to be very useful. If the investigator is using dummy variables (as discussed in the next section), then considering the missing values as a separate response to a question along with other responses can be an interesting option. Sometimes, in survey data, the fact that a respondent did not answer a particular question can be a useful piece of information to analyze.

The SPSS–X program also offers the option of computing each correlation, using cases with complete data on each pair of variables being correlated irrespective of whether or not the cases have missing data on other variables. This and other methods are also available in BMDP8D for computing correlation and covariances. These methods of using available pairs may result in computational problems (see the program manuals) and should be used with caution. Their use is only recommended when the number of completed cases is small and the investigator is trying to use as much of the data as possible.

The second type of missing data is called missing at random. Here the probability of response depends on $X$ but not on $Y$ (see Little and Rubin 1987). The observed values of $Y$ are not a random subsample of the desired sample of $Y$, but they are a random sample of the desired sample within subclasses defined by values of $X$. For example, in sampling it is common to obtain a better response from middle income families than from the poor. If income was an $X$ variable and the poor who responded were like the poor in the population, then the data would be missing at random. Methods for adjusting such data are available if data are obtainable on the population distribution of $X$. These methods involve giving greater weight to the cases that have a larger amount of missing data. A discussion of this can be found in Little and Rubin (1987).

The third type of missing data occurs when being missing is related to both $X$ and $Y$. For example, if we were predicting levels of depression and both being depressed ($Y$) and being poor ($X$) influenced whether or not a person responded, then the third type of missing data has occurred. Most methods for adjusting for missing values are not recommended in this case. It is more common to analyze the complete data and place caveats about the population to which inferences can be drawn.

In practice, missing values often occur for bizarre reasons that are difficult to fit into any statistical model. Even careful data screening can cause missing values, since observations that are declared outliers are commonly treated as missing values. A practical piece of advice is to use the various options available for analysis and then compare the results. If a consensus is obtained, then further credence is given to the results.

## 9.3 DUMMY VARIABLES

Often $X$ variables desired for inclusion in a regression model are not continuous. Such variables could be either nominal or ordinal. Ordinal measurements represent variables with an underlying scale. For example, the severity of a burn could be classified as mild, moderate, or severe. These burns are commonly called first-, second-, and third-degree burns, respectively. The $X$ variable representing these categories may be coded 1, 2, or 3. The data from ordinal variables could be analyzed by using the numerical coding for them and treated as though they were continuous (or interval data). This method takes advantage of the underlying order of the data. On the other hand, this analysis assumes equal intervals between successive values of the variable. Thus in the burn example, we would assume that the difference between a first- and a second-degree burn is equivalent to the difference between a second- and third-degree burn. As was discussed in Chapter 5, an appropriate choice of scale may be helpful in assigning numerical values to the outcome that would reflect the varying lengths of intervals between successive responses.

In contrast, the investigator may ignore the underlying order for ordinal variables and treat them as though they were nominal variables. In this section we discuss methods for using one or more nominal $X$ variables in regression analysis along with the role of interactions.

### A Single Binary Variable

We will begin with a simple example to illustrate the technique. Suppose that the dependent variable $Y$ is yearly income in dollars and the independent variable $X$ is the sex of the respondent (male or female). To represent sex, we create a *dummy* (or *indicator*) *variable* $D = 0$ if the respondent is male and $D = 1$ if the respondent is female. (All the major computer packages offer the option of recoding the data in this form.) The sample regression equation can then be written as $Y = A + BD$. The value of $Y$ is

$$Y = A \qquad \text{if} \qquad D = 0$$

and

$$Y = A + B \qquad \text{if} \qquad D = 1$$

Since our best estimate of $Y$ for a given group is that group's mean, $A$ is estimated as the average income for males ($D = 0$) and $A + B$ is the average income for females ($D = 1$). The regression coefficient $B$ is therefore $B = \bar{Y}_{\text{females}} - \bar{Y}_{\text{males}}$. In effect, males are considered the *reference group*; and females' income is measured by how much it differs from males' income.

## A Second Method of Coding Dummy Variables

An alternative way of coding the dummy variables is

$$D^* = -1 \qquad \text{for males}$$

and

$$D^* = +1 \qquad \text{for females}$$

In this case the regression equation would have the form

$$Y = A^* + B^*D^*$$

The average income for males is now

$$A^* - B^* \qquad (\text{when } D^* = -1)$$

and for females it is

$$A^* + B^* \qquad (\text{when } D^* = +1)$$

Thus

$$A^* = \tfrac{1}{2}(\bar{Y}_{\text{males}} + \bar{Y}_{\text{females}})$$

and

$$B^* = \bar{Y}_{\text{females}} - \tfrac{1}{2}(\bar{Y}_{\text{males}} + \bar{Y}_{\text{females}})$$

or

$$B^* = \tfrac{1}{2}(\bar{Y}_{\text{females}} - \bar{Y}_{\text{males}})$$

In this case neither males nor females are designated as the reference group.

## A Nominal Variable with Several Categories

For another example we consider an $X$ variable with $k > 2$ categories. Suppose income is now to be related to the religion of the respondent. Religion is classified as Catholic (C), Protestant (P), Jewish (J), and other (O). The religion variable can be represented by the dummy variables that follow:

| Religion | $D_1$ | $D_2$ | $D_3$ |
|----------|-------|-------|-------|
| C        | 1     | 0     | 0     |
| P        | 0     | 1     | 0     |
| J        | 0     | 0     | 1     |
| O        | 0     | 0     | 0     |

Note that to represent the four categories, we need only three dummy variables. In general, to represent $k$ categories, we need $k - 1$ dummy variables. Here the three variables represent C, P, and J, respectively. For example, $D_1 = 1$ if Catholic and zero otherwise; $D_2 = 1$ if Protestant and zero otherwise; and $D_3 = 1$ if Jewish and zero otherwise. The "other" group has a value of zero on each of the three dummy variables.

The estimated regression equation will have the form

$$Y = A + B_1 D_1 + B_2 D_2 + B_3 D_3$$

The average incomes for the four groups are

$$\bar{Y}_C = A + B_1$$
$$\bar{Y}_P = A + B_2$$
$$\bar{Y}_J = A + B_3$$
$$\bar{Y}_O = A$$

Therefore,

$$B_1 = \bar{Y}_C - A = \bar{Y}_C - \bar{Y}_O$$
$$B_2 = \bar{Y}_P - A = \bar{Y}_P - \bar{Y}_O$$
$$B_3 = \bar{Y}_J - A = \bar{Y}_J - \bar{Y}_O$$

Thus the group "other" is taken as the reference group to which all the others are compared. Although the analysis is independent of which group is chosen as the reference group, the investigator should select the group that makes the interpretation of the $B$'s most meaningful. The mean of the reference group is the constant $A$ in the equation. An example of program specification for dummy variables is given in Problem 9.2.

If none of the groups represent a natural choice of a reference group, an alternative choice of assigning values to the dummy variables is as follows:

| Religion | $D_1^*$ | $D_2^*$ | $D_3^*$ |
|----------|---------|---------|---------|
| C | 1 | 0 | 0 |
| P | 0 | 1 | 0 |
| J | 0 | 0 | 1 |
| O | −1 | −1 | −1 |

As before, Catholic has a value of 1 on $D_1^*$, Protestant a value of 1 on $D_2^*$, and Jewish a value of 1 on $D_3^*$. However, "other" has a value of $-1$ on each of the three dummy variables. Note that zero is also a possible value of these dummy variables.

The estimated regression equation is now

$$Y = A^* + B_1^* D_1^* + B_2^* D_2^* + B_3^* D_3^*$$

The group means are

$$\bar{Y}_\text{C} = A^* + B_1^*$$
$$\bar{Y}_\text{P} = A^* + B_2^*$$
$$\bar{Y}_\text{J} = A^* + B_3^*$$
$$\bar{Y}_\text{O} = A^* - B_1^* - B_2^* - B_3^*$$

In this case the constant

$$A^* = \tfrac{1}{4}(\bar{Y}_\text{C} + \bar{Y}_\text{P} + \bar{Y}_\text{J} + \bar{Y}_\text{O})$$

the unweighted average of the four group means. Thus

$$B_1^* = \bar{Y}_C - \tfrac{1}{4}(\bar{Y}_C + \bar{Y}_P + \bar{Y}_J + \bar{Y}_O)$$
$$B_2^* = \bar{Y}_P - \tfrac{1}{4}(\bar{Y}_C + \bar{Y}_P + \bar{Y}_J + \bar{Y}_O)$$

and

$$B_3^* = \bar{Y}_J - \tfrac{1}{4}(\bar{Y}_C + \bar{Y}_P + \bar{Y}_J + \bar{Y}_O)$$

Note that with this choice of dummy variables there is no reference group. Also, there is no slope coefficient corresponding to the "other" group. As before, the choice of the group to receive $-1$ values is arbitrary. That is, any of the four religious groups could have been selected for this purpose.

### *One Nominal and One Interval Variable*

Another example, which includes one dummy variable and one continuous variable, is the following: Suppose an investigator wants to relate vital capacity ($Y$) to height ($X$) for men and women ($D$) in a restricted age group. One model for the population regression equation is

$$Y = \alpha + \beta X + \delta D + e$$

where $D = 0$ for females and $D = 1$ for males. This equation is a multiple regression equation with $X = X_1$ and $D = X_2$. This equation breaks down to an equation for females,

$$Y = \alpha + \beta X + e$$

and one for males,

$$Y = \alpha + \beta X + \delta + e$$

or

$$Y = (\alpha + \delta) + \beta X + e$$

Figure 9.1a illustrates both equations.

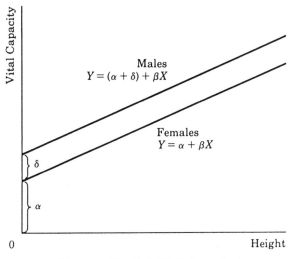

**a.** Lines are Parallel (*No* Interaction)

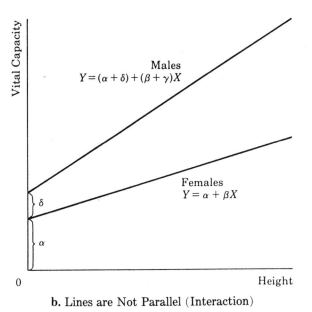

**b.** Lines are Not Parallel (Interaction)

*FIGURE 9.1.* Vital Capacity Versus Height for Males and Females

Note that this model forces the equation for males and females to have the same slope $\beta$. The only difference between the two equations is in the intercept: $\alpha$ for females and $(\alpha + \delta)$ for males.

## Interaction

A model that does not force the lines to be parallel is

$$Y = \alpha + \beta X + \delta D + \gamma(XD) + e$$

This equation is a multiple regression equation with $X = X_1$, $D = X_2$, and $XD = X_3$. The variable $X_3$ can be generated on the computer as the product of the other two variables. With this model the equation for females $(D = 0)$ is

$$Y = \alpha + \beta X + e$$

and for males it is

$$Y = \alpha + \beta X + \delta + \gamma X + e$$

or

$$Y = (\alpha + \delta) + (\beta + \gamma)X + e$$

The equations for this model are illustrated in Figure 9.1b for the case where $\gamma$ is a positive quantity.

Thus with this model we allow both the intercept and slope to be different for males and females. The quantity $\delta$ is the difference between the two intercepts, and $\gamma$ is the difference between the two slopes. Note that some investigators would call the product $XD$ the *interaction* between height and sex. With this model we can test the hypothesis of no interaction, i.e., $\gamma = 0$, and thus decide whether parallel lines are appropriate. This test can be done by using the methods discussed in Chapter 7, since this is an ordinary multiple regression model.

## Extensions

We can use these ideas and apply them to situations where there may be several $X$ variables and several $D$ (or dummy) variables. For example, we can estimate vital capacity by using age and height for males and females and for smokers and nonsmokers. The selection of the model—i.e., whether or not to include interaction—and the interpretation of the resulting equations must be done with caution. The previous methodology for variable selection applies here as well, with appropriate attention given to interpretation. For example, if a nominal variable such as religion is split into three new dummy variables, with "other" used originally as a reference group, a stepwise program may enter only one of these three dummy variables. Suppose the variable $D_1$ is entered, where $D_1 = 1$ signifies Catholic and $D_1 = 0$ signifies Protestant, Jewish, or other. In this case the reference group becomes Protestant, Jewish, or other. Care must be taken in the interpretation of the results to report the proper reference group. Sometimes, investigators will force in the three dummy variables $D_1$, $D_2$, $D_3$ if any one of them is entered in order to keep the reference group as it was originally chosen. This feature can be accomplished easily by using the subsets-of-variables option in BMDP2R.

The investigator is advised to write out the separate equation for each subgroup, as was shown in the previous examples. This technique will help clarify the interpretation of the regression coefficients and the implied reference group (if any). Also, it may be advisable to select a meaningful reference group prior to selecting the model equations rather than rely on the computer program to do it for you.

Another alternative to using dummy variables is to find separate regression equations for each level of the nominal or ordinal variables. For example, in the prediction of FEV1 from age and height, a better prediction can be achieved if an equation is found for males and a separate one for females. Females are not simply "short" men. If the sample size is adequate, it is a good procedure to check whether it is necessary to include an interaction term. If the slopes "seem" equal, no interaction term is necessary; otherwise, such terms should be included. Formal tests exist for this purpose, but they are beyond the scope of this book (see Graybill 1976).

## 9.4 CONSTRAINTS ON PARAMETERS

Some packaged programs, known as nonlinear regression programs, offer the user the option of restricting the range of possible values of the parameter estimates. In addition, some programs (e.g., BMDP3R,

BMDPAR, and SAS REG) offer the option of imposing *linear constraints* on the parameters. These constraints take the form

$$C_1\beta_1 + C_2\beta_2 + \cdots + C_P\beta_P = C$$

where $\beta_1, \beta_2, \ldots, \beta_P$ are the parameters in the regression equation and $C_1$, $C_2, \ldots, C_P$ and $C$ are constants supplied by the user. The program finds estimates of the parameters restricted to satisfy this constraint as well as any other supplied.

Although some of these programs are intended for nonlinear regression, they also provide a convenient method of performing a *linear* regression with constraints on the parameters. For example, suppose that the coefficient of the first variable in the regression equation was demonstrated from previous research to have a specified value, such as $B_1 = 2.0$. Then the constraint would simply be

$$C_1 = 1, \qquad C_2 = \cdots = C_P = 0$$

and

$$C = 2.0 \qquad \text{or} \qquad 1\beta_1 = 2.0$$

Another example of an inequality constraint is the situation when coefficients are required to be nonnegative. For example, if $\beta_2 \geq 0$, this constraint can also be supplied to the program.

The use of linear constraints offers a simple solution to the problem known as *spline regression* or *segmented-curve regression*. For instance, in economic applications we may want to relate the consumption function $Y$ to the level of aggregate disposable income $X$. A possible nonlinear relationship is a linear function up to some level $X_0$, i.e., for $X \leq X_0$, and another linear function for $X > X_0$. As illustrated in Figure 9.2, the equation for $X \leq X_0$ is

$$Y = \alpha_1 + \beta_1 X + e$$

and for $X > X_0$ it is

$$Y = \alpha_2 + \beta_2 X + e$$

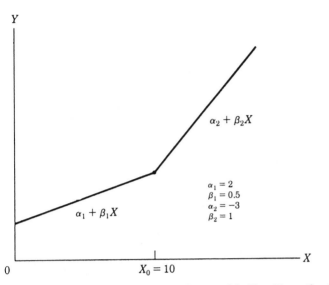

**FIGURE 9.2.** Segmented Regression Curve with $X = X_0$ as the Change Point

The two curves must meet at $X = X_0$. This condition produces the linear constraint

$$\alpha_1 + \beta_1 X_0 = \alpha_2 + \beta_2 X_0$$

Equivalently, this constraint can be written as

$$\alpha_1 + X_0 \beta_1 - \alpha_2 - X_0 \beta_2 = 0$$

To formulate this problem as a multiple linear regression equation, we first define a dummy variable $D$ such that

$$D = 0 \quad \text{if} \quad X \leq X_0$$
$$D = 1 \quad \text{if} \quad X > X_0$$

The segmented regression equation can be combined into one multiple linear equation as

$$Y = \gamma_1 + \gamma_2 D + \gamma_3 X + \gamma_4 DX + e$$

When $X \leq X_0$, then $D = 0$, and this equation becomes

$$Y = \gamma_1 + \gamma_3 X + e$$

Therefore $\gamma_1 = \alpha_1$ and $\gamma_3 = \beta_1$. When $X > X_0$, then $D = 1$, and the equation becomes

$$Y = (\gamma_1 + \gamma_2) + (\gamma_3 + \gamma_4)X + e$$

Therefore $\gamma_1 + \gamma_2 = \alpha_2$ and $\gamma_3 + \gamma_4 = \beta_2$.

With this model a nonlinear regression program can be employed with the restriction

$$\gamma_1 + X_0\gamma_3 - (\gamma_1 + \gamma_2) - X_0(\gamma_3 + \gamma_4) = 0$$

or

$$-\gamma_2 - X_0\gamma_4 = 0$$

or

$$\gamma_2 + X_0\gamma_4 = 0$$

For example, suppose that $X_0$ is known to be 10 and that the fitted multiple regression equation is estimated to be

$$Y = 2 - 5D + 0.5X + 0.5DX + e$$

Then we estimate $\alpha_1 = \gamma_1$ as 2, $\beta_1 = \gamma_3$ as 0.5, $\alpha_2 = \gamma_1 + \gamma_2$ as $2 - 5 = -3$, and $\beta_2 = \gamma_3 + \gamma_4$ as $0.5 + 0.5 = 1$. These results are pictured in Figure 9.2. Further examples can be found in Draper and Smith (1981).

Where there are more than two segments with known values of $X$ at which the segments intersect, a similar procedure using dummy variables could be employed. When these points of intersection are unknown, more complicated estimation procedures are necessary. Quandt (1972) presents a method for estimating two regression equations when an unknown number of the points belong to the first and second equation. The method involves numerical maximization techniques and is beyond the scope of this book.

Another kind of restriction occurs when the value of the dependent variable is constrained to be above (or below) a certain limit. For example, it may be physically impossible for $Y$ to be negative. Tobin (1958) derived a procedure for obtaining a multiple linear regression equation satisfying such a constraint.

## 9.5 ADDITIONAL METHODS FOR CHECKING MULTICOLLINEARITY

In Section 7.9, we discussed the problem of multicollinearity (i.e., when the independent variables are highly intercorrelated) and presented several solutions. In that section it was mentioned that most computer programs print out tolerance or its inverse, which is called the variance inflation factor. The tolerance of a variable, $X_i$, was defined as one minus the square multiple correlation of that variable with all the other $X$ variables already included in the model. A numerical value of tolerance less than 0.01 will result in a warning from the BMDP2R stepwise program.

If the numerical value of the tolerance coefficient is around 0.01 (the variance inflation factor about 100), then the investigator should check whether the slope coefficients are stable. Since the tolerance coefficient itself can be quite unstable, sometimes reliable slope coefficients can be obtained even if the tolerance is 0.01 or less. A useful method for checking stability is to standardize the data, compute the regular slope coefficients from the standardized data, and then compare them to the standardized coefficients obtained from the original regression equation (see Dillon and Goldstein 1984). If there are no appreciable differences between the standardized slope coefficients from the original equation and the regular slope coefficients using the standardized data, then there is less apt to be a problem due to multicollinearity.

To standardize the data, new variables are created whereby the sample mean of a variable is subtracted from each observation and the resulting difference is divided by the standard deviation. This is done for each independent and dependent variable. This will cause each variable to have a mean of zero and a standard deviation of one. The lack of large observations may make the computations of the covariance matrix somewhat more accurate. If there are no multicollinearity problems, the covariance matrix of the standardized variables should equal the correlation matrix of the original variables. Thus, another option to avoid multicollinearity is to start the regression computations using the correlation matrix as the original input. This also is an option in most of the major packaged programs. Note that the intercept term is zero if standardized variables or the correlation matrix is used.

Another indication of lack of stability is a change in the coefficients when the sample size changes. It is possible to take a random sample of the data using features provided by the packaged program. If the investigator randomly splits the sample in half and obtains widely different slope coefficients in the two halves, this may indicate a multicollinearity problem. If the coefficients are unstable, alternative estimation methods can be used as discussed in the next section.

## 9.6 RIDGE REGRESSION

In Section 7.9, it was mentioned that the concept of tolerance offers the basis for a method of removing redundant variables. Other methods for excluding variables were described in Chapter 8 (variable selection).

If a successful variable selection process is used, the problem of multicollinearity is often implicitly overcome. However, there are situations where the investigator wishes to use several independent variables that are highly intercorrelated. For example, such will be the case when the independent variables are the prices of related commodities or highly related physiological variables on animals. There are two major methods for performing regression analysis when multicollinearity exists. The first method uses a technique called ridge regression that will be explained next. The second technique uses a principal-components regression which will be given in Chapter 14. A fully worked-out example using both these techniques can be found in Chatterjee and Price (1977).

### Theoretical Background

One solution to the problem of multicollinearity is the so-called *ridge regression procedure* (Marquardt and Snee 1975; Hoerl and Kennard 1970; Gunst and Mason 1980). In effect, ridge regression artificially reduces the correlations among the independent variables in order to obtain more stable estimates of the regression coefficients. Note that although such estimates are biased, they may, in fact, produce a smaller mean square error for the estimates.

To explain the concept of ridge regression, we must follow through a theoretical presentation. Readers not interested in this theory may skip to the discussion of how to perform the analysis, using ordinary least squares regression programs.

We will restrict our presentation to the standardized form of the observations, i.e., where the mean of each variable is subtracted from the observation and the difference is divided by the standard deviation of the variable. The resulting least squares regression equation will automatically have a zero intercept and standardized regression coefficients. These coefficients are functions of only the correlation matrix among the $X$ variables, namely,

$$\begin{bmatrix} 1 & r_{12} & r_{13} & \cdots & r_{1P} \\ r_{12} & 1 & r_{23} & \cdots & \\ r_{13} & r_{23} & 1 & \cdots & \vdots \\ \vdots & \vdots & \vdots & & \\ r_{1P} & r_{2P} & r_{3P} & \cdots & 1 \end{bmatrix}$$

The instability of the least squares estimates stems from the fact that some of the independent variables can be predicted accurately by the other independent variables. (For those familiar with matrix algebra, note that this feature results in the correlation matrix having a nearly zero determinant.) The ridge regression method artificially inflates the diagonal elements of the correlation matrix by adding a positive amount $k$ to each of them. The correlation matrix is modified to

$$\begin{bmatrix} (1+k) & r_{12} & r_{13} & \cdots & r_{1P} \\ r_{12} & (1+k) & r_{23} & \cdots & \\ r_{13} & r_{23} & (1+k) & \cdots & \vdots \\ \vdots & \vdots & \vdots & & \\ r_{1P} & r_{2P} & r_{3P} & \cdots & (1+k) \end{bmatrix}$$

The value of $k$ can be any positive number and is usually determined empirically, as will be described later.

## *Example of Ridge Regression*

For $P = 2$—i.e., with two independent variables $X_1$ and $X_2$—the ordinary least squares *standardized* coefficients are computed as

$$b_1 = \frac{r_{1Y} - r_{12}r_{2Y}}{1 - r_{12}^2}$$

and

$$b_2 = \frac{r_{2Y} - r_{12}r_{1Y}}{1 - r_{12}^2}$$

The ridge estimators turn out to be

$$b_1^* = \frac{r_{1Y} - [r_{12}/(1 + k)]r_{2Y}}{1 - [r_{12}/(1 + k)]^2}\left(\frac{1}{1 + k}\right)$$

and

$$b_2^* = \frac{r_{2Y} - [r_{12}/(1 + k)]r_{1Y}}{1 - [r_{12}/(1 + k)]^2}\left(\frac{1}{1 + k}\right)$$

Note that the main difference between the ridge and least squares coefficients is that $r_{12}$ is replaced by $r_{12}/(1 + k)$, thus artificially reducing the correlation between $X_1$ and $X_2$.

For a numerical example, suppose that $r_{12} = 0.9$, $r_{1Y} = 0.3$, and $r_{2Y} = 0.5$. Then the standardized least squares estimates are

$$b_1 = \frac{0.3 - (0.9)(0.5)}{1 - (0.9)^2} = -0.79$$

and

$$b_2 = \frac{0.5 - (0.9)(0.3)}{1 - (0.9)^2} = 1.21$$

For a value of $k = 0.4$ the ridge estimates are

$$b_1^* = \frac{0.3 - [0.9/(1 + 0.4)](0.5)}{1 - [0.9/(1 + 0.4)]^2}\left(\frac{1}{1 + 0.4}\right) = -0.026$$

and

$$b_2^* = \frac{0.5 - [0.9/(1 + 0.4)](0.3)}{1 - [0.9/(1 + 0.4)]^2} \left(\frac{1}{1 + 0.4}\right) = 0.374$$

Note that both coefficients are reduced in magnitude from the original slope coefficients. We computed several other ridge estimates for $k = 0.05$ to 0.5. In practice, these estimates are plotted as a function of $k$, and the resulting graph, as shown in Figure 9.3, is called the *ridge trace*.

The ridge trace should be supplemented by a plot of the residual mean square of the regression equation. From the ridge trace together with the

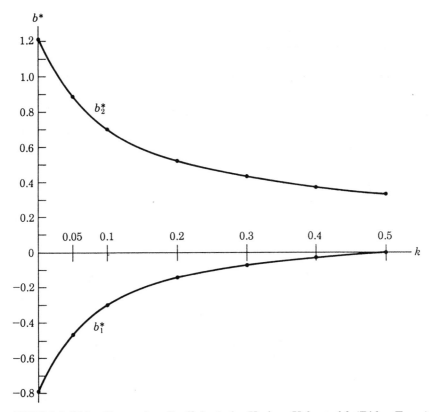

**FIGURE 9.3.** Ridge Regression Coefficients for Various Values of $k$ (Ridge Trace)

residual mean square graph, it becomes apparent when the coefficients begin to stabilize. That is, an increasing value of $k$ produces a small effect on the coefficients. The value of $k$ at the point of stabilization (say $k^*$) is selected to compute the final ridge coefficients. The final ridge estimates are thus a compromise between bias and inflated coefficients.

For the above example, as seen from Figure 9.3, the value of $k^* = 0.2$ seems to represent that compromise. Values of $k$ greater than 0.2 do not produce appreciable changes in the coefficients, although this decision is a subjective one on our part. Unfortunately, no objective criterion has been developed to determine $k^*$ from the *sample*. Proponents of ridge regression agree that the ridge estimates will give better predictions in the *population*, although they do not fit the *sample* as well as the least squares estimates. Further discussion of the properties of ridge estimators can be found in Gruber (1989).

## Implementation on the Computer

In practice, ordinary least squares regression programs such as those discussed in Chapter 7 can be used to obtain ridge estimates. As usual, we denote the dependent variable by $Y$ and the $P$ independent variables by $X_1, X_2, \ldots, X_P$. First, the $Y$ and $X$ variables are standardized by creating new variables in which we subtract the mean and divide the difference by the standard deviation for each variable. Then, $P$ additional dummy observations are added to the data set. In these dummy observations the value of $Y$ is always set equal to zero.

For a specified value of $k$ the values of the $X$ variables are defined as follows:

**1.** Compute $\sqrt{k(N-1)}$. Suppose $N = 20$ and $k = 0.2$; then $\sqrt{0.2(19)} = 1.95$.

**2.** The *first* dummy observation has $X_1 = \sqrt{k(N-1)}$, $X_2 = 0, \ldots, X_P = 0$, and $Y = 0$.

**3.** The *second* dummy observation has $X_1 = 0$, $X_2 = \sqrt{k(N-1)}$, $X_3 = 0, \ldots, X_P = 0$, and $Y = 0$, etc.

**4.** The $P$th dummy observation has $X_1 = 0$, $X_2 = 0$, $X_3 = 0, \ldots,$ $X_{P-1} = 0$, $X_P = \sqrt{k(N-1)}$, and $Y = 0$.

For a numerical example, suppose $P = 2$, $N = 20$, and $k = 0.2$. The two additional dummy observations are as follows:

| $X_1$ | $X_2$ | $Y$ |
|-------|-------|-----|
| 1.95  | 0     | 0   |
| 0     | 1.95  | 0   |

With the $N$ regular observations and these $P$ dummy observations, we use an ordinary least squares regression program in which the intercept is *forced* to be *zero*. The regression coefficients obtained from the output are automatically the ridge coefficients for the specified $k$. The resulting residual mean square should also be noted. Repeating this process for various values of $k$ will provide the information necessary to plot the ridge trace of the coefficients, similar to Figure 9.3, and the residual mean square. The usual range of interest for $k$ is between zero and one. Most computer packages allow the user to perform the above procedure for all desired values of $k$ in one run. An example of the control statements for such a run using BMDP2R is given in Problem 9.4.

The BMDP4R program has been written to perform both ridge regression and principal-components regression mentioned earlier. Plots such as that given in Figure 9.3, as well as several others, are printed. The output was designed to assist in the choice of $k^*$ in regression analysis. The SAS RIDGEREG procedure can also compute either ridge regression or principal-components regression. This procedure can be used on CMS and OS operating systems but not on personal computers. It is described in the *SAS SUGI Supplemental Library User's Guide*.

## SUMMARY

In this chapter we reviewed methods for handling some special problems in regression analysis and introduced some recent advances in the subject. We reviewed the literature on missing values and recommended that the largest possible complete sample be selected from the data set and then used in the analysis. In certain special circumstances it may be useful to replace missing values with estimates from the sample, and methods for this technique were referenced.

We gave a detailed discussion of the use of dummy variables in regression analysis. Several situations exist where dummy variables are very useful. These include incorporating nominal or ordinal variables in equations. The ideas explained here can also be used in other multivariate analyses. One reason we went into detail in this chapter is to enable you to adapt these methods to other multivariate techniques.

Two methods were discussed for producing "biased" parameter estimates, i.e., estimates that do not average out to the true parameter value. The first method is placing constraints on the parameters. This method is useful when you wish to restrict the estimates to a certain range or when natural constraints on some function of the parameters must be satisfied. The second biased regression technique is ridge regression. This method is used only when you must employ variables that are known to be highly intercorrelated. You should also consider some of the methods we discussed in Chapter 7, such as removing one of the intercorrelated variables. In addition, the tolerance option available in many programs will often prevent an intercorrelated variable from entering the equation if a variable selection method is used.

## BIBLIOGRAPHY

### Missing Data in Regression Analysis

Afifi, A. A., and Elashoff, R. M. 1966. Missing observations in multivariate statistics. I: Review of the literature. *Journal of the American Statistical Association* 61: 595–604.

————. 1969a. Missing observations in multivariate statistics. III: Large sample analysis of simple linear regression. *Journal of the American Statistical Association* 64:337–358.

————. 1969b. Missing observations in multivariate statistics. IV: A note on simple linear regression. *Journal of the American Statistical Association* 64:359–365.

*Beale, E. M. L., and Little, R. J. A. 1975. Missing values in multivariate analysis. *Journal of the Royal Statistical Society* 37:129–145.

Chatterjee, S., and Price, B. 1977. *Regression analysis by example.* New York: Wiley.

Dillon, W. R., and Goldstein, M. 1984. *Multivariate analysis: Methods and applications.* New York: Wiley.

Hartley, H. O., and Hocking, R. R. 1971. The analysis of incomplete data. *Biometrics* 27:783–824.

Little, R. J. A. 1982. Models for nonresponse in sample surveys. *Journal of the American Statistical Association* 77:237–250.

Little, R. J. A., and Rubin, D. B. 1987. *Statistical analysis with missing data.* New York: Wiley.

*Orchard, T., and Woodbury, M. A. 1972. A missing information principle: Theory and application. *Proceedings of the 6th Berkeley Symposium on Mathematical Statistics and Probability* 1:697–715.

## Ridge Regression

Fennessey, J., and D'Amico, R. 1980. Collinearity, ridge regression, and investigator judgement. *Social Methods and Research* 8:309–340.

*Gruber, M. 1989. *Regression estimators: A comparative study.* San Diego: Academic Press.

Gunst, R. F., and Mason, R. L. 1980. *Regression analysis and its application.* New York: Dekker.

Hoerl, A. E., and Kennard, R. W. 1970. Ridge regression: Application to non-orthogonal problems. *Technometrics* 12:55–82.

Marquardt, D. W., and Snee, R. D. 1975. Ridge regression in practice. *American Statistician* 29:3–20.

Myers, R. H. 1986. *Classical and modern regression with applications.* Boston: Duxbury Press.

## Segmented (Spline) Regression

Quandt, R. E. 1958. The estimation of the parameters of a linear regression system obeying two separate regimes. *Journal of the American Statistical Association* 53:873–880.

*————. 1960. Tests of the hypothesis that a linear regression system obeys two separate regimes. *Journal of the American Statistical Association* 55:324–331.

*————. 1972. New approaches to estimating switching regressions. *Journal of the American Statistical Association* 67:306–310.

## Other Regression

Box, G. E. P. 1966. Use and abuse of regression. *Technometrics* 8:625–629.

*Draper, N. R., and Smith, H. 1981. *Applied regression analysis.* 2nd ed. New York: Wiley.

*Graybill, F. A. 1976. *Theory and application of the linear model.* N. Scituate, Mass.: Duxbury Press.
*Tobin, J. 1958. Estimation of relationships for limited dependent variables. *Econometrica* 26:24–36.

## PROBLEMS

9.1 In the depression data set described in Chapter 3, data on educational level, age, sex, and income are presented for a sample of adults from Los Angeles County. Fit a regression plane with income as the dependent variable and the other variables as independent variables. Use a dummy variable for the variable "sex" that was originally coded 1,2 by stating sex = sex − 1. Which sex is the reference group?

9.2 Repeat Problem 9.1, but now use a dummy variable for education. Divide the education level into three categories: did not complete high school, completed at least high school, and completed at least a bachelor's degree. An example of a program specification using BMDP2R follows:

```
/problem  title is 'dummy edu'.
/input    format is free.
          variables are 37.
          cases are 294.
          file is depress.
/variable names are id,sex,age,marital,educ,employ,
          income,relig,c1,c2,c3,c4,c5,c6,c7,c8,c9,c10,
          c11,c12,c13,c14,c15,c16,c17,c18,c19,c20,
          cesd,cases,drink,health,regdoc,treat,
          beddays,acuteill,chronill,d1,d2.
          add = 2.
          use = sex,age,educ,income,d1,d2.
/transformation sex = sex − 1.
          if (educ eq 1 or educ eq 2) then (d1 = 1. d2 = 0.).
          if (educ eq 3 or educ eq 4) then (d1 = 0. d2 = 1.).
          if (educ gt 4) then (d1 = 0. d2 = 0.).
/regress  depend = income. indep = age,sex,d1,d2.
          enter = .001 .remove = .000.
/end
```

Compare the interpretation you would make of the effects of education on income in this problem and in Problem 9.1.

9.3 In the depression data set, determine whether religion has an effect on income when used as an independent variable along with age, sex, and educational level.

9.4 Draw a ridge trace for the accompanying data.

| | Variable | | | |
|---|---|---|---|---|
| Case | X1 | X2 | X3 | Y |
| 1 | 0.46 | 0.96 | 6.42 | 3.46 |
| 2 | 0.06 | 0.53 | 5.53 | 2.25 |
| 3 | 1.49 | 1.87 | 8.37 | 5.69 |
| 4 | 1.02 | 0.27 | 5.37 | 2.36 |
| 5 | 1.39 | 0.04 | 5.44 | 2.65 |
| 6 | 0.91 | 0.37 | 6.28 | 3.31 |
| 7 | 1.18 | 0.70 | 6.88 | 3.89 |
| 8 | 1.00 | 0.43 | 6.43 | 3.27 |
| Mean | 0.939 | 0.646 | 6.340 | 3.360 |
| Standard deviation | 0.475 | 0.566 | 0.988 | 1.100 |

The following control statements can be used for BMDP2R. Very low $F$-to-enter and -remove values have been chosen so that all variables will surely enter. The sentence "type $= 0$" is included to obtain the zero intercept. To run multiple problems with different values of $k$, include a "for" statement with the desired values of $k$, and provide a % sign ahead of the input paragraph to indicate that everything beyond is repeated.

```
/problem       title = 'ridge'.
for k = 0,0.1,0.3,0.5,1.0,2.0,3.0.
%/input        file = 'ridgedata'.
               format = free.
               cases = 11.
               var = 4.
/var           names = x1,x2,x3,y.
/transf        if (kase lt 9) then (x1 = (x1 − .939)/.475.
                   x2 = (x2 − .646)/0.566.
                   x3 = (x3 − 6.340)/0.988.
               y = (y − 3.360)/ 1.100).
               if (kase eq 9) then (x1 = (7 * k) ** 0.5).
               if (kase eq 10) then (x2 = (7 * k)** 0.5).
               if (kase eq 11) then (x3 = (7 * k)** 0.5).
```

```
/regress        dependent = y.
                enter = 0.001.remove = 0.0.
                type = 0.
/end
```

The details of the transform paragraph are given in the program manual.

9.5  Use the data in Appendix B. For the parents we wish to relate $Y$ = weight to $X$ = height for both men and women in a single equation. Using dummy variables, write an equation for this purpose, including an interaction term. Interpret the parameters. Run a regression analysis, and test whether the rate of change of weight versus height is the same for men and women. Interpret the results with the aid of appropriate graphs.

9.6  Continuation of Problem 9.5: Do a similar analysis for the first boy and girl. Include age and age squared in the regression equation.

9.7  Unlike the real data used in Problem 9.5, the accompanying data are "ideal" weights published by the Metropolitan Life Insurance Company for American men and women. Compute $Y$ = midpoint of weight range for medium-framed men and women for the various heights shown in the table. Pretending that the results represent a real sample, repeat the analysis requested in Problem 9.5, and compare the results of the two analyses.

| | | Men | |
|---|---|---|---|
| Height | Small Frame | Medium Frame | Large Frame |
| 5 ft 2 in. | 128–134 | 131–141 | 138–150 |
| 5 ft 3 in. | 130–136 | 133–143 | 140–153 |
| 5 ft 4 in. | 132–138 | 135–145 | 142–156 |
| 5 ft 5 in. | 134–140 | 137–148 | 144–160 |
| 5 ft 6 in. | 136–142 | 139–151 | 146–164 |
| 5 ft 7 in. | 138–145 | 142–154 | 149–168 |
| 5 ft 8 in. | 140–148 | 145–157 | 152–172 |
| 5 ft 9 in. | 142–151 | 148–160 | 155–176 |
| 5 ft 10 in. | 144–154 | 151–163 | 158–180 |
| 5 ft 11 in. | 146–157 | 154–166 | 161–184 |
| 6 ft 0 in. | 149–160 | 157–170 | 164–188 |
| 6 ft 1 in. | 152–164 | 160–174 | 168–192 |
| 6 ft 2 in. | 155–168 | 164–178 | 172–197 |
| 6 ft 3 in. | 158–172 | 167–182 | 176–202 |
| 6 ft 4 in. | 162–176 | 171–187 | 181–207 |

|  | **Women** | | |
|---|---|---|---|
| **Height** | **Small Frame** | **Medium Frame** | **Large Frame** |
| 4 ft 10 in. | 102–111 | 109–121 | 118–131 |
| 4 ft 11 in. | 103–113 | 111–123 | 120–134 |
| 5 ft 0 in. | 104–115 | 113–126 | 122–137 |
| 5 ft 1 in. | 106–118 | 115–129 | 125–140 |
| 5 ft 2 in. | 108–121 | 118–132 | 128–143 |
| 5 ft 3 in. | 111–124 | 121–135 | 131–147 |
| 5 ft 4 in. | 114–127 | 124–138 | 134–151 |
| 5 ft 5 in. | 117–130 | 127–141 | 137–155 |
| 5 ft 6 in. | 120–133 | 130–144 | 140–159 |
| 5 ft 7 in. | 123–136 | 133–147 | 143–163 |
| 5 ft 8 in. | 126–139 | 136–150 | 146–167 |
| 5 ft 9 in. | 129–142 | 139–153 | 149–170 |
| 5 ft 10 in. | 132–145 | 142–156 | 152–173 |
| 5 ft 11 in. | 135–148 | 145–159 | 155–176 |
| 6 ft 0 in. | 138–151 | 148–162 | 158–179 |

*Note*: Figures include 5 lb of clothing for men, 3 lb for women, and shoes with 1-in heels for both.

9.8  Use the data described in Problem 7.7. Since some of the X variables are intercorrelated, it may be useful to do a ridge regression analysis of Y on X1 to X9. Perform such an analysis, and compare the results to those of Problems 7.10 and 8.7.

9.9  Continuation of Problem 9.7: Using the data in the table given in Problem 9.7, compute the midpoints of weight range for all frame sizes for men and women separately. Pretending that the results represent a real sample, so that each height has three Y values associated with it instead of one, repeat the analysis of Problem 9.7. Produce the appropriate plots. Now repeat with indicator variables for the different frame sizes, choosing one as a reference group. Include any necessary interaction terms.

9.10  Take the family lung function data in Appendix B and delete (label as missing) the height of the middle child for every family with ID divisible by 6, that is, families 6, 12, 18, etc. (To find these, look for those ID's with ID/6 = integer part of (ID/6).) Delete the FEV1 of the middle child for families with ID divisible by 10. These data are missing at random. Use the various missing data options in SPSS REGRESSION to find the regression of FEV1 on height and weight for the middle child. Point out any differences in the results.

9.11  a. Using the family lung function data, relate FEV1 of the fathers to the area they live in. What is your reference group?

b. Repeat, including height in the regression. Do your coefficients for area change?

9.12 Using the family lung function data, relate FEV1 to height for the oldest child in the three ways: simple linear regression (see Problem 6.9), regression of FEV1 on height squared, and spline regression (split at HEI = 64). Which method is preferable?

9.13 Another way to answer the question of interaction between the independent variables in Problem 7.13 is to define a dummy variable that indicates whether an observation is above the median weight, and an equivalent variable for height. Relate FEV1 for the fathers to these dummy variables, including an interaction term. Test the hypothesis that there is no interaction. Compare your results with those of Problem 7.13.

9.14 Perform a ridge regression analysis of the family lung function data using FEV1 of the oldest child as the dependent variable and height, weight, and age of the oldest child as the independent variables.

9.15 Using the family lung function data, find the regression of height for the oldest child on mother's and father's height. Include a dummy variable for the sex of the child and any necessary interaction terms.

9.16 Using dummy variables, run a regression analysis that relates CESD as the dependent variable to marital status in the depression data set given in Chapter 3. Do it separately for males and females. Repeat using the combined group, but including a dummy variable for sex and any necessary interaction terms. Compare and interpret the results of the two regressions.

# Part Three

# *MULTIVARIATE ANALYSIS*

## Chapter Ten

# CANONICAL CORRELATION ANALYSIS

## 10.1 WHAT WILL YOU LEARN FROM THIS CHAPTER?

From this chapter you will learn how to analyze data when you have a set of $X$ variables and more than one $Y$ variable. In particular, you will learn:

- When to use canonical correlation analysis (10.2, 10.3).
- About the basic concepts used in analyzing linear relationships between two sets of data (10.4).
- How to interpret the linear combinations obtained in canonical correlation analysis (10.4).
- How to decide how many canonical correlations to obtain (10.4).
- About other methods for analyzing this type of problem (10.5).
- About the options available in the three statistical packages (10.6).
- What to watch out for in canonical correlation analysis (10.7).

## 10.2 WHEN IS CANONICAL CORRELATION ANALYSIS USED?

The technique of *canonical correlation analysis* is best understood by considering it as an extension of multiple regression and correlation analysis. In multiple regression analysis we find the best linear combination of $P$ variables, $X_1, X_2, \ldots, X_P$, to predict one variable $Y$. The multiple correlation coefficient is the simple correlation between $Y$ and its predicted value $\hat{Y}$. In multiple regression and correlation analysis our concern was therefore to examine the relationship between the $X$ variables and the single $Y$ variable.

In canonical correlation analysis we examine the linear relationships between a set of $X$ variables and a set of more than one $Y$ variable. The technique consists of finding several linear combinations of the $X$ variables and the same number of linear combinations of the $Y$ variables in such a way that these linear combinations best express the correlations between the two sets. Those linear combinations are called the *canonical variables*, and the correlations between corresponding pairs of canonical variables are called *canonical correlations*.

In a common application of this technique the $Y$'s are interpreted as outcome or dependent variables, while the $X$'s represent independent or predictive variables. The $Y$ variables may be harder to measure than the $X$ variables, as in the calibration situation discussed in Section 6.11.

Canonical correlation analysis applies to situations in which regression techniques are appropriate and where there exists more than one dependent variable. It is especially useful when the dependent or criterion variables are moderately intercorrelated; so it does not make sense to treat them separately, ignoring the interdependence. Another useful application is for testing independence between the sets of $Y$ and $X$ variables. This application will be discussed further in Section 10.4.

An example of an application is given by Waugh (1942); he studied the relationship between characteristics of certain varieties of wheat and characteristics of the resulting flour. Waugh was able to conclude that desirable wheat was high in texture, density, and protein content, and low on damaged kernels and foreign materials. Similarly, good flour should have high crude protein content and low scores on wheat per barrel of flour and ash in flour. Canonical correlation has also been used in psychology by Meredith (1964) to calibrate two sets of intelligence tests given to the same individuals. In addition, it has been used to relate linear combinations of personality scales to linear combinations of achievement tests (Tatsuoka 1988). Hopkins (1969) discusses several health-related applications of canonical correlation analysis, including, for example, a relationship between illness and housing quality.

Canonical correlation analysis is one of the less commonly used multivariate techniques. Its limited use may be due, in part, to the difficulty often encountered in trying to interpret the results. Also, prior to the advent of computers the calculations seemed forbidding.

## 10.3 DATA EXAMPLE

The depression data set presented in previous chapters is used again here to illustrate canonical correlation analysis. We select two dependent variables, CESD and health. The variable CESD is the sum of the scores on the 20 depression scale items; thus a high score indicates likelihood of depression. Likewise, "health" is a rating scale going from 1 to 4, where 4 signifies poor health and 1 signifies excellent health. The set of independent variables includes "sex", transformed so that 0 = male and 1 = female; "age" in years; "education," from 1 = less than high school up to 7 = doctorate; and "income" in thousands of dollars per year.

The summary statistics for the data are given in Tables 10.1 and 10.2. Note that the average score on the depression scale (CESD) is 8.9 in a possible range of 0 to 60. The average on the health variables is 1.8, indicating an average perceived health level falling between excellent and

**TABLE 10.1.** Means and Standard Deviations for Depression Data Set

| Variable | Mean | Standard Deviation |
|---|---|---|
| CESD | 8.88 | 8.82 |
| Health | 1.77 | 0.84 |
| Sex | 0.62 | 0.49 |
| Age | 44.41 | 18.09 |
| Education | 3.48 | 1.31 |
| Income | 20.57 | 15.29 |

**TABLE 10.2.** Correlation Matrix for Depression Data Set

|  | CESD | Health | Sex | Age | Education | Income |
|---|---|---|---|---|---|---|
| CESD | 1 | 0.212 | 0.124 | −0.164 | −0.101 | −0.158 |
| Health | | 1 | 0.098 | 0.308 | −0.270 | −0.183 |
| Sex | | | 1 | 0.044 | −0.106 | −0.180 |
| Age | | | | 1 | −0.208 | −0.192 |
| Education | | | | | 1 | 0.492 |
| Income | | | | | | 1 |

good. The average educational level of 3.5 shows that an average person has finished high school and perhaps attended some college.

Examination of the correlation matrix in Table 10.2 shows that neither CESD nor health is highly correlated with any of the independent variables. In fact, the highest correlation in this matrix is between education and income. Also, CESD is negatively correlated with age, education, and income (the younger, less educated, and lower-income person tends to be more depressed). The positive correlation between CESD and sex shows that females tend to be more depressed than males. Persons who perceived their health as good are more apt to be high on income and education and low on age.

In the following sections we will examine the relationship between the dependent variables (perceived health and depression) and the set of independent variables.

## 10.4 BASIC CONCEPTS OF CANONICAL CORRELATION

Suppose we wish to study the relationship between a set of variables $x_1$, $x_2, \ldots, x_P$ and another set $y_1, y_2, \ldots, y_Q$. The $x$ variables can be viewed as the independent or predictor variables, while the $y$'s are considered dependent or outcome variables. We assume that in any given sample the mean of each variable has been subtracted from the original data so that the sample means of all $x$ and $y$ variables are zero. In this section we discuss how the degree of association between the two sets of variables is assessed, and we present some related tests of hypotheses.

## First Canonical Correlation

The basic idea of canonical correlation analysis begins with finding one linear combination of the $y$'s, say

$$U_1 = a_1 y_1 + a_2 y_2 + \cdots + a_Q y_Q$$

and one linear combination of the $x$'s, say

$$V_1 = b_1 x_1 + b_2 x_2 + \cdots + b_P x_P$$

For any particular choice of the coefficients, the $a$'s and the $b$'s, we can compute values of $U_1$ and $V_1$ for each individual in the sample. From the $N$ individuals in the sample we can then compute the simple correlation between the $N$ pairs of $U_1$ and $V_1$ values in the usual manner. The resulting correlation depends on the choice of the $a$'s and the $b$'s.

In canonical correlation analysis we select values of $a$ and $b$ coefficients so as to *maximize* the correlation between $U_1$ and $V_1$. With this particular choice the resulting linear combination $U_1$ is called the *first canonical variable* of the $y$'s and $V_1$ is called the *first canonical variable* of the $x$'s. Note that both $U_1$ and $V_1$ have a mean of zero. The resulting correlation between $U_1$ and $V_1$ is called the *first canonical correlation*. The square of the first canonical correlation is often called the first *eigenvalue*.

The first canonical correlation is thus the highest possible correlation between a linear combination of the $x$'s and a linear combination of the $y$'s. In this sense it is the maximum linear correlation between the set of $x$ variables and the set of $y$ variables. The first canonical correlation is analogous to the multiple correlation coefficient between a single $Y$ variable and the set of $X$ variables. The difference is that in canonical correlation analysis we have several $y$ variables and we must find a linear combination of them also.

The BMDP6M program computed the coefficients ($a$'s and $b$'s) as shown in Table 10.3. Note that the first set of coefficients is used to compute the values of the canonical variables $U_1$ and $V_1$.

Table 10.4 shows the process used to compute the canonical correlation. For each individual we compute $V_1$ from the $b$ coefficients and the individual's $X$ variable values after subtracting the means. We do the same for $U_1$. These computations are shown for the first three individuals.

**TABLE 10.3.** Canonical Correlation Coefficients for First Correlation (Depression Data Set)

| Coefficients | Standardized Coefficients |
|---|---|
| $b_1 = 0.051$ (sex)  $a_1 = -0.055$ (CESD) | $b_1 = 0.025$ (sex)  $a_1 = -0.490$ (CESD) |
| $b_2 = 0.048$ (age)  $a_2 = 1.17$ (health) | $b_2 = 0.871$ (age)  $a_2 = +0.982$ (health) |
| $b_3 = -0.29$ (education) | $b_3 = -0.383$ (education) |
| $b_4 = +0.005$ (income) | $b_4 = 0.082$ (income) |

The correlation coefficient is then computed from the 294 values of $U_1$ and $V_1$. Note that the variances of $U_1$ and $V_1$ are each equal to 1.

The standardized coefficients are also shown in Table 10.3, and they are to be used with the standardized variables. The standardized coefficients can be obtained by multiplying each unstandardized coefficient by the standard deviation of the corresponding variable. For example, the unstandardized coefficient of $y_1$ (CESD) is $a_1 = -0.0555$, and from Table 10.1 the standard deviation of $y_1$ is 8.82. Therefore the standardized coefficient of $y_1$ is $-0.490$.

In this example the resulting canonical correlation is 0.405. This value represents the highest possible correlation between any linear combination of the independent variables and any linear combination of the dependent variables. In particular, it is larger than any simple correlation between an $x$ variable and a $y$ variable (see Table 10.2). One method for interpreting the linear combination is by examining the standardized coefficients. For the $x$ variables the canonical variable is determined largely by age and education. Thus a person who is relatively old and relatively uneducated would score high on canonical variable $V_1$. The canonical variable based on the $y$'s gives a large positive weight to the perceived health variables and a negative weight to CESD. Thus a person with a high health value (perceived poor health) and a low depression score would score high on canonical variable $U_1$. In contrast, a young person with relatively high education would score low on $V_1$, and a person in good perceived health but relatively depressed would score low on $U_1$. Sometimes, because of high intercorrelations between two variables in the same set, one variable may result in another having a small coefficient (see Levine 1977) and thus make the interpretation difficult. No very high correlations within a set existed in the present example.

**TABLE 10.4.** Computation of Correlation Between $U_1$ and $V_1$

| Individual | Sex $b_1(X_1 - \bar{X}_1)$ | Age $+ b_2(X_2 - \bar{X}_2)$ | Education $+ b_3(X_3 - \bar{X}_3)$ | Income $+ b_4(X_4 - \bar{X}_4)$ | $\rightarrow V_1$ | $U_1 \leftarrow$ | CESD $a_1(Y_1 - \bar{Y}_1)$ | Health $a_2(Y_2 - \bar{Y}_2)$ |
|---|---|---|---|---|---|---|---|---|
| 1 | 0.051(1 − 0.62) | + 0.048(68 − 44.4) | − 0.29(2 − 3.48) | + 0.0054(4 − 20.57) | = 1.49 | 0.76 = | −0.055(0 − 8.88) | + 1.17(2 − 1.77) |
| 2 | 0.051(0 − 0.62) | + 0.048(58 − 44.4) | − 0.29(4 − 3.48) | + 0.0054(15 − 20.57) | = 0.44 | −0.64 = | −0.055(4 − 8.88) | + 1.17(1 − 1.77) |
| 3 | 0.051(1 − 0.62) | + 0.048(45 − 44.4) | − 0.29(3 − 3.48) | + 0.0054(28 − 20.57) | = 0.23 | 0.54 = | −0.055(4 − 8.88) | + 1.17(2 − 1.77) |
| . | | | | | | | | |
| . | | | | | | | | |
| 294 | | | | | | | | |

Correlation between $U_1$ and $V_1$ = canonical correlation = 0.405

In summary, we conclude that older but uneducated people tend to be relatively undepressed although they perceive their health as relatively poor. Because the first canonical correlation is the largest possible, this impression is the strongest conclusion we can make from this analysis of the data. However, there may be other important conclusions to be drawn from the data, which will be discussed next.

## *Other Canonical Correlations*

Additional interpretation of the relationship between the $x$'s and the $y$'s is obtained by deriving other sets of canonical variables and their corresponding canonical correlations. Specifically, we derive a second canonical variable $V_2$ (linear combination of the $x$'s) and a corresponding canonical variable $U_2$ (linear combination of the $y$'s). The coefficients for these linear combinations are chosen so that the following conditions are met:

**1.** $V_2$ is uncorrelated with $V_1$ and $U_1$.
**2.** $U_2$ is uncorrelated with $V_1$ and $U_1$.
**3.** Subject to conditions 1 and 2, $U_2$ and $V_2$ have the maximum possible correlation.

The correlation between $U_2$ and $V_2$ is called the *second canonical correlation* and will necessarily be less than or equal to the first canonical correlation.

In our example the second set of canonical variables expressed in terms of the standardized coefficients is

$$V_2 = 0.396(\text{sex}) - 0.443(\text{age}) - 0.448(\text{education}) - 0.555(\text{income})$$

and

$$U_2 = 0.899(\text{CESD}) + 0.288(\text{health})$$

Note that $U_2$ gives a high positive weight to CESD and a low positive weight to health. In contrast, $V_2$ gives approximately the same moderate weight to all four variables, with sex having the only positive coefficient. A large value of $V_2$ is associated with young, poor, uneducated females. A large value of $U_2$ is associated with a high value of CESD (depressed) and to a lesser degree with a high value of health (poor perceived health). The value of the second canonical correlation is 0.266.

In general, this process can be continued to obtain other sets of canonical variables $U_3$, $V_3$; $U_4$, $V_4$; etc. The maximum number of canonical correlations and their corresponding sets of canonical variables is equal to the minimum of $P$ (the number of $x$ variables) and $Q$ (the number of $y$ variables). In our data example $P = 4$ and $Q = 2$, so the maximum number of canonical correlations is two.

## Tests of Hypotheses

Most packaged computer programs print the coefficients for all of the canonical variables, the values of the canonical correlations, and the values of the canonical variables for each individual in the sample. Some programs also print the computed values of Bartlett's chi-square test statistic. This test is an approximate test of the null hypothesis that the $k$ smallest population canonical correlations are zero. A large chi-square is an indication that not all of those $k$ correlations are zero. This test was derived with the assumption that the $x$'s and the $y$'s are jointly distributed according to a multivariate normal distribution (see Bartlett 1941 and Lawley 1959). In the SAS CANCORR and SPSS MANOVA procedures, a different test statistic is used but the same null hypothesis is tested.

In our data example the Bartlett chi-square for the hypothesis that both canonical correlations are zero was computed by BMDP6M to be 73.04 with eight degrees of freedom. Since the $P$ value is less than 0.00001, we conclude that at least one canonical correlation is nonzero and proceed to test the hypothesis that the smallest one is zero. The test statistic chi-square equals 21.17 with three degrees of freedom. The $P$ value is 0.001, and we conclude that both canonical correlations are significantly different from zero.

In data sets with more variables Bartlett's test can be a useful guide for selecting the number of significant canonical correlations. The test results are examined to determine at which step the remaining canonical correlations can be considered zero. In this case, as in stepwise regression, the significance levels should not be interpreted literally.

## 10.5 OTHER TOPICS RELATED TO CANONICAL CORRELATION

In this section we discuss some useful optional output available from packaged programs.

## Plots of Canonical Variable Scores

A useful option available in some programs is a plot of the canonical variable scores $U_i$ versus $V_i$. In Figure 10.1 we show a scatter diagram of $U_1$ versus $V_1$ for the depression data. The first individual from Table 10.4 is indicated on the graph. The degree of scatter gives the impression of a

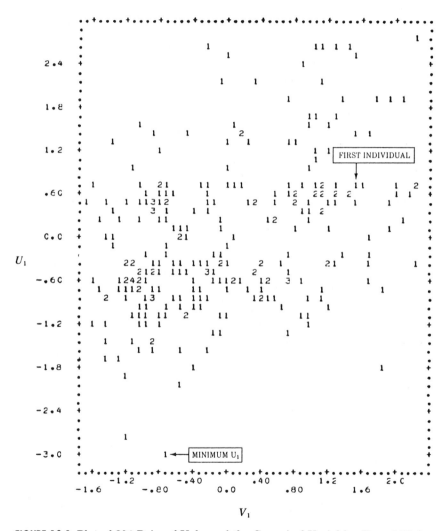

**FIGURE 10.1.** Plot of 294 Pairs of Values of the Canonical Variables $U_1$ and $V_1$ for the Depression Data Set (Canonical Correlation = 0.405)

somewhat weak but significant canonical correlation (0.405). For multivariate normal data the graph would approximate an ellipse of concentration. Such a plot can be useful in highlighting unusual cases in the sample as possible outliers or blunders. For example, the individual with the lowest value on $U_1$ is case number 289. This individual is a 19-year-old female with some high school education and with \$28,000-per-year income. These scores produce a value of $V_1 = -0.73$. Also, this woman perceives her health as excellent (1) and is very depressed (CESD = 47), resulting in $U_1 = -3.02$. This individual, then, represents an extreme case in that she is uneducated and young but has a good income. In spite of the fact that she perceives her health as excellent, she is extremely depressed. Although this case gives an unusual combination, it is not necessarily a blunder.

## Another Interpretation of Canonical Variables

Another useful optional output is the set of correlations between the canonical variables and the original variables used in deriving them. This output provides a way of interpreting the canonical variables when some of the variables within either the set of independent or the set of dependent variables are highly intercorrelated with each other (see Cooley and Lohnes 1971 or Levine 1977). For the depression data example these correlations are as shown in Table 10.5. These correlations are sometimes called *canonical variable loadings*. Other terms are *canonical loadings* and *canonical structural coefficients* (see Dillon and Goldstein 1984 or Thompson 1984 for additional interpretations of these coefficients).

Since the canonical variable loadings can be interpreted as simple correlations between each variable and the canonical variable, they are useful in understanding the relationship between the original variables and the canonical variables. When the set of variables used in one canonical variable are uncorrelated, the canonical variable loadings are equal to the standardized canonical variable coefficients. When some of the original variables are highly intercorrelated, the loadings and the coefficients can be quite different. It is in these cases that some statisticians find it simpler to try to interpret the canonical variable loadings rather than the canonical variable coefficients. For example, suppose that there are two $X$ variables that are highly positively correlated, and that each is positively correlated with the canonical variable. Then it is possible that

**TABLE 10.5.** Correlations Between Canonical Variables and Corresponding Variables (Depression Data Set)

|           | $U_1$   | $U_2$  |
|-----------|---------|--------|
| CESD      | −0.281  | 0.960  |
| Health    | 0.878   | 0.478  |

|           | $V_1$   | $V_2$  |
|-----------|---------|--------|
| Sex       | 0.089   | 0.525  |
| Age       | 0.936   | −0.225 |
| Education | −0.532  | −0.636 |
| Income    | −0.254  | −0.737 |

one canonical variable coefficient will be positive and one negative, while the canonical variable loadings are both positive, the result one expects.

In the present data set, the intercorrelations among the variables are neither zero nor strong. Comparing the results of Tables 10.5 and 10.3 for the first canonical variable shows that the standardized coefficients have the same signs as the canonical variable loadings, but with somewhat different magnitudes.

## Redundancy Analysis

The average of the squared canonical variable loadings for the first canonical variate, $V_1$, gives the proportion of the variance in the $X$ variables explained by the first canonical variate. The same is true for $U_1$ and $Y$. Similar results hold for each of the other canonical variates. For example, for $U_1$ we have $((-0.281)^2 + 0.878^2)/2 = 0.425$, or less than half of the variance in the $Y$'s explained by the first canonical variate. Sometimes the proportion of variance explained is quite low, even though there is a large canonical correlation. This may be due to only one or two variables having a major influence on the canonical variate.

The above computations provide one aspect of what is known as *redundancy analysis*. In addition, SAS CANCORR and BMDP6M can compute a quantity called the *redundancy coefficient* which is also useful in evaluating the adequacy of the prediction from the canonical analysis (see Muller 1981). The coefficient is a measure of the average proportion of variance in the $Y$ set that is accounted for by the $V$ set. It is comparable to the squared multiple correlation in multiple linear regression analysis. It is also possible to obtain the proportion of variance in the $X$ variables that is explained by the $U$ variables, but this is usually of less interest.

## 10.6 DISCUSSION OF COMPUTER PROGRAMS

The BMDP and SAS packages each contain a canonical correlation program. The SPSS–X MANOVA program will also perform canonical correlation analysis. The various options for these programs are summarized in Table 10.6.

For the data example described in Section 10.3 we want to relate reported health and depression levels to several typical demographic variables. The BMDP6M program was used to obtain the results reported in this chapter. The input and variable paragraphs are the usual ones. Because our data are divided into two sets of variables, we have to indicate which variables are in each set. This indication is given in the canonical paragraph. The "first = CESD,health" sentence defines which variables are the dependent variables. The second sentence defines the independent variables. Unlike the regression programs, this program does not need enter or remove sentences because the program will automatically compute up to a minimum of $P$ or $Q$ (the number of independent and

**TABLE 10.6.** Summary of Computer Output for Canonical Correlation

| Output | BMDP | SAS | SPSS–X |
|---|---|---|---|
| Means and standard deviations or variances of data | 6M | CANCORR | MANOVA |
| Correlation matrix | 6M | CANCORR | MANOVA |
| Canonical correlations | 6M | CANCORR | MANOVA |
| Canonical coefficients | 6M | CANCORR | MANOVA |
| Standardized canonical coefficients | 6M | CANCORR | MANOVA |
| Canonical variable loadings | 6M | CANCORR | MANOVA |
| Canonical variable scores | 6M | CANCORR | |
| Plots of canonical variable scores | 6M | PLOT | |
| Bartlett's test and $P$ value | 6M | | |
| Wilk's lambda and $F$ approximation | | CANCORR | MANOVA |
| Redundancy analysis | 6M | CANCORR | MANOVA |

dependent variables). The results of the Bartlett's test are used to decide how many significant correlations exist between the two sets of variables.

We ask for the coefficients of the canonical variables ($a$'s and $b$'s) and for the canonical variable loadings (the correlation between each variable and the canonical variable score; see Section 10.5).

Finally, we ask to see the plots of the first and second canonical variables ($U_1$ versus $V_1$ and $U_2$ versus $V_2$). These plots enable us to search for outliers and to interpret the correlation between the canonical variables. For example, if the relationship between $U_1$ and $V_1$ is not linear, then this plot enables us to assess the nonlinearity.

The paragraphs for our example are as follows:

```
/input       file is 'depress'.
             variables are 6.
             format is free.
/variable    names are (2)sex,(3)age,(5)educat,
             (7)income,(29)cesd,(32)health.
/canonical   first = cesd,health.
             second = sex,age,educat,income.
/print       matr = coef.
/plot        xvar = cnvrs1,cnvrs2.
             yvar = cnvrf1,cnvrf2.
/end
```

The program lists the first five cases so that we can check that it used the correct data; then it gives us simple summary statistics for each variable. It then prints the correlation matrix (see Table 10.2), and we readily see that although most of the variables are somewhat correlated, no very high correlations exist. Thus one variable cannot be considered a direct substitute for another. The program also prints the squared multiple correlation of each variable with all other variables in its set, which provides another method of checking multicollinearity. If a very high multiple correlation exists, then one strategy would be to discard a variable since it is not needed.

Next, the program prints out the canonical correlations between the two sets of canonical variables (0.405 and 0.266); these values are listed in order of magnitude. The square of these correlations is also printed and labeled "eigenvalue." Bartlett's test is also presented, as discussed in the last part of Section 10.4.

The unstandardized and standardized coefficients are printed for both sets of variables for the two canonical correlations. Also, the correlations of the canonical variables with the original variables are given in a separate table. In the case of the first canonical correlation a comparison of (1) the standardized coefficients and (2) the correlations of the canonical variables with the original variables shows no striking differences, although the sign is different for income and the two measures do not have a constant ratio from variable to variable, as shown below:

|  | *First Canonical Correlation* | |
|---|---|---|
| *Variable* | *Standardized Coefficients* | *Correlations* |
| CESD | −0.490 | −0.281 |
| Health | 0.982 | 0.878 |
| Sex | 0.025 | 0.089 |
| Age | 0.871 | 0.936 |
| Education | −0.383 | −0.532 |
| Income | 0.082 | −0.254 |

The correlation of age and health with their canonical variable is very high, indicating the close association between the outcome of those variables and the first canonical variable.

The plot of $U_1$ versus $V_1$ does not result in a nonlinear-appearing scatter diagram, nor does it look like a bivariate normal distribution (ellipse in shape), as can be seen in Figure 10.1. It may be that the skewness present in the CESD distribution has resulted in a somewhat skewed pattern for $U_1$ even though health has a greater overall effect on the first canonical variable. If this pattern were more extreme, it might be worthwhile to consider transformations on some of the variables such as CESD.

The SAS CANCORR procedure is very straightforward to run. You simply call the procedure and specify the following:

```
CANCORR DATA = DEPRESS ALL
VAR CESD HEALTH;
WITH SEX AGE EDUCAT INCOME;
```

Here the name of the data set is DEPRESS. The ALL statement produces all the optional output. The first set of variables represents the dependent variables, and the second set the independent variables.

The SAS CANCORR procedure will also perform the canonical correlation analysis starting with the partial correlations. Thus the linear effects of a variable or variables can be removed prior to obtaining the canonical correlation.

For the SPSS MANOVA program the instructions are as follows:

```
MANOVA   CESD HEALTH
         WITH SEX AGE EDUCAT INCOME/
         PRINT = SIGNIF (EIGEN)/
         DISCRIM (RAW, STAN, CORR)
```

These statements instruct the computer to perform the same analysis and print out the same statistics obtained from the other two packages, as discussed above.

## 10.7 WHAT TO WATCH OUT FOR

Because canonical correlation analysis can be viewed as an extension of multiple linear regression analysis, many of the precautionary remarks made at the ends of Chapters 6 through 8 apply here. The user should be aware of the following points:

1. The sample should be representative of the population to which the investigator wishes to make inferences. A simple random sample has this property. If this is not attainable, the investigator should at least make sure that the cases are selected in such a way that the full range of observations occurring in the population can occur in the sample. If the range is artificially restricted, the estimates of the correlations will be affected.

2. Poor reliability of the measurements can result in lower estimates of the correlations among the $X$'s and among the $Y$'s.

**3.** A search for outliers should be made by obtaining histograms and scatter diagrams of pairs of variables.

**4.** Stepwise procedures are not available in the programs described in this chapter. Variables that contribute little and are not needed for theoretical models should be candidates for removal. It may be necessary to run the programs several times to arrive at a reasonable choice of variables.

**5.** The investigator should check that the canonical correlation is large enough to make examination of the coefficients worthwhile. In particular, it is important that the correlation not be due to just one dependent variable and the independent variable. The proportion of variance should be examined, and if it is small then it may be sensible to reduce the number of variables in the model.

**6.** If the sample size is large enough, it is advisable to split it, run a canonical analysis on both halves, and compare the results to see if they are similar.

**7.** If the canonical coefficients and the canonical variable loadings differ considerably, (i.e., if they have different signs), then both should be examined carefully to aid in interpreting the results. Problems of interpretation are often more difficult in the second or third canonical variates than in the first. The condition that subsequent linear combinations of the variables be independent of those already obtained places restrictions on the results that may be difficult to understand.

**8.** Tests of hypotheses regarding canonical correlations assume that the joint distribution of the $X$'s and $Y$'s is multivariate normal. This assumption should be checked if such tests are to be reported.

## SUMMARY

In this chapter we presented the basic concepts of canonical correlation analysis, an extension of multiple regression and correlation analysis. The extension is that the dependent variable is replaced by two or more dependent variables. If $Q$, the number of dependent variables, equals 1, then canonical correlation reduces to multiple regression analysis.

In general, the resulting canonical correlations quantify the strength of the association between the dependent and independent sets of variables. The derived canonical variables show which combinations of the original variables best exhibit this association. The canonical variables can be interpreted in a manner similar to the interpretation of principal components or factors, as will be explained in Chapters 14 and 15.

## BIBLIOGRAPHY

*Bartlett, M. S. 1941. The statistical significance of canonical correlations. *Biometrika* 32:29–38.

*Cooley, W. W., and Lohnes, P. R. 1971. *Multivariate data analysis.* New York: Wiley.

Dillon, W. R., and Goldstein, M. 1984. *Multivariate analysis: Methods and applications.* New York: Wiley.

Hopkins, C. E. 1969. Statistical analysis by canonical correlation: A computer application. *Health Services Research* (Winter); 4:304–312.

*Hotelling, H. 1936. Relations between two sets of variables. *Biometrika* 28:321–377.

*Lawley, D. N. 1959. Tests of significance in canonical analysis. *Biometrika* 46:59–66.

Levine, M. S. 1977. *Canonical analysis and factor comparison.* Sage University Paper. Beverly Hills: Sage.

Meredith, W. 1964. Canonical correlation with fallible data. *Psychometrika* 29:55–65.

*Morrison, D. F. 1976. *Multivariate statistical methods.* 2nd ed. New York: McGraw-Hill.

Muller, K. E. 1981. Relationships between redundancy analysis, canonical correlation, and multivariate regression. *Psychometrika* 46:139–142.

Muller, K. E. 1982. Understanding canonical correlation through the general linear model and principal components. *The American Statistician* 36:342–354.

*Tatsuoka, M. M. 1988. *Multivariate analysis: Techniques for educational and psychological research.* 2nd ed. New York: Wiley.

Thompson, B. 1984. *Canonical correlation analysis.* Beverly Hills: Sage.

Thorndike, R. M. 1978. *Correlational procedures for research.* New York: Gardner Press.

Waugh, F. V. 1942. Regressions between sets of variables. *Econometrica* 10:290–310.

## PROBLEMS

10.1 For the depression data set, perform a canonical correlation analysis between the following:

- Set 1: AGE, MARITAL (married versus other), EDUCAT (high school or less versus other), EMPLOY (full-time versus other), and INCOME.
- Set 2: The last seven variables.

Perform separate analyses for men and women. Interpret the results.

10.2 For the data set given in Appendix B, do a canonical correlation analysis on height, weight, FVC, and FEV1 for fathers versus the same variables for mothers. Interpret.

10.3 For the Chemical Companies data given in Table 8.1, perform a canonical correlation analysis using P/E and EPS5 as dependent variables and the remaining variables as independent variables. Write the interpretations of the significant canonical correlations in terms of their variables.

10.4 Using the data given in Appendix B, perform a canonical correlation analysis using height, weight, FVC, and FEV1 of the oldest child as the dependent variables and the same measurements for the parents as the independent variables. Interpret the results.

10.5 Generate the sample data for X1, X2, ..., X9, Y as in Problem 7.7. For each of the following pairs of sets of variables, do a canonical correlation analysis between the variables in Set 1 and those in Set 2.

|     | **Set 1**      | **Set 2**    |
| --- | -------------- | ------------ |
| a.  | X1, X2, X3     | X4–X9        |
| b.  | X4, X5, X6     | X7, X8, X9   |
| c.  | Y, X9          | X1, X2, X3   |

Interpret the results in light of what you know about the population.

# Chapter Eleven

# DISCRIMINANT ANALYSIS

## 11.1 WHAT WILL YOU LEARN FROM THIS CHAPTER?

From this chapter you will learn how to classify an individual into one of two or more populations on the basis of the values of one or more variables. In particular, you will learn:

- When discriminant analysis is used (11.2, 11.3).
- About the basic concepts underlying classification of individuals (11.4).
- The meaning of the classical method of classification, the Fisher discriminant function, and how to obtain it (11.5).
- How to interpret discriminant function programs (11.6).
- How to incorporate prior information into the classification procedure (11.7).
- How to evaluate the degree of success of the classification procedure (11.8).
- How to evaluate the contribution of variables (11.9).
- How to select variables for use in classification (11.10).
- About classification into more than two groups (11.11).

- How to use canonical correlation in discriminant function analysis (11.12).
- How to choose the appropriate computer program and options (11.13).
- What to watch out for in discriminant analysis (11.14).

## 11.2 WHEN IS DISCRIMINANT ANALYSIS USED?

*Discriminant analysis* techniques are used to classify individuals into one of two or more alternative groups (or populations) on the basis of a set of measurements. The populations are known to be distinct, and each individual belongs to one of them. These techniques can also be used to identify which variables contribute to making the classification. Thus, as in regression analysis, we have two uses, prediction and description.

As an example, consider an archeologist who wishes to determine which of two possible tribes created a particular statue found in a dig. The archeologist takes measurements on several characteristics of the statue and must decide whether these measurements are more likely to have come from the distribution characterizing the statues of one tribe or from the other tribe's distribution. These distributions are based on data from statues known to have been created by members of one tribe or the other. The problem of classification is therefore to guess who made the newly found statue on the basis of measurements obtained from statues whose identities are certain.

The measurements on the new statue may consist of a single observation, such as its height. However, we would then expect a low degree of accuracy in classifying the new statue since there may be quite a bit of overlap in the distribution of heights of statues from the two tribes. If, on the other hand, the classification is based on several characteristics, we would have more confidence in the prediction. The discriminant analysis methods described in this chapter are multivariate techniques in the sense that they employ several measurements.

As another example, consider a loan officer at a bank who wishes to decide whether to approve an applicant's automobile loan. This decision is made by determining whether the applicant's characteristics are more similar to those of persons who in the past repaid loans successfully or to those of persons who defaulted. Information on these two groups, available from past records, would include factors such as age, income, marital status, outstanding debt, and home ownership.

A third example, which is described in detail in the next section, comes from the depression data set (Chapters 1 and 3). We wish to predict whether an individual living in the community is more or less likely to be depressed on the basis of readily available information on the individual.

## 11.3 DATA EXAMPLE

As described in Chapter 1, the depression data set was collected for individuals residing in Los Angeles County. To illustrate the ideas described in this chapter, we will develop a method for estimating whether an individual is likely to be depressed. For the purposes of this example "depression" is defined by a score of 16 or greater on the CESD scale (see the code book given in Table 3.2). This information is given in the variable called "cases." We will base the estimation on demographic and other characteristics of the individual. The variables used are education and income. We may also wish to determine whether we can improve our prediction by including information on illness, sex, or age. Additional variables are an overall health rating, number of bed days in the past two months (0 if less than eight days, 1 if eight or more), acute illness (1 if yes in the past two months, 0 if no), and chronic illness (0 if none, 1 if one or more).

The first step in examining the data is to obtain descriptive measures of each of the groups. Table 11.1 lists the means and standard deviations for each variable in both groups. Note that in the depressed group, group II, we have a higher percentage of females, a lower average age, a lower educational level, and lower incomes. The standard deviations in the two groups are similar except for income, where they are slightly different. Note also that the health characteristics of the depressed group are generally worse than those of the nondepressed, even though the members of the depressed group tend to be younger on the average. Because sex is coded males = 1 and females = 2, the average sex of 1.80 indicates that 80% of the depressed group are females. Similarly, 59% of the non-depressed individuals are female.

Suppose that we wish to predict whether or not individuals are depressed, on the basis of their incomes. Examination of Table 11.1 shows

**TABLE 11.1.** Means and Standard Deviations for Nondepressed and Depressed Adults in Los Angeles County

| Variable | Group I, Nondepressed (N = 244) | | Group II, Depressed (N = 50) | |
|---|---|---|---|---|
| | Mean | Standard Deviation | Mean | Standard Deviation |
| Sex (male = 1, female = 2) | 1.59* | 0.49 | 1.80* | 0.40 |
| Age (in years) | 45.2 | 18.1 | 40.4 | 17.4 |
| Education (1 to 7, 7 high) | 3.55 | 1.33 | 3.16 | 1.17 |
| Income (thousands of dollars per year) | 21.68* | 15.98 | 15.20* | 9.84 |
| Health index (1 to 4, 1 = excellent) | 1.71* | 0.80 | 2.06* | 0.98 |
| Bed days (0 : less than 8 days per year; 1 : 8 or more days) | 0.17* | 0.39 | 0.42* | 0.50 |
| Acute conditions (0 = no, 1 = yes) | 0.28 | 0.45 | 0.38 | 0.49 |
| Chronic conditions (0 = none, 1 = one or more) | 0.48 | 0.50 | 0.62 | 0.49 |

* For a test of equal means, $P$ less than 0.01, assuming a normal distribution.

that the mean value for depressed individuals is significantly lower than that for the nondepressed. Thus, intuitively, we would classify those with lower incomes as depressed and those with higher incomes as non-depressed. Similarly, we may classify the individuals on the basis of age alone, or sex alone, etc. However, as in the case of regression analysis, the use of several variables simultaneously can be superior to the use of any one variable. The methodology for achieving this result will be explained in the next sections.

## 11.4 BASIC CONCEPTS OF CLASSIFICATION

In this section, we present the underlying concepts of classification as given by Fisher (1936) and give an example illustrating its use. We also briefly discuss the coefficients from the Fisher discriminant function.

Statisticians have formulated different ways of performing and evaluating discriminant function analysis. One method of evaluating the results uses what are called *classification functions*. This approach will be described next. Necessary computations are given in Sections 11.6, 11.11, and 11.13. In addition, discriminant function analysis can be viewed as a special case of canonical correlation analysis presented in Chapter 10. Many of the programs print out canonical coefficients and graphical output based upon evaluation of canonical variates. This approach will be discussed in Section 11.12.

## Principal Ideas

Suppose that an individual may belong to one of two populations. We begin by considering how an individual can be classified into one of these populations on the basis of a measurement of one characteristic, say $X$. Suppose that we have a representative sample from each population, enabling us to estimate the distributions of $X$ and their means. Typically, these distributions can be represented as in Figure 11.1.

From the figure it is intuitively obvious that a low value of $X$ would lead us to classify an individual into population II and a high value would

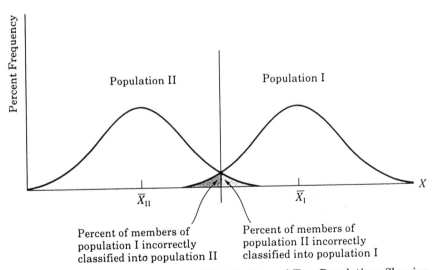

**FIGURE 11.1.** Hypothetical Frequency Distributions of Two Populations Showing Percentage of Cases Incorrectly Classified

lead us to classify an individual into population I. To define what is meant by *low* or *high*, we must select a dividing point. If we denote this dividing point by $C$, then we would classify an individual into population I if $X \geq C$. For any given value of $C$ we would be incurring a certain percentage of error. If the individual came from population I but the measured $X$ were less than $C$, we would incorrectly classify the individual into population II, and vice versa. These two types of errors are illustrated in Figure 11.1. If we can assume that the two populations have the same variance, then the usual value of $C$ is

$$C = \frac{\bar{X}_{\mathrm{I}} + \bar{X}_{\mathrm{II}}}{2}$$

This value ensures that the two probabilities of error are equal.

The idealized situation illustrated in Figure 11.1 is rarely found in practice. In real-life situations the degree of overlap of the two distributions is frequently large, and the variances are rarely precisely equal. For example, in the depression data the income distributions for the depressed and nondepressed individuals do overlap to a large degree, as illustrated in Figure 11.2. The usual dividing point is

$$C = \frac{15.20 + 21.68}{2} = 18.44$$

As can be seen from Figure 11.2, the percentage errors are rather large. The exact data on the errors are shown in Table 11.2. These numbers

**TABLE 11.2.** Classification of Individuals as Depressed or Not Depressed on the Basis of Income Alone

| | Classified as | | |
|---|---|---|---|
| **Actual Status** | **Not Depressed** | **Depressed** | **% Correct** |
| Not depressed ($N = 244$) | 123 | 121 | 50.4 |
| Depressed ($N = 50$) | 19 | 31 | 62.0 |
| Total $N = 294$ | 142 | 152 | 52.4 |

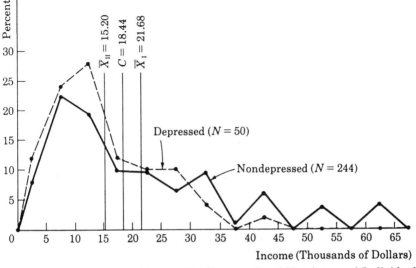

**FIGURE 11.2.** Distribution of Income for Depressed and Nondepressed Individuals Showing Effects of a Dividing Point at an Income of $18,440

were obtained by first checking whether each individual's income was greater than or equal to $18.44 \times 10^3$ and then determining whether the individual was correctly classified. For example, of the 244 nondepressed individuals, 123 had incomes greater than or equal to $18.44 \times 10^3$ and were therefore correctly classified as not depressed (see Table 11.2). Similarly, of the 50 depressed individuals, 31 were correctly classified. The total number of correctly classified individuals is $123 + 31 = 154$, amounting to 52.4% of the total sample of 294 individuals, as shown in Table 11.2. Thus, although the mean incomes were significantly different from each other $(P < 0.01)$, income alone is not very successful in identifying whether an individual is depressed.

Combining two or more variables may provide better classification. Note that the number of variables used must be less than $N_I$ plus $N_{II}$ minus 1. For two variables $X_1$ and $X_2$ concentration ellipses may be illustrated as shown in Figure 11.3 (see Section 7.5 for an explanation of concentration ellipses). Figure 11.3 also illustrates the univariate distributions of $X_1$ and $X_2$ separately. The univariate distribution of $X_1$ is what is obtained if the values of $X_2$ are ignored. On the basis of $X_1$ alone, and its corresponding dividing point $C_1$, a relatively large amount of error of misclassification

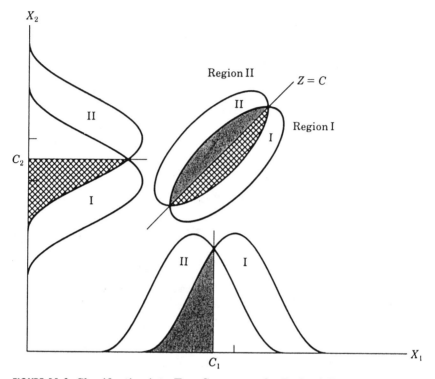

**FIGURE 11.3.** Classification into Two Groups on the Basis of Two Variables

would be encountered. Similar results occur for $X_2$ and its corresponding dividing point $C_2$. To use both variables simultaneously, we need to divide the plane of $X_1$ and $X_2$ into two regions, each corresponding to one population, and classify the individuals accordingly. A simple way of defining the two regions is to draw a straight line through the points of intersection of the two concentration ellipses, as shown in Figure 11.3.

The percentage of individuals from population II incorrectly classified is shown in the crosshatched areas. The shaded areas show the percentage of individuals from population I who are misclassified. The errors incurred by using two variables are often much smaller than those incurred by using either variable alone. In the illustration in Figure 11.3 this result is, in fact, the case.

The dividing line was represented by R. A. Fisher (1936) as an equation $Z = C$, where $Z$ is a linear combination of $X_1$ and $X_2$ and $C$ is a constant

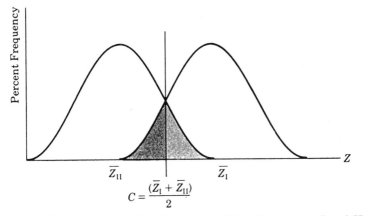

**FIGURE 11.4.** Frequency Distributions of $Z$ for Populations I and II

defined as follows:

$$C = \frac{\bar{Z}_\mathrm{I} + \bar{Z}_\mathrm{II}}{2}$$

where $\bar{Z}_\mathrm{I}$ is the average value of $Z$ in population I and $\bar{Z}_\mathrm{II}$ is the average value of $Z$ for population II.

In this book we will call $Z$ the *Fisher discriminant function*, written as

$$Z = a_1 X_1 + a_2 X_2$$

for the two-variable case. The formulas for computing the coefficients $a_1$ and $a_2$ can be found in Fisher (1936), Lachenbruch (1975), or Afifi and Azen (1979).

For each individual from each population, the value of $Z$ is calculated. When the frequency distributions of $Z$ are graphed separately for each population, the result is as illustrated in Figure 11.4. In this case, the bivariate classification problem with $X_1$ and $X_2$ is reduced to a univariate situation using the single variable $Z$.

## Example

As an example of this technique for the depression data, it may be better to use both income and age to classify depressed individuals. Program

BMDP7M was used to obtain the Fisher discriminant function. Unfortunately, this equation cannot be obtained directly from the output, and some intermediate computations must be made, as explained in Section 11.6. The result is

$$Z = 0.0209(\text{age}) + 0.0336(\text{income})$$

The mean $Z$ value for each group can be obtained as follows, using the means from Table 11.1:

$$\text{mean } Z = 0.0209(\text{mean age}) + 0.0336(\text{mean income})$$

Thus

$$\bar{Z}_{\text{not depressed}} = 0.0209(45.2) + 0.0336(21.68) = 1.67$$

and

$$\bar{Z}_{\text{depressed}} = 0.0209(40.4) + 0.0336(15.20) = 1.36$$

The dividing point is therefore

$$C = \frac{1.67 + 1.36}{2} = 1.515$$

An individual is then classified as depressed if his or her $Z$ value is less than 1.52.

For two variables it is possible to illustrate the classification procedure as shown in Figure 11.5. This figure was obtained from the output of a BMDP6D program (see Chapter 6). In Figure 11.5 each A denotes the income and age of a nondepressed person, and each B denotes the same information for a depressed person. The dividing line is a graph of the equation $Z = C$, i.e.,

$$0.0209(\text{age}) + 0.0336(\text{income}) = 1.515$$

An individual falling in the region above the dividing line is classified as not depressed. Note that, indeed, very few depressed persons fall far above the dividing line.

To measure the degree of success of the classification procedure for this sample, we must count how many of each group are correctly classified.

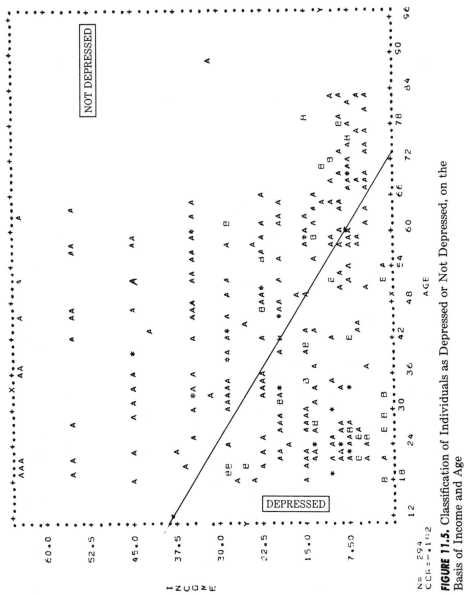

**FIGURE 11.5.** Classification of Individuals as Depressed or Not Depressed, on the Basis of Income and Age

281

**TABLE 11.3.** Classification of Individuals as Depressed or Not Depressed on the Basis of Income and Age

| | Classified As | | |
|---|---|---|---|
| *Actual Status* | *Not Depressed* | *Depressed* | *% Correct* |
| Not depressed ($N = 244$) | 154 | 90 | 63.1 |
| Depressed ($N = 50$) | 20 | 30 | 60.0 |
| Total $N = 294$ | 174 | 120 | 62.6 |

The computer program produces these counts automatically; they are shown in Table 11.3. Note that 63.1% of the nondepressed are correctly classified. This value is compared with 50.4%, which results when income alone is used (Table 11.2). The percentage of depressed correctly classified is comparable in both tables. Combining age with income improved the overall percentage of correct classification from 52.4% to 62.6%.

## Interpretation of Coefficients

In addition to its use for classification, the Fisher discriminant function is helpful in indicating the direction and degree to which each variable contributes to the classification. The first thing to examine is the sign of each coefficient: If it is positive, the individuals with larger values of the corresponding variable tend to belong to population I, and vice versa. In the depression data example, both coefficients are positive, indicating that large values of both variables are associated with a lack of depression. In more complex examples, comparisons of those variables having positive coefficients with those having negative coefficients can be revealing. To quantify the magnitude of the distribution, the investigator may find standardized coefficients helpful, as explained in Section 11.6.

The concept of discriminant functions applies as well to situations where there are more than two variables, say $X_1, X_2, \ldots, X_P$. As in multiple linear regression, it is often sufficient to select a small number of variables. Variable selection will be discussed in Section 11.10. In the next section we present some necessary theoretical background for discriminant function analysis.

## 11.5 THEORETICAL BACKGROUND

In deriving his linear discriminant function, R. A. Fisher (1936) did not have to make any distributional assumptions for the variables used in classification. Fisher denoted the discriminant function by

$$Z = a_1 X_1 + a_2 X_2 + \cdots + a_P X_P$$

As in the previous section, we denote the two mean values of $Z$ by $\bar{Z}_I$ and $\bar{Z}_{II}$. We also denote the pooled sample variance of $Z$ by $S_Z^2$ (this statistic is similar to the pooled variance used in the standard two-sample $t$ test; e.g., see Dixon and Massey 1983). To measure how "far apart" the two groups are in terms of values of $Z$, we compute

$$D^2 = \frac{(\bar{Z}_I - \bar{Z}_{II})^2}{S_Z^2}$$

Fisher selected the coefficients $a_1, a_2, \ldots, a_P$ so that $D^2$ has the maximum possible value.

The term $D^2$ can be interpreted as the squared distance between the means of the standardized value of $Z$. A larger value of $D^2$ indicates that it is easier to discriminate between the two groups. The quantity $D^2$ is called the *Mahalanobis distance*. Both $a_i$ and $D^2$ are functions of the group means and the pooled variances and covariances of the variables. The manuals for the statistical package programs make frequent use of these formulas, and you will find a readable presentation of them in Lachenbruch (1975) or Klecka (1980).

Some distributional assumptions make it possible to develop further statistical procedures relating to the problem of classification. These procedures include tests of hypotheses for the usefulness of some or all of the variables and methods for estimating errors of classification.

The variables used for classification are denoted by $X_1, X_2, \ldots, X_P$. The standard model makes the assumption that for each of the two populations these variables have a multivariate normal distribution. It further assumes that the covariance matrix is the same in both populations. However, the mean values for a given variable may be different in the two populations. We further assume that we have a random sample from each of the populations. The sample sizes are denoted by $N_I$ and $N_{II}$.

Alternatively, we may think of the two populations as subpopulations of a single population. For example, in the depression data the original population consists of all adults over 18 years old in Los Angeles County. Its two subpopulations are the depressed and nondepressed. A single sample was collected and later diagnosed to form two subsamples.

## 11.6 INTERPRETATION

In this section we present various methods for interpreting discriminant functions. Specifically, we discuss the regression analogy, computations of the coefficients, standardized coefficients, and posterior probabilities.

### Regression Analogy

A useful connection exists between regression and discriminant analyses. For the regression interpretation we think of the classification variables $X_1, X_2 \ldots, X_P$ as the independent variables. The dependent variable is a dummy variable indicating the population from which each observation comes. Specifically,

$$Y = \frac{N_{\mathrm{II}}}{N_{\mathrm{I}} + N_{\mathrm{II}}}$$

if the observation comes from population I, and

$$Y = -\frac{N_{\mathrm{I}}}{N_{\mathrm{I}} + N_{\mathrm{II}}}$$

if the observation comes from population II. For instance, for the depression data $Y = 50/(244 + 50)$ if the individual is not depressed and $Y = -244/(244 + 50)$ if the individual is depressed.

When the usual multiple regression analysis is performed, the resulting regression coefficients are proportional to the discriminant function coefficients $a_1, a_2, \ldots, a_P$ (see Lachenbruch 1975). The value of the resulting multiple correlation coefficient $R$ is related to the Mahalanobis $D^2$ by the following formula:

$$D^2 = \frac{R^2}{1 - R^2} \frac{[N_{\text{I}} + N_{\text{II}}][N_{\text{I}} + N_{\text{II}} - 2]}{N_{\text{I}} \times N_{\text{II}}}$$

Hence from a multiple regression program it is possible to obtain the coefficients of the discriminant function and the value of $D^2$. The $\bar{Z}$'s for each group can be obtained by multiplying each coefficient by the corresponding variable's sample mean and adding the results. The dividing point $C$ can then be computed as

$$C = \frac{\bar{Z}_{\text{I}} + \bar{Z}_{\text{II}}}{2}$$

As in regression analysis, some of the independent variables (or classification variables) may be dummy variables (see Section 9.3). In the depression example we may, for instance, use sex as one of the classification variables by treating it as a dummy variable. Research has shown that even though such variables do not follow a normal distribution, their use in linear discriminant analysis can still help improve the classification.

## Computing the Fisher Discriminant Function

In the discriminant analysis programs that will be discussed in Section 11.13, some computations must be performed to obtain the values of the discriminant coefficients. Some programs (such as BMDP7M and the SPSS–X DISCRIMINANT procedure) print what is called a "classification function" for each group. Other programs (such as the SAS DISCRIM procedure) call these functions the "linearized discriminant functions." For each population the coefficients are printed for each variable. The discriminant function coefficients $a_1, a_2, \ldots, a_P$ are then obtained by subtraction.

As an example, we again consider the depression data using age and income. The classification functions are shown in Table 11.4.

The coefficient $a_1$ for age is $0.1634 - 0.1425 = 0.0209$. For income, $a_2 = 0.1360 - 0.1024 = 0.0336$. The dividing point $C$ is also obtained by

**TABLE 11.4.** Classification Function and Discriminant Coefficients for Age and Income from BMDP7M

| | Classification Function | | |
|---|---|---|---|
| Variables | Group I,<br>Not Depressed | Group II,<br>Depressed | Discriminant Function |
| Age | 0.1634 | 0.1425 | $0.0209 = a_1$ |
| Income | 0.1360 | 0.1024 | $0.0336 = a_2$ |
| Constant | −5.8641 | −4.3483 | $1.5158 = C$ |

subtraction, but in *reverse order*. Thus $C = -4.3483 - (-5.8641) = 1.5158$. (This agrees closely with the previously computed value $C = 1.515$, which we use throughout this chapter.) For more than two variables the same procedure is used to obtain $a_1, a_2, \ldots, a_P$, and $C$.

## Renaming the Groups

If we wish to put the depressed persons into group I and the nondepressed into group II, we can do so by reversing the zero and one values for the "cases" variable. In the data used for our example, "cases" equals 1 if a person is depressed and zero if not depressed. However, we can make cases equal zero if a person is depressed and one if a person is not depressed. The BMDP TRANSFORM paragraph for this conversion is as follows:

```
/TRANSFORM
X = CASES.
IF (X EQ 0) THEN CASES = 1.
IF (X EQ 1) THEN CASES = 0.
```

Note that this reversal does not change the classification functions but simply changes their order so that *all* the signs in the linear discriminant function are changed. The new constant and discriminant function are

$-1.515$ and $-0.0209(\text{age}) - 0.0336(\text{income})$, respectively.

The ability to discriminate is exactly the same, and the number of individuals correctly classified is the same.

## Standardized Coefficients

As in the case of regression analysis, the values of $a_1, a_2, \ldots, a_P$ are not directly comparable. However, an impression of the relative effect of each variable on the discriminant function can be obtained from the *standardized discriminant coefficients*. This technique involves the use of the pooled (or within-group) covariance matrix from the computer output. In the original example this covariance matrix is as follows:

|        | Age    | Income |
|--------|--------|--------|
| Age    | 324.8  | −57.7  |
| Income | −57.7  | 228.6  |

Thus the pooled standard deviations are $\sqrt{324.8} = 18.02$ for age and $\sqrt{228.6} = 15.10$ for income. The standardized coefficients are obtained by multiplying the $a_i$'s by the corresponding pooled standard deviations. Hence the standardized discriminant coefficients are

$$(0.0209)(18.02) = 0.377 \quad \text{for age}$$

and

$$(0.0336)(15.10) = 0.505 \quad \text{for income}$$

It is therefore seen that income has a slightly larger effect on the discriminant function than age.

## Posterior Probabilities

Thus far the classification procedure assigned an individual to either group I or group II. Since there is always a possibility of making the wrong classification, we may wish to compute the probability that the individual has come from one group or the other. We can compute such a probability under the multivariate normal model discussed in Section 11.5. The formula is

$$\text{probability of belonging to population I} = \frac{1}{1 + \exp(-Z + C)}$$

where $\exp(-Z + C)$ indicates $e$ raised to the power $(-Z + C)$, as discussed in Truett, Cornfield, and Kannell (1967). The probability of belonging to population II is one minus the probability of belonging to population I.

For example, suppose that an individual from the depression study is 42 years old and earns $24 \times 10^3$ income per year. For that individual the value of the discriminant function is

$$Z = 0.0209(42) + 0.0336(24) = 1.718$$

Since $C = 1.515$—and therefore $Z$ is greater than $C$—we classify the individual as not depressed (in population I). To determine how likely this person is to be not depressed, we compute the probability

$$\frac{1}{1 + \exp(-1.718 + 1.515)} = 0.55$$

The probability of being depressed is $1 - 0.55 = 0.45$. Thus this individual is only slightly more likely to be not depressed than to be depressed.

Several packaged programs compute the probabilities of belonging to both groups for each individual in the sample. In some programs these probabilities are called the *posterior probabilities* since they express the probability of belonging to a particular population posterior to (i.e., after) performing the analysis.

Posterior probabilities offer a valuable method of interpreting classification results. The investigator may wish to classify only those individuals whose probabilities clearly favor one group over the other. Judgment could be withheld for individuals whose posterior probabilities are close to 0.5. In the next section another type of probability, called prior probability, will be defined and used to modify the dividing point.

Finally, we note that the discriminant function presented here is a sample estimate of the population discriminant function. We would compute the latter if we had the actual values of the population parameters. If the populations were both multivariate normal with equal covariance matrices, then the population discriminant classification procedure would be optimal; i.e., no other classification procedure would produce a smaller total classification error (see Anderson 1984).

## 11.7 ADJUSTING THE VALUE OF THE DIVIDING POINT

In this section we indicate how prior probabilities and costs of misclassification can be incorporated into the choice of the dividing point $C$.

### Incorporating Prior Probabilities into the Choice of C

Thus far, the dividing point $C$ was used as the point producing an equal percentage of errors of both types, i.e., the probability of misclassifying an individual from population I into population II, or vice versa. This use can be seen in Figure 11.4. But the choice of the value of $C$ can be made to produce any desired ratio of these probabilities of errors. To explain how this choice is made, we must introduce the concept of *prior probability*. Since the two populations constitute an overall population, it is of interest to examine their relative size. The prior probability of population I is the probability that an individual selected at random actually comes from population I. In other words, it is the proportion of individuals in the overall population who fall in population I. This proportion is denoted by $q_I$.

In the depression data the definition of a depressed person was originally designed so that 20% of the population would be designated as depressed and 80% nondepressed. Therefore the prior probability of not being depressed (population I) is $q_I = 0.8$. Likewise, $q_{II} = 1 - q_I = 0.2$. Without knowing any of the characteristics of a given individual, we would thus be inclined to classify him or her as nondepressed, since 80% were in that group. In this case we would be correct 80% of the time. This example offers an intuitive interpretation of prior probabilities. Note, however, that we would be always wrong in identifying depressed individuals.

The theoretical choice of the dividing point $C$ is made so that the total probability of misclassification is minimized. This total probability is defined as $q_I \cdot$ (probability of misclassifying an individual from population I into population II) plus $q_{II} \cdot$ (probability of misclassifying an individual from population II into population I), or

$$q_I \cdot \text{Prob(II given I)} + q_{II} \cdot \text{Prob(I given II)}$$

Under the multivariate normal model mentioned in Section 11.5 the optimal choice of the dividing point $C$ is

$$C = \frac{\bar{Z}_I + \bar{Z}_{II}}{2} + \ln \frac{q_{II}}{q_I}$$

where ln stands for the natural logarithm. Note that if $q_I = q_{II} = \frac{1}{2}$, then $q_{II}/q_I = 1$ and $\ln q_{II}/q_I = 0$. In this case $C$ is

$$C = \frac{\bar{Z}_I + \bar{Z}_{II}}{2} \qquad \text{if} \qquad q_I = q_{II}$$

Thus in the previous sections we have been implicitly assuming that $q_I = q_{II} = \frac{1}{2}$.

For the depression data we have seen that $q_I = 0.8$, and therefore the theoretical dividing point should be

$$C = 1.515 + \ln(0.25) = 1.515 - 1.386 = 0.129$$

In examining the data, we see that using this dividing point classifies all of the nondepressed individuals correctly and all of the depressed individuals incorrectly. Therefore the probability of classifying a nondepressed individual (population I) as depressed (population II) is zero. On the other hand, the probability of classifying a depressed individual (population II) as nondepressed (population I) is 1. Therefore the total probability of misclassification is $(0.8)(0) + (0.2)(1) = 0.2$. When $C = 1.515$ was used, the two probabilities of misclassification were 0.369 and 0.400, respectively (see Table 11.3). In that case the total probability of misclassification is $(0.8)(0.379) + (0.2)(0.400) = 0.383$. This result verifies that the theoretical dividing point did produce a smaller value of this total probability of misclassification.

In practice, however, it is not appealing to identify none of the depressed individuals. If the purpose of classification were preliminary screening, we would be willing to incorrectly label some individuals as depressed in order to avoid missing too many of those who are truly depressed. In practice, we would choose various values of $C$ and for each value determine the two probabilities of misclassification. The desired choice of $C$ would be made when some balance of these two is achieved.

## Incorporating Costs into the Choice of C

One method of weighting the errors is to determine the relative costs of the two types of misclassification. For example, suppose that it is four times as serious to falsely label a depressed individual as nondepressed as it is to label a nondepressed individual as depressed. These costs can be denoted as

$$\text{cost(II given I)} = 1$$

and

$$\text{cost(I given II)} = 4$$

The dividing point $C$ can then be chosen to minimize the total cost of misclassification, namely

$$q_\text{I} \cdot \text{Prob(II given I)} \cdot \text{cost(II given I)}$$
$$+ q_\text{II} \cdot \text{Prob(I given II)} \cdot \text{cost(I given II)}$$

The choice of $C$ that achieves this minimization is

$$C = \frac{\bar{Z}_\text{I} + \bar{Z}_\text{II}}{2} + K$$

where

$$K = \ln \frac{q_\text{II} \cdot \text{cost(I given II)}}{q_\text{I} \cdot \text{cost(II given I)}}$$

In the depression example the value of $K$ is

$$K = \ln \frac{0.2(4)}{0.8(1)} = \ln 1 = 0$$

In other words, this numerical choice of cost of misclassification and the use of prior probabilities counteract each other so that $C = 1.515$, the same value obtained without incorporating costs and prior probabilities.

Finally, it is important to note that incorporating the prior probabilities and costs of misclassification alters only the choice of the dividing point $C$. It does not affect the computation of the coefficients $a_1, a_2, \ldots, a_P$ in the discriminant function. If the computer program does not allow the option of incorporating those quantities, you can easily modify the dividing point as was done in the above example. In Section 11.13, we show how costs of misclassification can be incorporated into a program that allows only prior probabilities to be specified.

## 11.8 HOW GOOD IS THE DISCRIMINANT FUNCTION?

A *measure of goodness* for the classification procedure consists of the two probabilities of misclassification, probability(II given I) and probability(I given II). Various methods exist for estimating these probabilities. One method, called the *empirical method*, was used in the previous examples. That is, we applied the discriminant function to the same samples used for deriving it and computed the proportion incorrectly classified from each group (see Tables 11.2 and 11.3). This process is a form of validation of the discriminant function. Although this method is intuitively appealing, it does produce biased estimates. In fact, the resulting proportions under-estimate the true probabilities of misclassification, because the same sample is used for deriving and validating the discriminant function.

Ideally, we would like to derive the function from one sample and apply it to another sample to estimate the proportion misclassified. This procedure is called *cross-validation*, and it produces unbiased estimates. The investigator can achieve cross-validation by randomly splitting the original sample from each group into two subsamples: one for deriving the discriminant function and one for cross-validating it.

The investigator may be hesitant to split the sample if it is small. An alternative method sometimes used in this case, which imitates splitting the samples, is called the *jackknife procedure*. In this method we exclude one observation from the first group and compute the discriminant function on the basis of the remaining observations. We then classify the excluded observation. This procedure is repeated for each observation in the first sample. The proportion of misclassified individuals is the jack-knife estimate of Prob(II given I). A similar procedure is used to estimate Prob(I given II). This method produces nearly unbiased estimators. Some programs offer this option.

If we accept the multivariate normal model, theoretical estimates of the probabilities are also available and require only an estimate of $D^2$. The formulas are

$$\text{estimated Prob(II given I)} = \text{area to left of } \left( \frac{K - 1/2D^2}{D} \right) \text{ under standard}$$

$$\text{normal curve}$$

and

$$\text{estimated Prob(I given II)} = \text{area to left of } \left( \frac{-K - 1/2D^2}{D} \right) \text{ under}$$

$$\text{standard normal curve}$$

where

$$K = \ln \frac{q_{II} \cdot \text{cost(I given II)}}{q_1 \cdot \text{cost(II given I)}}$$

If $K = 0$, these two estimates are each equal to the area to the left of $(-D/2)$ under the standard normal curve. For example, in the depression example $D^2 = 0.319$ and $K = 0$. Therefore $D/2 = 0.282$, and the area to the left of $-0.282$ is $0.389$. From this method we estimate both Prob(II given I) and Prob(I given II) as $0.39$. This method is particularly useful if the discriminant function is derived from a regression program, since $D^2$ can be easily computed from $R^2$ (see Section 11.6).

Unfortunately, this last method also underestimates the true probabilities of misclassification. An *unbiased estimator* of the population Mahalanobis $D^2$ is

$$\text{unbiased } D^2 = \frac{N_I + N_{II} - P - 3}{N_I + N_{II} - 2} D^2 - P \left( \frac{1}{N_I} + \frac{1}{N_{II}} \right)$$

In the depression example we have

$$\text{unbiased } D^2 = \frac{50 + 244 - 2 - 3}{50 + 244 - 2} (0.319) - 2 \left( \frac{1}{50} + \frac{1}{244} \right) = 0.316 - 0.048$$

$$= 0.268$$

The resulting area is computed in a similar fashion to the last method. Since unbiased $D/2 = 0.259$, the resulting area to the left of $-D/2$ is $0.398$. In comparing this result with the estimate based on the biased $D^2$, we note that the difference is small because (1) only two variables are used and (2) the sample sizes are fairly large. On the other hand, if the number of variables $P$ were close to the total sample size $(N_\mathrm{I} + N_\mathrm{II})$, the two estimates could be very different from each other.

Whenever possible, it is recommended that the investigator obtain at least some of the above estimates of errors of misclassification and the corresponding probabilities of correct prediction.

To evaluate how well a particular discriminant functions is performing, the investigator may also find it useful to compute the probability of correct prediction based on pure *guessing*. The procedure is as follows: Suppose that prior probability of belonging to population I is known to be $q_\mathrm{I}$. Then $q_\mathrm{II} = 1 - q_\mathrm{I}$. One way to classify individuals using these probabilities alone is to imagine a coin that comes up heads with probability $q_\mathrm{I}$ and tails with probability $q_\mathrm{II}$. Every time an individual is to be classified, the coin is tossed. The individual is classified into population I if the coin comes up heads and into population II if it is tails. Overall, a proportion $q_\mathrm{I}$ of all individuals will be classified into population I.

Next, the investigator computes the total probability of correct classification. Recall that the probability that a person comes from population I is $q_\mathrm{I}$, and the probability that any individual is classified into population I is $q_\mathrm{I}$. Therefore the probability that a person comes from population I and is *correctly* classified into population I is $q_\mathrm{I}^2$. Similarly, $q_\mathrm{II}^2$ is the probability that an individual comes from population II and is correctly classified into population II. Thus the total probability of correct classification using only knowledge of the prior probabilities is $q_\mathrm{I}^2 + q_\mathrm{II}^2$. Note that the lowest possible value of this probability occurs when $q_\mathrm{I} = q_\mathrm{II} = 0.5$, i.e., when the individual is equally likely to come from either population. In that case $q_\mathrm{I}^2 + q_\mathrm{II}^2 = 0.5$.

Using this method for the depression example, with $q_\mathrm{I} = 0.8$ and $q_\mathrm{II} = 0.2$, gives us $q_\mathrm{I}^2 + q_\mathrm{II}^2 = 0.68$. Thus we would expect more than two-thirds of the individuals to be correctly classified if we simply flipped a coin that comes up heads 80% of the time. Note, however, that we would be wrong on 80% of the depressed individuals, a situation we may not be willing to tolerate. (In this context you might recall the role of costs of misclassification discussed in Section 11.7.)

## 11.9  TESTING  FOR  THE  CONTRIBUTIONS  OF  CLASSIFICATION VARIABLES

Can we classify individuals by using variables available to us better than we can by chance alone? One answer to this question assumes the multivariate normal model presented in Section 11.4. The question can be formulated as a hypothesis-testing problem. The null hypothesis being tested is that none of the variables improve the classification based on chance alone. Equivalent null hypotheses are that the two population means for each variable are identical, or that the population $D^2$ is zero. The test statistic for the null hypothesis is

$$F = \frac{N_{\mathrm{I}} + N_{\mathrm{II}} - P - 1}{P(N_{\mathrm{I}} + N_{\mathrm{II}} - 2)} \times \frac{N_{\mathrm{I}}N_{\mathrm{II}}}{N_{\mathrm{I}} + N_{\mathrm{II}}} \times D^2$$

with degrees of freedom of $P$ and $N_{\mathrm{I}} + N_{\mathrm{II}} - P - 1$ (see Rao 1973). The $P$ value is the tail area to the right of the computed test statistic. We point out that $N_{\mathrm{I}}N_{\mathrm{II}}D^2/(N_{\mathrm{I}} + N_{\mathrm{II}})$ is known as the two-sample Hotelling $T^2$, derived originally for testing the equality of two sets of means (see Morrison 1976).

For the depression example using age and income, the computed $F$ value is

$$F = \frac{244 + 50 - 2 - 1}{2(244 + 50 - 2)} \times \frac{244 \times 50}{244 + 50} \times 0.319 = 6.60$$

with 2 and 291 degrees of freedom. The $P$ value for this test is less than 0.005. Thus these two variables together significantly improve the prediction based on chance alone. Equivalently, there is statistical evidence that the population means are not identical in both groups. It should be noted that most existing computer programs do not print the value of $D^2$. However, from the printed value of the above $F$ statistic we can compute $D^2$ as follows:

$$D^2 = \frac{P(N_{\mathrm{I}} + N_{\mathrm{II}})(N_{\mathrm{I}} + N_{\mathrm{II}} - 2)}{(N_{\mathrm{I}}N_{\mathrm{II}})(N_{\mathrm{I}} + N_{\mathrm{II}} - P - 1)} F$$

Another useful test is whether one additional variable improves the discrimination. Suppose that the population $D^2$ based on $X_1, X_2, \ldots, X_P$

variables is denoted by pop $D_P^2$. We wish to test whether an additional variable $X_{P+1}$ will significantly increase the pop $D^2$, i.e., we test the hypothesis that pop $D_{p+1}^2 = $ pop $D_P^2$. The test statistic under the multivariate normal model is also an $F$ statistic and is given by

$$F = \frac{(N_{\mathrm{I}} + N_{\mathrm{II}} - P - 2)(N_{\mathrm{I}}N_{\mathrm{II}})(D_{P+1}^2 - D_P^2)}{(N_{\mathrm{I}} + N_{\mathrm{II}})(N_{\mathrm{I}} + N_{\mathrm{II}} - 2) + N_{\mathrm{I}}N_{\mathrm{II}}D_P^2}$$

with 1 and $(N_{\mathrm{I}} + N_{\mathrm{II}} - P - 2)$ degrees of freedom (see Rao 1965).

For example, in the depression data we wish to test the hypothesis that age improves the discriminant function when combined with income. The $D_1^2$ for income alone is 0.183, and $D_2^2$ for income and age is 0.319. Thus with $P = 1$

$$F = \frac{(50 + 244 - 1 - 2)(50 \times 244)(0.319 - 0.183)}{(294)(292) + 50 \times 244 \times 0.183} = 5.48$$

with 1 and 291 degrees of freedom. The $P$ value for this test is equal to 0.02. Thus age significantly improves the classification when combined with income.

A generalization of this last test allows for the checking of the contribution of several additional variables simultaneously. Specifically, if we start with $X_1, X_2, \ldots, X_P$ variables, we can test whether $X_{P+1}, \ldots, X_{P+Q}$ variables improve the prediction. We test the hypothesis that pop $D_{P+Q}^2 = $ pop $D_P^2$. For the multivariate normal model the test statistic is

$$F = \frac{(N_{\mathrm{I}} + N_{\mathrm{II}} - P - Q - 1)}{Q} \times \frac{N_{\mathrm{I}}N_{\mathrm{II}}(D_{P+Q}^2 - D_P^2)}{(N_{\mathrm{I}} + N_{\mathrm{II}})(N_{\mathrm{I}} + N_{\mathrm{II}} - 2) + N_{\mathrm{I}}N_{\mathrm{II}}D_P^2}$$

with $Q$ and $(N_{\mathrm{I}} + N_{\mathrm{II}} - P - Q - 1)$ degrees of freedom.

The last two formulas for $F$ are useful in variable selection, as will be shown in the next section.

## 11.10 VARIABLE SELECTION

Recall that there is an analogy between regression analysis and discriminant function analysis. Therefore much of the discussion of variable selection given in Chapter 8 applies to selecting variables for classification into two groups. In fact, the computer programs discussed in Chapter 8 may be used here as well. These include, in particular, stepwise regression programs and subset regression programs. In addition, some computer programs are available for performing stepwise discriminant analysis. They employ the same concepts discussed in connection with stepwise regression analysis.

In discriminant function analysis, instead of testing whether the value of multiple $R^2$ is altered by adding (or deleting) a variable, we test whether the value of pop $D^2$ is altered by adding or deleting variables. The $F$ statistic given in Section 11.9 is used for this purpose. As before, the user may specify a value for the $F$-to-enter and $F$-to-remove values. For $F$-to-enter Costanza and Afifi (1979) recommend using a value corresponding to a $P$ of 0.15. No recommended value from research can be given for the $F$-to-remove value, but a reasonable choice may be a $P$ of 0.30.

## 11.11 CLASSIFICATION INTO MORE THAN TWO GROUPS

A comprehensive discussion of classification into more than two groups is beyond the scope of this book. However, several books do include detailed discussion of this subject, including Tatsuoka (1988), Lachenbruch (1975), Morrison (1976), and Afifi and Azen (1979). In this section we summarize the classification procedure for a multivariate normal model and discuss an example.

We now assume that an individual is to be classified into one of $k$ populations, for $k \geq 2$, on the basis of the values of $P$ variables $X_1$, $X_2, \ldots, X_P$. We assume that in each of the $k$ populations the $P$ variables have a multivariate normal distribution, with the same covariance matrix. A typical packaged computer program will compute, from a sample from each population, a classification function. In applications, the variable values from a given individual are substituted in each classification function, and the individual is classified into the population corresponding to the highest classification function.

Now we will consider an example. In the depression data set values of CESD can vary from 0 to 60 (see Chapter 3). In the previous runs, persons were classified as depressed if their CESD scores were greater than or equal to 16; otherwise, they were classified as not depressed.

Another possibility is to divide the individuals into $k = 3$ groups: those who deny any symptoms of depression (CESD score = 0), those who have CESD scores between 1 and 15 inclusive, and those who have scores of 16 or greater. Arbitrarily, we call these three groups lowdep(1), meddep (2), and highdep (3).

The variables considered for entry in a stepwise fashion are sex, age, education, income, health, bed days, and chronic illness. These variables are also used in Section 11.13, where the control statements for $k = 2$ are given for BMDP7M. The method of entering variables used in this program is similar to that described for forward stepwise regression in Chapter 8.

Partial results for this example are given in Table 11.5. Note that not all the variables entered the discriminant function.

The approximate $F$ statistic given in Table 11.5 tests the null hypothesis that the means of the three groups are equal for all variables simultaneously. From Appendix Table A.4 we observe that the $P$ value is very small, indicating that the null hypothesis should be rejected.

The $F$ matrix part of Table 11.5 gives $F$ statistics for testing the equality of means for each pair of groups. The $F$ value (2.36) for the lowdep and meddep groups is barely significant at the 5% level. The other $F$ statistics indicate a significant difference between the highdep group and each of the other two. This result partially justifies our previous analysis of two groups, where highdep is the depressed group and lowdep plus meddep constitute the nondepressed group.

The classification functions are also shown in Table 11.5. To classify a new individual into one of the three groups, we first evaluate each of the three classification functions, using that individual's scores on all the variables that entered the function. Then the individual is assigned to the group for which the computed classification function is the highest. If we are interested in the groups taken two at a time, the corresponding pair of classification functions could be subtracted from each other to produce a discriminant function, as explained in Section 11.6.

In the example considered here we divided a discrete variable into three subsets to get the three groups. Often when we are working with nominal data, three or more separate groups exist; examples include classification by religion or by race of individuals. Thus the capability that discriminant function analysis has of handling more than two groups is a useful feature in some applications.

**TABLE 11.5.** Partial Printout from BMDP7M for Classification into More Than Two Groups, Using the Depression Data with $k = 3$ Groups

---

```
APPROXIMATE F-STATISTIC  4.347  DEGREES OF FREEDOM  12.00  572.00

          F-MATRIX      DEGREES OF FREEDOM = 6   286

          lowdep       meddep
meddep     2.36
hishdep    7.12          5.58

CLASSIFICATION FUNCTIONS

             GROUP = lowdep       meddep          hishdep
   VARIABLE
    2 sex          7.07529        7.49617          8.14165
    3 age           .16774         .13935           .11698
    5 educat       2.54993        2.82551          2.68116
    7 income        .10533         .09005           .06537
   32 health       2.13954        2.75024          3.10425
   35 beddays      -.97394        -.80246           .46685

   CONSTANT      -17.62107      -18.54811        -18.81630

   VARIABLE    COEFFICIENTS FOR CANONICAL VARIABLES

    2 sex         -.73103         -.01977
    3 age          .03167          .02531
    5 educat      -.00617         -.65481
    7 income       .02758         -.00067
   32 health      -.57524         -.68822
   35 beddays    -1.13644         1.13117

   CONSTANT       .49631          2.17769
```

---

# 11.12 USE OF CANONICAL CORRELATION IN DISCRIMINANT FUNCTION ANALYSIS

Recall that there is an analogy between regression and discriminant analysis for two groups. It is unfortunate that this analogy does *not* extend to the case of $k$ greater than two. There is, however, a correspondence between canonical correlation and classification into several populations.

We begin by defining a set of new variables called $Y_1, Y_2, \ldots, Y_{k-1}$. These are dummy or indicator variables that show which group each member of the sample came from. Note that as discussed in Chapter 9, we need $k - 1$ dummy variables to describe $k$ groups. For example, suppose that there are $k = 4$ groups. Then the dummy variables $Y_1$, $Y_2$, and $Y_3$ are formed as follows:

| Group | $Y_1$ | $Y_2$ | $Y_3$ |
|-------|-------|-------|-------|
| 1 | 1 | 0 | 0 |
| 2 | 0 | 1 | 0 |
| 3 | 0 | 0 | 1 |
| 4 | 0 | 0 | 0 |

Thus an individual coming from group 1 would be assigned a value of 1 on $Y_1$, 0 on $Y_2$, and 0 on $Y_3$, etc.

If we make $Q = k - 1$, then we have a sample with two sets of variables, $Y_1$, $Y_2, \ldots$, $Y_Q$ and $X_1$, $X_2, \ldots$, $X_P$. We now perform a canonical correlation analysis of these variables. This analysis will result in a set of $U_i$ variables and a set of $V_i$ variables. As explained in Chapter 10, the number of these pairs of variables is the smaller of $P$ and $Q$. Thus this number is the smaller of $P$ and $k - 1$.

The variable $V_1$ is the linear combination of the $X$ variables with the maximum correlation with $U_1$. In this sense $V_1$ maximizes the correlation with the dummy variables representing the groups, and therefore it exhibits the maximum difference among the groups. Similarly, $V_2$ exhibits this maximum difference with the condition that $V_2$ is uncorrelated with $V_1$; etc.

Once derived, the variables $V_i$ should be examined as discussed in Chapter 10 in an effort to give them a meaningful interpretation. Such an interpretation can be based on the magnitude of the standardized coefficients given in the outputs. The variables with the largest standardized coefficients can help give "names" to the canonical discriminant functions. To further assist the user in this interpretation, the SPSS–X DISCRIMINANT procedure offers the option of rotating the $V_i$ variables by the varimax method (explained in Chapter 15, for factor analysis).

The $V_i$ variables are called the *canonical discriminant functions* or *canonical variables* in program output. For the depression data example, the results are given in the bottom of Table 11.5. Since there are three groups, we obtained $k - 1 = 2$ canonical (discriminant function) variables. The numerical values that we would obtain had we used a canonical discriminant function program are proportional to those in Table 11.5. As noted earlier, $V_1$ is the linear combination of the $X$ variables with the maximum correlation with $U_1$, a linear function of the dummy variables representing group membership. In this sense, $V_1$ maximizes the correlation with the dummy variables representing the groups, and therefore it exhibits the maximum differences amongst the groups. Similarly, $V_2$ exhibits this maximum difference with the condition that $V_2$ is uncorrelated with $V_1$, etc.

As was true with canonical correlation analysis, often the first canonical discriminant function is easier to interpret than subsequent ones. For the depression data example, the variables sex, health, and bed days have large negative coefficients in the first canonical variable. Thus, patients who are female with perceived poor health and with a large number of bed days tend to be more depressed. The second canonical variable is more difficult to interpret.

When there are two or more canonical variables, some packaged programs produce a plot of the values of the first two canonical variables for each individual in the total sample. In this plot each group is identified by a different letter or number. Such a plot is a valuable tool in seeing how separate the groups are since this plot illustrates the maximum possible separation among the groups. The same plot can also be used to identify outliers or blunders.

Figure 11.6 shows a plot of the two canonical variables for this example. The symbols l, m, and h indicate the three groups. The figure also indicates the position of the mean values (1, 2, and 3) of the canonical variables for each of the three groups. These means show that the main variation is exhibited in canonical variable 1. The plot also shows a great deal of overlap between the individuals in the three groups. Finally, no extreme outliers are evident, although the case in the upper right-hand corner may merit further examination.

When there are only two groups, there is only one canonical variable. In this case, the coefficients of the canonical variable are proportional to those of the Fisher discriminant function. The resulting pair of histograms for the two groups should be examined, rather than the scatter diagram. In other words, there is no advantage to using the more general canonical

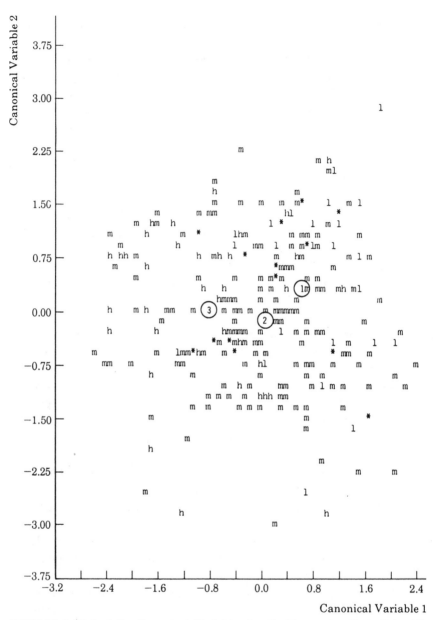

**FIGURE 11.6.** Plot of the Canonical Variables for the Depression Data Set with $k = 3$ Groups

procedure. The advantage of the canonical procedure is that it aids in interpreting the results when there are three or more groups.

Various options and extensions for classification are offered in standard packaged programs, some of which are presented in the next section. In particular, the stepwise procedure of variable selection is available with most standard programs. However, unless the investigator is familiar with the complexities of a given option, we recommend the use of the default options. For most standard packaged programs, the variable to be entered is selected to maximize a test statistic called Wilks' lambda. The one exception to the procedure is that we recommend modifying the values of the $F$-to-enter and $F$-to-remove as discussed in the previous section.

When the number of groups, $k$, is greater than two, the investigator may wish to examine the discrimination between the groups taken two at a time. This examination will serve to highlight specific differences between any two groups, and the results may be easier to interpret. Another possibility is to contrast each group with the remaining groups taken together.

## 11.13 DISCUSSION OF COMPUTER PROGRAMS

In this section, we present the features of packaged programs for discriminant analysis and we give an example of a program run for the depression data set.

### Features of Packaged Programs

Table 11.6 summarizes the items available in the output of standard packaged programs. BMDP has one program that can be used to run regular discriminant function analysis or stepwise discriminant function analysis. It also provides canonical coefficients and canonical scores for each individual. It is basically a stepwise program, so to enter all the variables the user can set the values for $F$ very low (as illustrated in the example later in this section) or specify FORCE = 2 with no level statement in the DISCRIMINANT paragraph.

SAS has three programs. DISCRIM is recommended for discriminant function analysis with two or more groups. It also includes an option for performing nonparametric discriminant function analysis which is not

**TABLE 11.6.** Summary of Computer Output from BMDP, SAS, and SPSS–X for Discriminant Function Analysis

| Output | BMDP | SAS | SPSS–X |
|---|---|---|---|
| Means and standard deviations by group | 7M | All | DISCRIMINANT |
| Pooled covariance and correlations | 7M | All | DISCRIMINANT |
| Classification function | 7M | DISCRIM | DISCRIMINANT |
| $D^2$ | 7M | DISCRIM, STEPDISC | |
| $F$ statistics | 7M | All | DISCRIMINANT |
| Wilks' lambda | 7M | All | DISCRIMINANT |
| Stepwise options | 7M | STEPDISC | DISCRIMINANT |
| Classification tables | 7M | DISCRIM | DISCRIMINANT |
| Jackknife classification tables | 7M | DISCRIM | |
| Cross-validation with subsample | 7M | | |
| Canonical coefficients | 7M | CANDISC | DISCRIMINANT |
| Standardized canonical coefficients | 7M | CANDISC | DISCRIMINANT |
| Canonical plots | 7M | CANDISC | DISCRIMINANT |
| Prior probabilities | 7M | DISCRIM | DISCRIMINANT |
| Posterior probabilities | 7M | DISCRIM | DISCRIMINANT |

discussed in this book (see Lachenbruch 1975; Hand 1981). The STEPDISC procedure is available for stepwise discriminant function analysis. The CANDISC procedure emphasizes output from the canonical correlation standpoint.

The SPSS procedure DISCRIMINANT performs discriminant function analysis by default and stepwise analysis by use of the optional METHOD subcommand. It prints and plots the canonical scores, which the SPSS manual calls "discriminant scores."

An examination of the variable means is useful in obtaining a feel for how the groups differ from each other. When there are only two groups, it is useful to compute the Mahalanobis $D^2$ for a single variable or a group of variables. Unfortunately, most programs do not print these values of $D^2$. For a single variable $X$, say, the value of $D^2$ is easily computed from the group means $\bar{X}_I$ and $\bar{X}_{II}$ and the pooled variance $S^2$ as

$$D^2 \text{ (for single variable } X) = \frac{(\bar{X}_I - \bar{X}_{II})^2}{S^2}$$

These grouped means and pooled variances are available from all programs. For a subset of variables each program prints an $F$ statistic to test the equality of the group means. The value of this $F$ can be used to compute $D^2$, as shown in Section 11.9. Note that BMDP7M prints the Mahalanobis $D^2$ for each individual. This value is a measure of the distance between the point representing an individual and the point representing the estimated population mean. A small value of $D^2$ indicates that the individual probably belongs to that population.

In Section 11.7, we discussed how prior probabilities and costs of misclassification can be used to adjust the value of the dividing point $C$ for two groups. If the costs are assumed equal, we may have the program do this adjustment automatically by supplying the prior probabilities as input. If the costs are not equal, we may trick the program into incorporating them into the prior probabilities, as the following example illustrates. Suppose $q_I = 0.4$, $q_{II} = 0.6$, cost(II given I) = 5, and cost(I given II) = 1. Then

$$\text{adjusted } q_I = q_I \cdot \text{cost(II given I)} = (0.4)(5) = 2$$

and

$$\text{adjusted } q_{II} = q_{II} \cdot \text{cost(I given II)} = (0.6)(1) = 0.6$$

Since the prior probabilities must add up to one, we further adjust $q_I$ and $q_{II}$ such that their sum is one, i.e.,

$$\text{adjusted } q_I = \frac{2}{2.6} \quad \text{and} \quad \text{adjusted } q_{II} = \frac{0.6}{2.6}$$

The packaged programs BMDP7M, SAS CANDISC, and SPSS–X DISCRIMINANT compute canonical variables. As mentioned in Section 11.12, these variables are linear combinations of the variables chosen to represent the maximum separation possible among the groups. When the number of groups is two, only one canonical variable exists, and its value for a given individual is proportional to the value of the Fisher discriminant function. The program may then produce a histogram of this variable for each of the two groups.

For the two-groups case we remind you of the analogy between regression and discrimination. Thus it is possible to perform discriminant analysis for two groups using any of the regression programs discussed in Part 2 of this book. In particular, an all-subsets regression, such as BMDP9R or SAS REG, can be used to perform variable selection for discriminant function analysis. Note that the discriminant function coefficients resulting from two different programs may be different, but they will be proportional to each other.

## Example

An example of the input statements used to run the stepwise discriminant function program BMDP7M is presented here for the depression data set. In this example eight variables are considered as possible variables for the discriminant function: sex, age, education, income, health status, bed days, acute illness, and chronic illness. A description of these variables is given in the code book in Table 3.2. Note that the first four variables are typical demographic data and the last four are measures of health and illness. On the basis of income and age alone, we are able to classify 62.6% correctly (see Table 11.3), and this run will enable us to see whether we can improve on that percentage.

One confusing use of terminology emerges in this run. In BMDP programs the word *case* is used to indicate the number of observations. But note that we have used the same word to signify whether or not a person is depressed, following common medical terminology use of the word *case* as someone who is ill.

The program input statements for BMDP7M are as follows:

```
/input      file is 'depress'.
            format is '(8x,f1.0,f2.0,1x,f1.0,1x,
            f2.0,23x,f1.0,1x,f1.0,2x,f1.0,f1.0,f1.0)'.
            variables are 9.
/variable   names are sex,age,educat,income,cases,health,
            beddays,acuteill,chronill.
            group is cases.
/group      code(5) = 0,1.
            names(5) = notdep,depress.
/disc       enter = 1,1.
            remove = 0,0.
/end
```

The "group" sentence must be used to signify how the two separate groups are defined. This is the fifth variable and takes on the value 0 if normal and 1 if depressed. The $F$-to-enter and $F$-to-remove values have been set very low in order to see what will enter. It is also possible to obtain additional printout by using the "print" paragraph, but here we will just use the default printing option.

The program first prints the means and standard deviations for both groups (50 depressed and 244 not-depressed persons). These results are shown in Table 11.7.

Note from the table that the standard deviations are not grossly different between the two groups but that the data cannot be considered to be multivariate normal. Some of the variables only take on two values, 1 and 2, or 0 and 1. Also, income appears to be skewed, as indicated by the large standard deviation relative to the mean and the lack of negative incomes. Simply looking at the mean values indicates that depressed people are more apt to be female, to be younger, to have less income and education, and to be in poorer health.

At step 0 the largest $F$-to-enter is for bed days, so it enters at step 1. Income enters next at step 2, sex at step 3, age at step 4, and health at step 5. Then no more variables enter since all of the $F$-to-enter levels are less than 1. At the last step the $F$-to-remove levels are 3.91 for sex, 6.02 for age, 6.95 for income, 4.20 for health, and 8.65 for bed days. It is not possible to attach precise $P$ values to these $F$-to-remove levels because of the lack of

**TABLE 11.7.** Means and Standard Deviations for Depression Data

| | Not Depressed | | Depressed | |
|---|---|---|---|---|
| Variable | Mean | Standard Deviation | Mean | Standard Deviation |
| Sex | 1.59 | 0.49 | 1.80 | 0.41 |
| Age | 45.24 | 18.15 | 40.38 | 17.40 |
| Education | 3.55 | 1.33 | 3.16 | 1.17 |
| Income | 21.68 | 15.98 | 15.20 | 9.84 |
| Health | 1.71 | 0.80 | 2.06 | 0.98 |
| Bed days | 0.17 | 0.38 | 0.42 | 0.50 |
| Acute illness | 0.29 | 0.45 | 0.38 | 0.49 |
| Chronic illness | 0.48 | 0.50 | 0.62 | 0.49 |

**TABLE 11.8.** Classification and Discriminant Functions

| Variables | Classification Function | | Discriminant Function |
| | Not Depressed | Depressed | |
|---|---|---|---|
| Sex | 7.268 | 7.962 | −0.694 |
| Age | 0.132 | 0.108 | 0.024 |
| Income | 0.181 | 0.151 | 0.030 |
| Health | 1.793 | 2.240 | −0.447 |
| Bed days | −0.140 | 1.119 | −1.259 |
| Constant | −12.932 | −13.724 | −0.792 |

normality and the use of the stepwise procedure, but the $P$ values appear to be high enough so that these variables should be left in.

The classification function is given in Table 11.8 from the printout, and the discriminant function is obtained by subtraction (see Section 11.6).

Note that the magnitudes of the coefficients for age and income are quite similar to those in Table 11.4 even with the addition of three other variables.

The value of $F$ at the last step is given as 7.60 with 5 and 288 degrees of freedom. This value tests the hypothesis that the population $D^2 = 0$ and is highly significant ($P < 0.001$). We can compute $D^2$ (see Section 11.9) for five variables with $N_I = 244$ and $N_{II} = 50$ as

$$D^2 = \frac{(5)(294)(292)}{(50)(244)(294 - 5 - 1)}(7.60) = 0.9285$$

Thus $-D/2$ is $-0.4818$, which results in an area of about 0.31 to the left of this value in Appendix Table A.1. The unbiased estimate of $D^2$ is

$$\frac{244 + 50 - 5 - 3}{244 + 50 - 2}(0.9285) - 5\left(\frac{1}{50} + \frac{1}{244}\right) = 0.7889$$

(see Section 11.8). This estimate modifies the above area to 0.33, rather close to 0.31.

The program prints the posterior probability of belonging to the not-depressed and the depressed group for each individual. For the first individual in the sample, who is a 68-year-old woman with a $4000-per-year reported income, a health status of 2, and no bed days (see Table 3.3),

$$\text{Prob(not depressed)} = \frac{1}{1 + \exp(-Z + C)} = \frac{1}{1 + \exp(0.530 - 0.792)}$$

$$= 0.565$$

where $C = -0.792$ and

$$Z = -0.694(2) + 0.024(68) + 0.030(4) - 0.447(2) - 1.259(0) = -0.530$$

The probability of being not depressed is close to one-half, but the program classifies this individual as not depressed. The first individual classified as depressed is the fifth person in Table 3.3, who is a 33-year-old female with an income of $35,000, a health status of 1, and bed days of 1. This woman actually had a CESD score of only 6, so she was not depressed. Overall, the program classified correctly 71.1% of the time, as Table 11.9 indicates. These results appear to be an improvement over the results listed in Table 11.3.

Using the method described in Section 11.9, we can test whether the population $D^2$ for all five variables is equal to the population $D^2$ for two

**TABLE 11.9.** Classification of Individuals as Depressed or Not Depressed on the Basis of Sex, Age, Income, Health, and Bed Days

| Actual Status | N | Classified As | | % Correct |
| | | Not depressed | Depressed | |
|---|---|---|---|---|
| Not depressed | 244 | 175 | 69 | 71.7 |
| Depressed | 50 | 16 | 34 | 68.0 |
| Total | 294 | 191 | 103 | 71.1 |

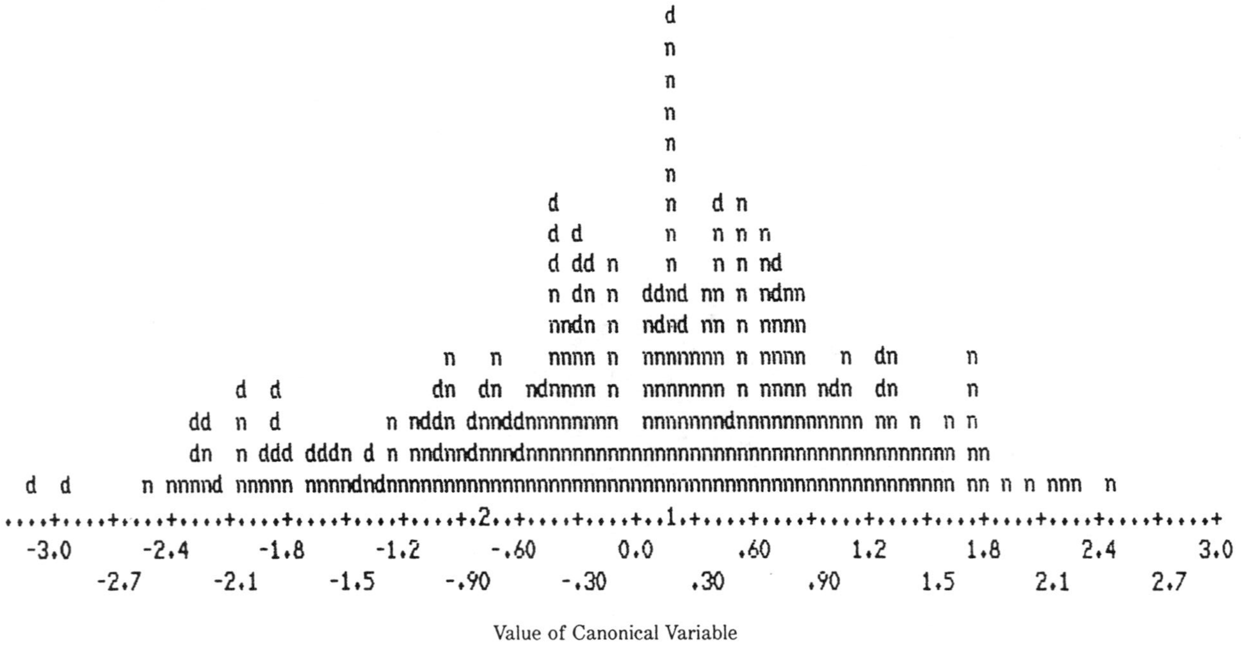

```
                                                          d
                                                          n
                                                          n
                                                          n
                                                          n
                                                          n
                                           d              n    d n
                                           d d            n    n n n
                                           d dd n         n    n n n nd
                                           n dn n         ddnd m n n ndm
                                           ndn n          ndnd m n n mmm
                          n n               mmn n         mmmmm n nmm   n dn        n
                          dn dn             dn nmm n      nmmmm n mmm ndn dn        n
             dd d         n nddn dnnddnmmmmm               nmnnmmdmmmmmmmm mm n n n m
 dd  n  d    dn n ddd dddn d n nmdndndmmmmmmmmmmmmmmmmmmmmmmmmmmmmmmmmmmmmmmm mm n n nm n
 d d    n mmmd nmmm mmmmdndmmmmmmmmmmmmmmmmmmmmmmmmmmmmmmmmmmmmmmmmmmmmmmmmmm mm n n nmm  n
 ....+....+....+....+....+....+....+....+2..+....+....+....1.+....+....+....+....+....+....+....+....+
 -3.0   -2.4   -1.8   -1.2   -.90   -.60   -.30    0.0    .30    .60    .90    1.2    1.5    1.8    2.4    3.0
    -2.7   -2.1   -1.5             -1.8        -.60              .30         .60        1.5        2.1    2.7
```

Value of Canonical Variable

**FIGURE 11.7.** Histogram of Canonical Variable from BMDP7M

310

variables (age and income). With $P = 2$ and $Q = 3$ we compute, using $D_5^2 = 0.929$ and $D_2^2 = 0.319$.

$$F = \frac{244 + 50 - 2 - 3 - 1}{3}$$
$$\times \frac{(244)(50)(0.929 - 0.319)}{(244 + 50)(244 + 50 - 2) + (244)(50)(0.319)}$$
$$= 7.96$$

with 3 and 288 degrees of freedom. Since this result corresponds to a $P$ value less than 0.005, we conclude that sex, health, and bed days significantly improve the prediction based on age and income alone.

The program next lists the canonical variable for each individual and makes a histogram of the values, with the not-depressed labeled n and the depressed labeled d, as shown in Figure 11.7. Note that the upper end of the graph is dominated by the letter n, as should be the case. When better discrimination is possible, such a figure would consist of two overlapping but clearly discernable histograms.

The person the program classifies as most depressed according to the value of the canonical variable and the posterior probability is an 18-year-old female with a reported income of $2000 per year, a health status of 3, and one reported bed day. This woman had a CESD score of 39, and so the program correctly classified her as depressed.

Further runs could be made by introducing other variables, considering transformations on some variables, such as the logarithm of income, and setting costs and prior probabilities equal to desired values.

## 11.14 WHAT TO WATCH OUT FOR

Partly because there is more than one group, discriminant function analysis has additional aspects to watch out for beyond those discussed in regression analysis. A thorough discussion of possible problems is given in Lachenbruch (1977). A list of important trouble areas is as follows:

1. Theoretically, a simple random sample from each population is assumed. As this is often not feasible, the sample taken should be examined for possible biasing factors.

**2.** It is critical that the correct group identification be made. For example, in the example in this chapter it was assumed that all persons had a correct score on the CESD scale so that they could be correctly identified as normal or depressed. Had we identified some individuals incorrectly, this would increase the reported error rates given in the computer output. It is particularly troublesome if one group contains more misidentifications than another.

**3.** The choice of variables is also important. Analogies can be made here to regression analysis. Similar to regression analysis, it is important to remove outliers, make necessary transformations on the variables, and check independence of the cases. It is also a matter of concern if there are considerably more missing values in one group than another.

**4.** Multivariate normality is assumed in discriminant function analysis when computing posterior probabilities or performing statistical tests. The use of dummy variables has been shown not to cause undue problems, but very skewed or long-tailed distributions for some variables can increase the total error rate.

**5.** Another assumption is equal covariance matrices in the groups. If one covariance matrix is very different from the other, then quadratic discriminant function analysis should be considered for two groups (see Lachenbruch 1975) or transformations should be made on the variables that have the greatest difference in their variances.

**6.** If the sample size is sufficient, researchers sometimes obtain a discriminant function from one-half or two-thirds of their data and apply it to the remaining data to see if the same proportion of cases are classified correctly in the two subsamples. Often when the results are applied to a different sample, the proportion classified correctly is smaller. If the discriminant function is to be used to classify individuals, it is important that the original sample come from a population that is similar to the one it will be applied to in the future.

**7.** If some of the variables are dichotomous and one of the outcomes rarely occurs, then logistic regression analysis (discussed in the next chapter) should be considered.

## SUMMARY

In this chapter we discussed discriminant function analysis, a technique dating back to at least 1936. However, its popularity began with the introduction of large-scale computers in the 1960s. The method's original concern was to classify an individual into one of several populations. It is also used for explanatory purposes to identify the relative contributions of a single variable or a group of variables to the classification.

In this chapter our main emphasis was on the case of two populations. We gave some theoretical background and presented an example of the use of a packaged program for this situation. We also discussed the case of more than two groups.

Interested readers may pursue the subject of classification further by consulting the references given in the Bibliography. In particular, Lachenbruch (1975) and James (1985) present a comprehensive discussion of the subject.

## BIBLIOGRAPHY

Afifi, A. A., and Azen, S. P. 1979. *Statistical analysis: A computer oriented approach*. 2nd ed. New York: Academic Press.

*Anderson, T. W. 1984. *An introduction to multivariate statistical analysis*. 2nd ed. New York: Wiley.

Costanza, M. C., and Afifi, A. A. 1979. Comparison of stopping rules for forward stepwise discriminant analysis. *Journal of the American Statistical Association* 74:777–785.

Dixon, W. J., and Massey, F. J. 1983. *Introduction to statistical analysis*. 4th ed. New York: McGraw-Hill.

Fisher, R. A. 1936. The use of multiple measurements in taxonomic problems. *Annals of Eugenics* 7:179–188.

Hand, D. J. 1981. *Discrimination and classification*. New York: Wiley.

James, M. 1985. *Classification algorithms*. New York: Wiley.

Klecka, W. R. 1980. *Discriminant analysis*. Beverly Hills: Sage.

Lachenbruch, P. A. 1975. *Discriminant analysis*. New York: Hafner Press.

Lachenbruch, P. A. 1977. Some misuses of discriminant analysis. *Methods of information in medicine* 16:255–258.

*Morrison, D. F. 1976. *Multivariate statistical methods*. 2nd ed. New York: McGraw-Hill.

*Rao, C. R. 1973. *Linear inference and its application.* 2nd ed. New York: Wiley.

*Tatsuoka, M. M. 1988. *Multivariate analysis: Techniques for educational and psychological research.* 2nd ed. New York: Wiley.

Truett, J., Cornfield, J., and Kannell, W. 1967. Multivariate analysis of the risk of coronary heart disease in Framingham. *Journal of Chronic Diseases* 20:511–524.

## PROBLEMS

11.1 Using the depression data set, perform a stepwise discriminant function analysis with age, sex, log(income), bed days, and health as possible variables. Compare the results with those given in Section 11.13.

11.2 For the data shown in Table 8.1, divide the chemical companies into two groups: group I consists of those companies with a P/E less than 9, and group II consists of those companies with a P/E greater than or equal to 9. Group I should be considered mature or troubled firms, and group II should be considered growth firms. Perform a discriminant function analysis, using ROR5, D/E, SALESGR5, EPS5, NPM1, and PAYOUTR1. Assume equal prior probabilities and costs of misclassification. Test the hypothesis that the population $D^2 = 0$. Produce a graph of the posterior probability of belonging to group I versus the value of the discriminant function. Use of the sentence "POST." in the PRINT paragraph of BMDP7M will result in both these quantities being printed. Estimate the probabilities of misclassification by several methods.

11.3 Continuation of Problem 11.2: Test whether D/E alone does as good a classification job as all six variables.

11.4 Continuation of Problem 11.2: Choose a different set of prior probabilities and costs of misclassification that seem reasonable to you and repeat the analysis.

11.5 Continuation of Problem 11.2: Perform a variable selection analysis, using stepwise and best-subset programs. Compare the results with those of the variable selection analysis given in Chapter 8.

11.6 Continuation of Problem 11.2: Now divide the companies into three groups: group I consists of those companies with a P/E of 7 or less, group II consists of those companies with a P/E of 8 to 10, and group III consists of those companies with a P/E greater than or equal to 11. Perform a stepwise discriminant function analysis, using these three groups and the same variables as in Problem 11.2. Comment.

11.7 In this problem you will modify the data set created in Problem 7.7 to make it suitable for the theoretical exercises in discriminant analysis. Generate the sample data for X1, X2, ..., X9 as in Problem 7.7 (Y is not used here). Then for the first 50 cases, add 6 to X1, add 3 to X2, add 5 to X3, and leave the values for X4 to X9 as they are. For the last 50 cases, leave all the data as they are. Thus the first 50 cases represent a random sample from a multivariate normal population called population I with the following means: 6 for X1, 3 for X2, 5 for X3, and zero for X4 to X9. The last 50 observations represent a random sample from a multivariate normal population (called population II) whose mean is zero for each variable. The population Mahalanobis $D^2$'s for each variable separately are as follows: 1.44 for X1, 1 for X2, 0.5 for X3, and zero for each of X4 to X9. It can be shown that the population $D^2$ is as follows: 3.44 for X1 to X9; 3.44 for X1, X2, and X3; and zero for X4 to X9. For all nine variables the population discriminant function has the following coefficients; 0.49 for X1, 0.5833 for X2, $-0.25$ for X3, and zero for each of X4 to X9. The population errors of misclassification are

$$\text{Prob(I given II)} = \text{Prob(II given I)} = 0.177$$

Now perform a discriminant function analysis on the data you constructed, using all nine variables. Compare the results of the sample with what you know about the populations.

11.8 Continuation of Problem 11.7: Perform a similar analysis, using only X1, X2, and X3. Test the hypothesis that these three variables do as well as all nine in classifying the observations. Comment.

11.9 Continuation of Problem 11.7: Do a variable selection analysis for all nine variables. Comment.

11.10 Continuation of Problem 11.7: Do a variable selection analysis, using variables X4 to X9 only. Comment.

11.11 From the family lung function data in Appendix B, create a data set containing only those families from Burbank and Long Beach (AREA = 1 or 3). The observations now belong to one of two AREA-defined groups.
a. Assuming equal prior probabilities and costs, perform a discriminant function analysis for the fathers using FEV1 and FVC.
b. Now choose prior probabilities based on the population of each of these two California cities (at the time the study was conducted, the population of Burbank was 84,625 and that of Long Beach was 361,334), and assume the cost of Long Beach given Burbank is twice that of Burbank given Long Beach. Repeat the analysis of part a with these assumptions. Compare and comment.

11.12 At the time the study was conducted, the population of Lancaster was 48,027 while Glendora had 38,654 residents. Using prior probabilities based on these population figures and those given in Problem 11.11, and the entire lung function data set (with four AREA-defined groups), perform a discriminant function analysis on the fathers. Use FEV1 and FVC as classifying variables. Do you think the lung function measurements are useful in distinguishing among the four areas?

11.13 a. In the family lung function data in Appendix B, divide the fathers into two groups: group I with FEV1 less than or equal to 4.09, and group II with

FEV1 greater than 4.09. Assuming equal prior probabilities and costs, perform a stepwise discriminant function analysis using the variables height, weight, age, and FVC. What are the probabilities of misclassification?

b. Use the classification function you found in part a to classify first the mothers and then the oldest children. What are the probabilities of misclassification? Why is the assumption of equal prior probabilities not realistic?

11.14 Divide the oldest children in the family lung function data set into two groups based on weight: less than or equal to 101 versus greater than 101. Perform a stepwise discriminant function analysis using the variables OCHEIGHT, OCAGE, MHEIGHT, MWEIGHT, FHEIGHT, and FWEIGHT. Now temporarily remove from the data set all observations with OCWEIGHT between 87 and 115 inclusive. Repeat the analysis and compare the results.

11.15 Is it possible to distinguish between men and women in the depression data set on the basis of income and level of depression? What is the classification function? What are your prior probabilities? Test whether the following variables help discriminate: EDUCAT, EMPLOY, HEALTH.

11.16 Refer to the table of ideal weights given in Problem 9.7 and calculate the midpoint of each weight range for men and women. Pretending these represent a real sample, perform a discriminant function analysis to classify observations as male or female on the basis of height and weight. How could you include frame size in the analysis? What happens if you do?

# Chapter Twelve

# LOGISTIC REGRESSION

## 12.1 WHAT WILL YOU LEARN FROM THIS CHAPTER?

In Chapter 11 we presented a method of classifying individuals into one of two possible populations, a method developed originally for continuous variables. From this chapter, you will learn another method, which applies to discrete or continuous variables. In particular, you will learn:

- When logistic regression is used (12.2, 12.3).
- The meaning of the multiple logistic regression model (12.4).
- How to choose between logistic regression and discriminant function analysis (12.4).
- How to interpret the results of a logistic regression analysis (12.5, 12.6).
- How to evaluate the resulting equation (12.7).
- How to obtain and adjust risk functions (12.8).
- About what computer programs to use to obtain various output for the logistic regression model (12.9).
- What to watch out for when performing logistic regression analysis (12.10).

## 12.2 WHEN IS LOGISTIC REGRESSION USED?

Logistic regression can be used whenever an individual is to be classified into one of *two* populations. Thus it is an alternative to the discriminant analysis presented in Chapter 11. In the past most of the applications of logistic regression were in the medical field. It has been used, for example, to calculate the risk of developing heart disease as a function of certain personal and behavioral characteristics. Similarly, logistic regression can be used to decide which characteristics are predictive of teenage pregnancy. Different variables such as grade point average in elementary school, religion, or family size could be used to predict whether or not a teenager will become pregnant. In industry, operation units could be classified as successful or not successful according to some objective criteria. Then several characteristics of the units could be measured and logistic regression analysis could be used to determine which characteristics best predict success.

The linear discriminant function gives rise to the logistic posterior probability when the multivariate normal model is assumed. Logistic regression represents an alternative method of classification when the multivariate normal model is not justified. As we will discuss in this chapter, the logistic regression analysis is applicable for any combination of discrete and continuous variables. (If the multivariate normal model is applicable, the methods discussed in Chapter 11 will result in a better classification procedure and will require less computer time to analyze.)

## 12.3 DATA EXAMPLE

The same depression data set described in Chapter 11 will be used in this chapter. However, the first group will consist of people who are depressed rather than not depressed. Based on the discussion given in Section 11.6, this redefinition of the groups implies that the discriminant function based on age and income is

$$Z = -0.0209(\text{age}) - 0.0336(\text{income})$$

with a dividing point $C = -1.515$. Assuming equal prior probabilities, the posterior probability of being depressed is

$$\text{Prob(depressed)} = \frac{1}{1 + \exp[-1.515 + 0.0209(\text{age}) + 0.0336(\text{income})]}$$

For a given individual with a discriminant function value of $Z$, we can write this posterior probability as

$$P_Z = \frac{1}{1 + e^{C-Z}}$$

As a function of $Z$, the probability $P_Z$ has the logistic form shown in Figure 12.1. Note that $P_Z$ is always positive; in fact, it must lie between zero and 1 because it is a probability. The minimum age is 18 years, and the minimum income is $\$2 \times 10^3$. These minimums result in a $Z$ value of $-0.443$ and a probability $P_Z = 0.745$, as graphed. When $Z = -1.515$, the dividing point $C$, then $P_Z = 0.5$. Larger values of $Z$ occur when age is younger and/or income is lower. For an older person with a higher income, the probability of being depressed is low.

The same data and other data from this depression set will be used later in the chapter to illustrate a different method of computing the probability of being depressed.

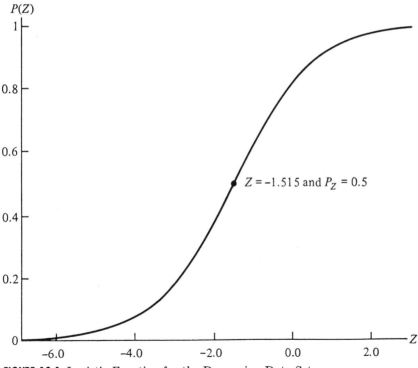

**FIGURE 12.1.** Logistic Function for the Depression Data Set

## 12.4 BASIC CONCEPTS OF LOGISTIC REGRESSION

The *logistic function*

$$P_Z = \frac{1}{1 + e^{C-Z}}$$

may be transformed to produce a new interpretation. Specifically, we define the *odds* as the following ratio:

$$\text{odds} = \frac{P_Z}{1 - P_Z}$$

Computing the odds is a commonly used technique of interpreting probabilities (see Fleiss 1981). For example, in sports we may say that the odds are 3 to 1 that one team will defeat another in a game. This statement means that the favored team has a probability of $3/(3 + 1)$ of winning, since $0.75/(1 - 0.75) = 3/1$.

Note that as the value of $P_Z$ varies from 0 to 1, the odds vary from 0 to $\infty$. When $P_Z = 0.5$, the odds are 1. On the odds scale the values from 0 to 1 correspond to values of $P_Z$ from 0 to 0.5. On the other hand, values of $P_Z$ from 0.5 to 1.0 result in odds of 1 to $\infty$. Taking the logarithm of the odds will cure this asymmetry. When $P_Z = 0$, ln odds $= -\infty$; when $P_Z = 0.5$, ln odds $= 0.0$; and when $P_Z = 1.0$, ln odds $= +\infty$. The term *logit* is sometimes used instead of ln(odds).

By taking the natural logarithm of the odds and performing some algebraic manipulation, we obtain

$$\ln\left(\frac{P_Z}{1 - P_Z}\right) = -C + Z$$

In words, the logarithm of the odds is linear in the discriminant function $Z$. Since $Z = a_1 X_1 + a_2 X_2 + \cdots + a_P X_P$ (from Chapter 11), ln(odds) is seen to be linear in the original variables. If we rewrite $-C + Z$ as $a + b_1 X_1 + b_2 X_2 + \cdots + b_P X_P$, the equation relating ln(odds) to the discriminant function is

$$\ln(\text{odds}) = a + b_1 X_1 + b_2 X_2 + \cdots + b_P X_P$$

This equation is in the same form as the multiple linear regression equation (see Chapter 7), where $a = -C$ and $b_i = a_i$ for $i = 1$ to $P$. For this reason the logistic function has been called the *multiple logistic regression equation*, and the coefficients in the equation can be interpreted as regression coefficients.

The fundamental assumption in logistic regression analysis is that ln(odds) is linearly related to the independent variables. No assumptions are made regarding the distributions of the $X$ variables. In fact, one of the major advantages of this method is that the $X$ variables may be discrete or continuous.

The model assumed is

$$\ln(\text{odds}) = \alpha + \beta_1 X_1 + \beta_2 X_2 + \cdots + \beta_P X_P$$

In terms of the probability of belonging to population I, the equation can be written as

$$\frac{\text{probability of belonging}}{\text{to population I}} = \frac{1}{1 + \exp[-(\alpha + \beta_1 X_1 + \beta_2 X_2 + \cdots + \beta_P X_P)]}$$

This equation is called the *logistic regression equation*.

As mentioned earlier, the technique of linear discriminant analysis can be used to compute estimates of the parameters $\alpha, \beta_1, \beta_2, \ldots, \beta_p$. However, the method of *maximum likelihood* produces estimates that depend only on the logistic model. The maximum likelihood estimates should, therefore, be more robust than the linear discriminant function estimates. However, if the distribution of the $X$ variables is, in fact, multivariate normal, then the discriminant analysis method requires a smaller sample size to achieve the same precision as the maximum likelihood method (see Efron 1975). The maximum likelihood estimates of the probabilities of belonging to one population or the other are preferred to the discriminant function estimates when the $X$'s are nonnormal (see Halperin, Blackwelder, and Verter 1971 or Press and Wilson 1978). The estimates of the coefficients or the probabilities derived from the two methods will rarely be substantially different from each other, whether or not the multivariate normality assumption is satisfied. An exception to this statement is given in O'Hara et al. (1982); they demonstrate that the use of discriminant function analysis may lead to underestimation of the coefficients when all the $X$'s are categorical and the probability of the outcome is small.

Most logistic regression programs use the method of maximum likelihood to compute estimates of the parameters. The procedure is iterative, and a theoretical discussion of it is beyond the scope of this book. When the number of variables is large (say more than nine), the computer time required may be prohibitively long. Some programs, such as BMDPLR, allow the investigator to use an approximation to the maximum likelihood method.

## 12.5 INTERPRETATION: CATEGORICAL VARIABLES

One of the most important benefits of the logistic model is that it allows the use of categorical $X$ variables. Any number of such variables can be used in the model. The simplest situation is one in which we have a single $X$ variable with two possible values. For example, for the depression data we can attempt to predict who is depressed on the basis of the individual's sex. Table 12.1 shows the individuals classified by depression and sex.

If the individual is a female, then the odds of being depressed are 40/143. Similarly, for males the odds of being depressed are 10/101. The *ratio* of these odds is

$$\text{odds ratio} = \frac{40/143}{10/101} = \frac{40 \times 101}{10 \times 143} = 2.825$$

The odds of a female being depressed are 2.825 times that of a male. Note that we could just as well compute the odds ratio of *not* being

**TABLE 12.1.** Classification of Individuals by Depression Level and Sex

|            | Depression |     |       |
| ---------- | ---------- | --- | ----- |
| Sex        | Yes        | No  | Total |
| Female (1) | 40         | 143 | 183   |
| Male (0)   | 10         | 101 | 111   |
| Total      | 50         | 244 | 294   |

depressed. In this case we have

$$\text{odds ratio} = \frac{143/40}{101/10} = 0.354$$

The concept of odds ratio is used extensively in biomedical applications (see Fleiss 1981). It is a measure of association of a binary variable (risk factor) with the occurrence of a given event (disease)(see Reynolds 1977 for applications in behavioral science).

To represent a variable such as sex, we customarily use a dummy variable: $X = 0$ if male and $X = 1$ if female. The logistic regression equation can then be written as

$$\text{Prob(depressed)} = \frac{1}{1 + e^{-\alpha - \beta X}}$$

The sample estimates of the parameters are

$$a = \text{estimate of } \alpha = -2.313$$
$$b = \text{estimate of } \beta = 1.039$$

We note that the estimate of $\beta$ is the natural logarithm of the odds ratio of females to males; or

$$1.039 = \ln 2.825$$

Equivalently,

$$\text{odds ratio } e^b = e^{1.039} = 2.825$$

Also, the estimate of $\alpha$ is the natural logarithm of the odds for males $(X = 0)$; or

$$-2.313 = \ln \frac{10}{101}$$

When there is only a single dichotomous variable, it is not worthwhile to perform a logistic regression analysis. However, in a multivariate logistic equation the value of the coefficient of a dichotomous variable can be related to the odds ratio in a manner similar to that outlined above. For example, for the depression data, if we include age, sex, and income in the same logistic model, the estimated equation is

Prob(depressed)

$$= \frac{1}{1 + \exp\{-[-0.676 - 0.021(\text{age}) - 0.037(\text{income}) + 0.929(\text{sex})]\}}$$

Since sex is a 0, 1 variable, its coefficient can be given an interesting interpretation. The quantity $e^{0.929} = 2.582$ may be interpreted as the odds ratio of being depressed if female after adjusting for the linear effects of age and income. It is important to note that such an interpretation is valid only when we do *not* include the interaction of the dichotomous variable with any of the other variables. Further discussion on this subject may be found in Breslow and Day (1980), Schlesselman (1982), and Hosmer and Lemeshow (1989). In particular, the case of several dummy variables is discussed in some detail in these books.

Approximate confidence intervals for the odds ratio for a binary variable can be computed using the slope coefficient $b$ and its standard error listed in the output. For example, 95% approximate confidence limits for the odds ratio can be computed as $\exp(b \pm 1.96 \times$ standard error of $b$).

If a categorical variable takes on more than two values, we may create several dummy variables to represent it, as was discussed in Chapter 9. The BMDPLR program creates such variables automatically. However, the program assigns the values $-1$ and $+1$ to each dummy variable if the standard default is used and not the values 0 and 1. The user may force the computer to use any two values by reading the dummy variables as "interval" variables after transforming the variable to the desired values. Alternatively, using the sentence

DVAR = PART.

in the REGRESSION paragraph will result in the usual dummy variables, with the first category the reference group (see the program manual).

## 12.6 INTERPRETATION: CONTINUOUS AND MIXED VARIABLES

To show the similarity between discriminant function and logistic regression analysis, we performed a logistic regression on age and income in the depression data set, using the BMDPLR program. The program requires the user to state, for each variable, whether it is categorical or interval. Variables called interval are used as they are supplied to the program. The program, however, assigns dummy values to variables called categorical. For this example both age and income are called interval. The estimates obtained from the program are as follows:

| Term | Coefficient | Standard Error |
|------|-------------|----------------|
| Age | $-0.020$ | 0.009 |
| Income | $-0.041$ | 0.014 |
| Constant | 0.028 | 0.487 |

So the estimate of $\alpha$ is 0.028, of $\beta_1$ is $-0.020$, and of $\beta_2$ is $-0.041$. The equation for ln(odds), or logit, is estimated by $0.028 - 0.020(\text{age}) - 0.041(\text{income})$. The coefficients $-0.020$ and $-0.041$ are interpreted in the same manner they are interpreted in a multiple linear regression equation, where the dependent variable is $\ln[P_I/(1 - P_I)]$ and where $P_I$ is the logistic regression equation, estimated as

$$\begin{matrix} \text{probability of} \\ \text{being depressed} \end{matrix} = \frac{1}{1 + \exp\{-[0.028 - 0.020(\text{age}) - 0.041(\text{income})]\}}$$

Recall that the coefficients for age and income in the discriminant function are $-0.0209$ and $-0.0336$, respectively. These estimates are within one standard error of $-0.020$ and $-0.041$, respectively, the estimates obtained from the BMDPLR program. The constant 0.028 corresponds to a dividing point of $-0.028$. This value is different from the dividing point of $-1.515$ obtained by the discriminant analysis. The explanation for this discrepancy is that the logistic regression program implicitly uses prior probability estimates obtained from the sample. In this example these prior probabilities are 250/294 and 50/294, respectively. When these prior probabilities are used, the discriminant function dividing point is $-1.515 + \ln q_{II}/q_I = -1.515 + 1.585 = 0.070$. This value is closer to the value 0.028 obtained by the logistic regression program than is $-1.515$.

For a continuous variable $X$ with slope coefficient $b$, the quantity $\exp(b)$ is interpreted as the ratio of the odds for a person with value $(X + 1)$ relative to the odds for a person with value $X$. Therefore, $\exp(b)$ is the incremental odds ratio corresponding to an increase of one unit in the variable $X$, assuming that the values of all other $X$ variables remain unchanged. The incremental odds ratio corresponding to the change of $k$ units in $X$ is $\exp(kb)$. For example, for the depression data, the odds of depression can be written as

$$P_Z/(1 - P_Z) = \exp[0.028 - 0.020(\text{age}) - 0.041(\text{income})]$$

For the variable age, a reasonable increment is $k = 10$ years. The incremental odds ratio corresponding to an increase of 10 years in age can be computed as $\exp(10b)$ or $\exp(10x - 0.020) = \exp(-0.200) = 0.819$. In other words, the odds of depression is estimated as 0.819 of what it would be if a person were 10 years younger. An odds ratio of one signifies no effect, but in this case increasing the age has the effect of lowering the odds of depression. If the computed value of $\exp(10b)$ had been two, then the odds of depression would have doubled for that incremental change in a variable. This statistic can be called the ten-year odds ratio for age or, more generally, the $k$ incremental odds ratio for variable $X$ (see Hosmer and Lemeshow 1989).

Ninety-five percent confidence intervals for the above statistic can be computed from $\exp[kb \pm k(1.96)\text{se}(b)]$, where se stands for standard error. For example, for age

$$\exp(10 \times -0.020 \pm 10 \times 1.96 \times 0.009) \qquad \text{or}$$

$$\exp(-0.376 \text{ to } -0.024) \qquad \text{or}$$

$$0.687 < \text{odds ratio} < 0.976$$

Note that for each coefficient printed by the program, an *asymptotic standard error* is also printed. These errors can be used to test hypotheses and obtain confidence intervals. For example, to test the hypothesis that the coefficient for age is zero, we compute the test statistic

$$Z = \frac{-0.020 - 0}{0.009} = -2.22$$

For large samples an approximate $P$ value can be obtained by comparing this $Z$ statistic to percentiles of the standard normal distribution.

## 12.7 EVALUATING HOW WELL LOGISTIC REGRESSION PREDICTS OUTCOMES

If the purpose of performing the logistic regression is to obtain a regression equation that could predict in which of two groups a person or thing could be placed in the future, then the investigator would want to know how much to trust predictions from the logistic regression equation. In other words, can the equation predict correctly a high proportion of the time? This question is somewhat different from asking about the statistical significance, as it is possible to obtain results that are statistically significant but still do not predict very well. In discriminant function programs, output is available to assist in this evaluation. For example, the classification table where the number predicted correctly in each group is displayed. Also, the histogram and scatter diagram plots from such programs provide visual information on group separation. In this section, we will discuss what information is available from logistic regression programs and how it can be used to evaluate prediction.

In addition to determining how well it predicts, if the investigator wishes to actually classify persons, then a cutoff point on the probability of being depressed must be found. Information that is provided in the BMDPLR program to assist in determining how well the regression equation predicts can also be used to determine a cutoff point. This cutoff point is denoted by $P_c$. We would classify a person as depressed if the probability of depression is greater than or equal to $P_c$. First the program gives two histograms, one for the depressed and one for the normals, using the same values on the horizontal axes. This allows one to see the overlap in the two groups and to consider the effects of various cutpoints on the proportion incorrectly classified.

The program also provides a table in which the percent of people correctly classified in the nondepressed, depressed, and total group are listed along with other information for a series of cutpoints. Naturally, we wish the percent correctly classified in each group to be as close to one as possible. These percentages are shown in Figure 12.2. For example, for a cutoff point of about 0.175 each group is approximately 60% correctly

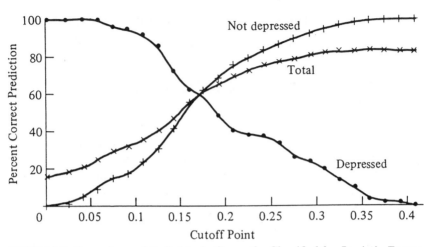

**FIGURE 12.2.** Percentage of Individuals Correctly Classified by Logistic Regression

classified. This may be a good choice of a cutoff point because it treats both groups equally. In contrast, a cutoff point of 0.002 would result in 98% of the depressed group classified correctly but only 7% of the normals. Sixty percent correctly classified appears to be better than flipping a coin, but it would lead one to consider other variables to see if the prediction could be improved.

One method of choosing a cutoff point (discussed in Chapter 11) depends on the relative cost of misclassification. The BMDPLR program allows the user to supply these costs as input, and then prints out the loss for each cutpoint in the same table mentioned above so the user can choose the minimum loss.

The table can also be used to provide the information needed to draw a receiver operating characteristic, or *ROC* curve. ROC was originally proposed for signal-detection work when the signals were not always correctly received. In an ROC curve, the proportion of depressed persons correctly classified as depressed is plotted on the vertical axis for the various cutpoints. This is often called the *true positive fraction*, or *sensitivity* in medical research. On the horizontal axis is the proportion of non-depressed persons classified as depressed (called the *false positive fraction*, or one minus *specificity* in medical studies). Swets (1973) gives a summary of its use in psychology and Metz (1978) presents a discussion of its use in medical studies (see also Kraemer 1988).

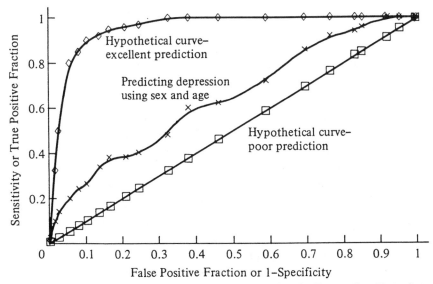

**FIGURE 12.3.** ROC Curve from Logistic Regression for the Depression Data Set

Three ROC curves are drawn in Figure 12.3. The top one represents a hypothetical curve which would be obtained from a logistic regression equation that resulted in excellent prediction. Even for small values of the proportion of normal persons incorrectly classified as depressed, it would be possible to get a large proportion of depressed persons correctly classified as depressed. The middle curve is what was actually obtained from the prediction of depression just using age and sex. The graph is produced by choosing many cutoff points, finding the true positive fraction and the false positive fraction for each point, and then plotting them. Here we can see that to obtain a high proportion, say 0.8, of depressed persons classified as depressed results in a proportion of about 0.65 of the normal persons classified as depressed, an unacceptable level. The lower hypothetical curve (straight line) represents the chance-alone assignment (i.e., flipping a coin).

If we believe that the prevalence of depression is low, then a cutpoint on the lower part of the ROC curve would be chosen since most of the population is normal (nondepressed) and we do not want too many normals classified as depressed. A case would be called depressed only if we were quite sure they were depressed. This is called a *strict threshold*. The downside of this approach is that many cases would be missed. If we thought we were taking persons from a population with a high rate of depression, then a cutpoint higher up the curve should be chosen. Here a person is called depressed if there is any indication that the person might fall in that group. Very few depressed persons would be missed, but many normals would be called depressed. This type of cutpoint is called a *lax threshold*.

Note that the curve must pass through the points (0,0) and (1,1). The maximum area under the curve is one. The SAS LOGIST procedure prints out the area under the ROC curve as a measure of the predictability of the logistic regression equation. The numerical value will be close to one if the prediction is excellent, and close to one-half if it is poor. The ROC curve can be useful in deciding which of two logistic functions to use. All else being equal, the one with the greatest area would be chosen. Otherwise, you might wish to choose the one with the greatest height to the ROC curve at the desired cutoff point.

Several approaches have been proposed for testing how well the derived multiple logistic regression equation fits the data. They all rely on the idea of comparing an observed number of individuals with the number expected if the model were valid. These observed ($O$) and expected ($E$) numbers are combined to form a $\chi^2$ statistic called the *goodness-of-fit* $\chi^2$. Large values of the test statistic indicate a poor fit of the model. Equivalently, small $P$ values are an indication of poor fit. In this section we present two different approaches to the goodness-of-fit test.

The *classical approach* begins with identifying different combinations of values of the variables used in the equation, called *patterns*. For example, if there are two dichotomous variables, sex and employment status, there are four distinct patterns: male employed, male unemployed, female employed, female unemployed. For each of these combinations we count the observed number of individuals ($O$) in populations I and II. Similarly, for each of these individuals we compute the probability of being in population I or II on the basis of the logistic equation. The sum of these probabilities for a given pattern is denoted by $E$. The goodness-of-fit test statistic is computed as

$$\text{goodness-of-fit } \chi^2 = \sum 2 \times O \times \ln\left(\frac{O}{E}\right)$$

where the summation is extended over all the distinct patterns. The BMDPLR program computes this statistic and the corresponding $P$

value. The LOGLINEAR procedure in SPSS–X also computes this statistic, calling it the likelihood ratio $\chi^2$. The investigator is warned though, that *this statistic may give a misleading impression when the number of distinct patterns is too small or too large.* A large number of distinct patterns occur when continuous variables are used.

Another goodness-of-fit approach is described by Lemeshow and Hosmer (1982). In this approach the probability of belonging to population I is calculated for every individual in the sample, and the resulting numbers are arranged in increasing order. The range of probability values is then divided into ten groups (deciles). For each decile the observed numbers of individuals in population I are computed $(O)$. Also, the expected numbers $(E)$ are calculated by adding the logistic probabilities for all individuals in each decile. The goodness-of-fit statistic is calculated as

$$\text{goodness-of-fit } \chi^2 = \sum \frac{(O - E)^2}{E}$$

where the summation extends over the two populations and the ten deciles. The BMDPLR program prints this statistic and its $P$ value as well. A similar goodness-of-fit statistic, called Pearson's $\chi^2$, is calculated by the SPSS–X procedures LOGLINEAR and PROBIT. These procedures calculate the observed and expected numbers for all individuals, or in each category defined by the independent variables if they are categorical. The above summation then extends over all individuals or over all categories, respectively.

All of these goodness-of-fit tests are approximate and require large sample sizes. A large goodness-of-fit chi-square (or a small $P$ value) indicates that the fit may not be good.

A different sort of statistic is also computed to test whether inclusion of one or more variables *improves* the prediction. The BMDPLR program prints an "improvement chi-square" that tests whether one new variable entered in a stepwise fashion significantly improves prediction. The LOGIST procedure in SAS computes this statistic, called the "*model chi-square*," for the total set of variables in the equation. This statistic may be used to test the hypothesis that the set of variables is useless in classifying individuals. A large value of this statistic (or a small $P$ value) is an indication that the variables are *useful* in classification. This test is analogous to testing the hypothesis that the population Mahalanobis $D^2$ is zero in discriminant function analysis (see Chapter 11).

The test for whether a *single* variable improves the prediction forms the basis for a stepwise logistic regression procedure. The principle is the same as that used in stepwise linear regression or stepwise discriminant function analysis. The chi-square value or an approximate $F$ value is used for entering or deleting variables. The BMDPLR program prints an improvement chi-square that tests whether a new variable entered in a stepwise fashion significantly improves prediction. A large value of chi-square (or a small $P$ value) indicates that the variable should be added.

## 12.8 APPLICATIONS OF LOGISTIC REGRESSION

Multiple logistic regression equations are often used to estimate the probability of a certain event occurring to a given individual. Examples of such events are failure to repay a loan, the occurrence of a heart attack, or death from lung cancer. In such applications the period of time in which such an event is to occur must be specified. For example, the event might be a heart attack occurring within ten years from the start of observation.

For estimation of the equation a sample is needed in which each individual has been observed for the specified period of time and values of a set of relevant variables have been obtained at or up to the start of the observation. Such a sample can be selected and used in the following two ways:

1. A sample is selected in a random manner and observed for the specified period of time. This sample is called a *cross-sectional sample*. From this single sample two subsamples result, namely, those who experience the event and those who do not. The methods described in this chapter are then used to obtain the estimated logistic regression equation. This equation can be applied directly to a new member of the population from which the original sample was obtained. This application assumes that the population is in a steady state; i.e., no major changes occur that alter the relationship between the variables and the occurrence of the event. Use of the equation with a different population may require an adjustment, as described in the next paragraph.

2. The second way of obtaining a sample is to select two random samples, one for which the event occurred and one for which the

event did not occur. This sample is called a *case-control sample*. Values of the predictive variables must be obtained in a retrospective fashion, i.e., from past records or recollection. The data can be used to estimate the logistic regression equation. This method has the advantage of enabling us to specify the number of individuals with or without the event. In application of the equation the regression coefficients $b_1, b_2, \ldots, b_P$ are valid. However, the constant $a$ must be adjusted to reflect the true population proportion of individuals with the event. To make the adjustment, we need an estimate of the probability $P$ of the event in the target population. The adjusted constant $a^*$ is then computed as follows:

$$a^* = a + \ln[\hat{P}n_0/(1 - \hat{P})n_1],$$

where

$a =$ the constant obtained from the program

$\hat{P} =$ estimate of $P$

$n_0 =$ the number of individuals in the sample for whom the event did not occur

$n_1 =$ the number of individuals in the sample for whom the event did occur

Further details regarding this subject can be found in Bock and Afifi (1988).

For example, in the depression data suppose that we wish to estimate the probability of being depressed on the basis of sex, age, and income. From the depression data we obtain an equation for the probability of being depressed as follows:

Prob(depressed)

$$= \frac{1}{1 + \exp\{-[-0.676 - 0.021(\text{age}) - 0.037(\text{income}) + 0.929(\text{sex})]\}}$$

This equation is derived from data on an urban population with a depression rate of approximately 20%. Since the sample has 50/294, or

17%, depressed individuals, we must adjust the constant. Since $a = -0.676$, $\hat{P} = 0.20$, $n_0 = 244$, and $n_1 = 50$, we compute the adjusted constant as:

$$a^* = -0.676 + \ln[(0.2)(244)/(0.8)(50)]$$
$$= -0.676 + \ln(1.22)$$
$$= -0.676 + 0.199$$
$$= -0.477$$

Therefore the equation used for estimating the probability of being depressed is

Prob(depressed)

$$= \frac{1}{1 + \exp\{-[-0.477 - 0.021(\text{age}) - 0.037(\text{income}) + 0.929(\text{sex})]\}}$$

This can be compared to the equation given in Section 12.5.

Another method of sampling is to select *pairs* of individuals *matched* on certain characteristics. One member of the pair has the event and the other does not. Data from such matched studies may be analyzed by the logistic regression programs described in the next section after making certain adjustments. Holford, White, and Kelsey (1978) describe these adjustments for one-to-one matching; Breslow and Day (1980) give a theoretical discussion of the subject. Woolson and Lachenbruch (1982) describe adjustments that allow a least-squares approach to be used for the same estimation, and Hosmer and Lemeshow (1989) discuss many-to-one matching.

As an example of one-to-one matching, consider a study of melanoma among workers in a large company. Since this disease is not common, all the new cases are sampled over a five-year period. These workers are then matched by gender and age intervals to other workers from the same company who did not have melanoma. The result is a matched-paired sample. The researchers may be concerned about a particular set of exposure variables, such as exposure to some chemicals or working in a particular part of the plant. There may be other variables whose effect the researcher also wants to control for, although they were not considered matching variables. All these variables, except those that were used in the matching, can be considered as independent variables in a logistic regression analysis. Three adjustments have to be made to running the usual logistic regression program in order to accommodate matched-pairs data:

1. For each of the $N$ pairs, compute the difference $X(\text{case}) - X(\text{control})$ for each $X$ variable. These differences will form the new variables for a single sample of size $N$ to be used in the logistic regression equation.
2. Create a new outcome variable $Y$ with a value of 1 for each of the $N$ pairs.
3. Set the constant term in the logistic regression equation equal to zero.

An example of this procedure is given in the BMDP manual. The BMDP 1990 personal computer logistic regression program provides this as one of its options. Breslow and Day (1980) give a listing of a FORTRAN program that allows for a variable number of controls per case. An example of the use of multiple number of controls per case is given in the new LE program in the 1990 BMDP manual.

Schlesselman (1982) compares the output that would be obtained from a paired-matched logistic regression analysis to that obtained if the matching was ignored. He explains that if the matching is ignored and the matching variables are associated with the exposure variables, then the estimates of the odds ratio of the exposure variables will be biased toward no effect (or an odds ratio of 1). If the matching variables are unassociated with the exposure variables, then it is not necessary to use the form of the analysis that accounts for matching. Furthermore, in the latter case, using the analysis that accounts for matching would result in larger estimates of the standard error of the slope coefficients.

## 12.9 DISCUSSION OF COMPUTER PROGRAMS

Of the three major packages, only SAS and BMDP include specific logistic regression programs (see Table 12.2). The SAS package has one program, regression program LOGIST, that is author-supported and available on

**TABLE 12.2.** Summary of Computer Output from BMDP, SAS, and SPSS–X for Logistic Regression Analysis

| Output | BMDP | SAS | SPSS-X |
|---|---|---|---|
| Overall means of variables | LR | LOGIST | DESCRIPTIVES |
| Beta coefficients | LR | LOGIST, CATMOD | PROBIT, LOGLINEAR |
| Standard error coefficients | LR | LOGIST, CATMOD | PROBIT, LOGLINEAR |
| Stepwise features | LR | LOGIST | |
| Model chi-square | | LOGIST | |
| Goodness-of-fit chi-square | LR | | PROBIT, LOGLINEAR |
| Classification table (one cutpoint) | | LOGIST | |
| Output table (multiple cutpoints) | LR | | |
| Correlation matrix coefficients | LR | LOGIST, CATMOD | PROBIT |
| Covariance matrix coefficients | | LOGIST, CATMOD | PROBIT |
| Cost matrix | LR | | |
| Probabilities of event for individuals | LR | LOGIST | PROBIT |

mainframe or CMS- and OS-supported computers. The LOGIST procedure can be used for forward stepwise or backward elimination. A classification table can be obtained, but this is somewhat arbitrary due to the choice of the cutpoint. In the LOGIST write-up, a description is given on how to test for goodness-of-fit of the model by specifying a more unrestricted model (see Cox 1970 for further discussion). In SAS, the CATMOD procedure can also be used to compute a logistic regression equation. Note that the signs for the coefficients in this procedure are the reverse of those discussed in this book. Neither of the SAS procedures is available in the personal computer version.

The BMDPLR program was used and output from it discussed throughout this chapter. One especially good feature of this program is the table which lists the various cutpoints and the percent of correct predictions in each group. This can be obtained by using the COST option in the PRINT paragraph. In addition, the 1990 manual includes a new PR program that performs logistic regression when there are more than two outcomes. The procedures LOGLINEAR and PROBIT in SPSS–X can be used to obtain estimates of the coefficients of the logistic equation.

LOGLINEAR can be used only when the independent variables are categorical. Neither procedure is available in SPSS/PC+.

It is important to note that the computational method for each of these programs is iterative, and hence the program can be expensive if the number of variables is large. It may be advisable for the investigator to perform some initial variable selection, using discriminant analysis programs. In BMDP, two computation options are offered. The default option, called the ACE method, can be much less expensive than the MLE option. Due to the importance of logistic regression, programs for the PC have been written to perform this analysis for epidemiological purposes. The reader may wish to check for local availability of such programs.

To illustrate the use of BMDPLR, we give below the control statements needed to predict nondepressed or depressed individuals, with age, sex, and income as independent variables. A "transform" paragraph is added to change sex to a 0,1 variable from a 1,2 variable. Since it is called by the same name, no new variable is considered to be added. In addition, the variable-named cases have been switched so that the depressed are being predicted as explained in Section 11.6. This program was run on a personal computer using a BMDP save file called depress.save with code called depress.

```
/input      files = 'depress.sav'. code = depress.
/var        use = sex,age,income,cases.
/group      code(cases) = 0, 1. name(cases) = depress,normal.
/transform  x = cases.
            if(x eq 0) then cases = 1.
            if(x eq 1) then cases = 0.
            sex = sex - 1.
/regr       depend = cases. interval = sex,age,income.
            enter = .2.
            remove = 1.
/end
```

A "regress" paragraph is needed to indicate which variable is the dependent variable. Also, the form of the independent variables are specified. Here we say that they are all interval so that we will not have to use the "design" variable sentence. The enter and remove values in this program are not $F$ levels. Instead, they are $P$ values. The $P$ values for enter must be less than the $P$ for remove. In this example we set the $P$ value for enter at 0.2, so that some variables are highly likely to enter, and the $P$ value for remove at 1, so that no variable is removed. In most cases we recommend that you use the default option.

The program prints the first ten observations so that you can check that it is reading the data you want it to use. It also prints simple summary statistics.

Using the default options results in a stepwise inclusion of variables. At step 0 the constant term is entered, and a goodness-of-fit chi-square is printed. A table of approximate $F$-to-enter values is printed, and variable "sex" that has the largest numerical value is entered. At step 2 income is entered, and finally at step 3 age enters. The printout includes the coefficients along with their standard errors and each coefficient divided by its standard error, as shown in Table 12.3 for steps 2 and 3.

In our run we noted that the coefficient for income at step 2 was $-0.031$ but became $-0.037$ when age entered at the third step. One question, then, is whether a significant age–income interaction or other interaction exists among the independent variables. This issue can be checked by adding the following sentence to the "regress" paragraph:

model = age*sex*income.

This sentence causes the interaction of age and sex and income, and all subsets of this interaction (age and sex; age and income; sex and income; age, sex, and income), to be considered as possible variables for entry by the stepwise procedure.

**TABLE 12.3.** Partial Output of BMDPLR for Steps 2 and 3

| Term | Coefficient | Standard Error | Coefficient ÷ Standard Error |
|------|-------------|----------------|------------------------------|
| **Step 2** | | | |
| Income | −0.031 | 0.014 | −2.286 |
| Sex | +0.903 | 0.382 | +2.360 |
| Constant | −1.660 | 0.417 | −3.975 |
| **Step 3** | | | |
| Age | −0.021 | 0.009 | −2.318 |
| Income | −0.037 | 0.014 | −2.595 |
| Sex | +0.929 | 0.386 | +2.409 |
| Constant | −0.676 | 0.579 | −1.169 |

The program also prints the number and the percent of correct and incorrect predictions, as was discussed in Section 12.7.

## 12.10 WHAT TO WATCH OUT FOR

When researchers first read that they do not have to assume a multivariate normal distribution to use logistic regression analysis, they often leap to the conclusion that no assumptions at all are needed for this technique. Unfortunately, as with any statistical analysis, we still need some assumptions: a simple random sample, correct knowledge of group membership, and data that are free of miscellaneous errors and outliers. Additional points to watch out for include:

1. The model assumes that ln(odds) is linearly related to the independent variables. This should be checked using goodness-of-fit measures or other means provided by the program. It may require transformations of the data to achieve this.

2. The computations use an iterative procedure that can be quite time-consuming on some personal computers. To reduce computer time, researchers sometimes limit the number of variables.

3. Logistic regression should not be used to evaluate risk factors in longitudinal studies in which the studies are of different durations (see Woodbury, Manton, and Stallard 1981 for an extensive discussion of problems in interpreting the logistic function in this context).

4. The coefficient for a variable in a logistic regression equation depends on the other variables included in the model. The coefficients for the same variable, when included with different sets of variables, could be quite different.

5. If a matched analysis is performed, any variable used for matching cannot also be used as an independent variable.

## SUMMARY

In this chapter we presented another method of classifying individuals into one of two populations. It is based on the assumption that the logarithm of the odds of belonging to one population is a linear function of the variables used for classification. The result is that the probability of belonging to the population is a multiple logistic function.

We described in some detail how this relatively new method can be implemented by packaged programs, and we made the comparison with the linear discriminant function. Thus logistic regression is appropriate when both categorical and continuous variables are used, while the linear discriminant function is preferable when the multivariate normal model can be assumed. We also described situations in which logistic regression is a useful method of analysis.

Aldrich and Nelson (1984) discuss logistic regression as one of a variety of models used in situations where the dependent variable is dichotomous. Hosmer and Lemeshow (1989) provide an extensive treatment of the subject of logistic regression.

## BIBLIOGRAPHY

Aldrich, J. H., and Nelson, F. D. 1984. *Linear probability, logit, and probit models.* Newbury Park, CA: Sage.

Bock, J., and Afifi, A. A. 1988. Estimation of probability using the logistic model in retrospective studies. *Computers and Biomedical Research* 21:449–470.

*Breslow, N. E., and Day, N. E. 1980. *Statistical methods in cancer research.* IARC Scientific Publications No. 32. Lyons, France: World Health Organization.

Brittain, E. 1980. *Probability of developing coronary heart disease.* Technical Report 54. Stanford, Calif: Stanford University, Division of Biostatistics.

*Cox, D. R. 1970. *Analysis of binary data.* London: Methuen.

Efron, B. 1975. The efficiency of logistic regression compared to normal discriminant analysis. *Journal of the American Statistical Association* 70:892–898.

Fleiss, J. L. 1981. *Statistical methods for rates and proportions.* New York: Wiley.

Halperin, M., Blackwelder, W. C., and Verter, J. I. 1971. Estimation of the multivariate logistic risk function: A comparison of the discriminant function and maximum likelihood approach. *Journal of Chronic Diseases* 24:125–158.

Holford, T. R., White, C., and Kelsey, J. L. 1978. Multivariate analyses for matched case-control studies. *American Journal of Epidemiology* 107:245–256.

Hosmer, D. W., Jr., and Lemeshow, S. 1989. *Applied logistic regression.* New York: Wiley.

Kraemer, H. C. 1988. Assessment of 2 × 2 associations: Generalization of signal-detection methodology. *The American Statistician* 42:37–49.

Lemeshow, S., and Hosmer, D. W. 1982. A review of goodness-of-fit statistics for use in the development of logistic regression models. *American Journal of Epidemiology* 115:92–106.

Metz, C. E. 1978. Basic principles of ROC analysis. *Seminars in Nuclear Medicine* 8:283–298.

O'Hara, T. F., Hosmer, D. W., Lemeshow, S., and Hartz, S. C. 1982. A comparison of discriminant function and maximum likelihood estimates of logistic coefficients for categorical data. University of Massachusetts, Amherst, Mass.

Press, S. J., and Wilson, S. 1978. Choosing between logistic regression and discriminant analysis. *Journal of the American Statistical Society* 73:699–705.

Reynolds, H. T. 1977. *Analysis of nominal data*. Beverly Hills: Sage.

Schlesselman, J. J. 1982. *Case-control studies*. New York: Oxford University Press.

Swets, J. A. 1973. The receiver operating characteristic in psychology. *Science* 182:990–1000.

Woodbury, M. A., Manton, K. G., and Stallard, E. 1981. Longitudinal models for chronic disease risk: An evaluation of logistic multiple regression and alternatives. *International Journal of Epidemiology* 10:187–197.

Woolson, R. F., and Lachenbruch, P. A. 1982. Regression analysis of matched case-control data. *American Journal of Epidemiology* 115:444–452.

## PROBLEMS

12.1 If the probability of an individual getting a hit in baseball is 0.20, then the odds of getting a hit are 0.25. Check to determine that the previous statement is true. Would you prefer to be told that your chances are one in five of a hit or that for every four hitless times at bat you can expect to get one hit?

12.2 Using the formula odds $= P/(1 - P)$, fill in the accompanying table.

| Odds | P |
|------|------|
| 0.25 | 0.20 |
| 0.5  |      |
| 1.0  | 0.5  |
| 1.5  |      |
| 2.0  |      |
| 2.5  |      |
| 3.0  | 0.75 |
| 5.0  |      |

12.3 The accompanying table presents the number of individuals by smoking and disease status. What are the odds that a smoker will get disease A? That a nonsmoker will get disease A? What is the odds ratio?

|         | Disease A |     |       |
|---------|-----------|-----|-------|
| Smoking | Yes       | No  | Total |
| Yes     | 80        | 120 | 200   |
| No      | 20        | 280 | 300   |
| Total   | 100       | 400 | 500   |

12.4 Perform a logistic regression analysis with the same variables and data used in the example in Section 11.13 and compare the results.

12.5 Perform a logistic regression analysis on the data described in Problem 11.2. Compare the two analyses.

12.6 Repeat Problem 11.5, using stepwise logistic regression. Compare the results.

12.7 Repeat Problem 11.7, using stepwise logistic regression. Compare the results.

12.8 Repeat Problem 11.10, using stepwise logistic regression. Compare the results.

12.9 a. Using the depression data set, fill in the following table:

|                 | Sex    |      |       |
|-----------------|--------|------|-------|
| Regular Drinker | Female | Male | Total |
| yes             |        |      |       |
| no              |        |      |       |
| total           |        |      |       |

What are the odds that a woman is a regular drinker? That a man is a regular drinker? What is the odds ratio?

b. Repeat the tabulation and calculations for part a separately now for people who are depressed and those who are not. Compare the odds ratios for the two groups.

c. Run a logistic regression analysis with DRINK as the dependent variable and CESD and SEX as the independent variables. Include an interaction term. Is it significant? How does this relate to part b above?

12.10 Using the depression data set and stepwise logistic regression, describe the probability of an acute illness as a function of age, education, income, depression, and regular drinking.

12.11 Repeat Problem 12.10, but for chronic rather than acute illness.

12.12 Repeat Problem 11.13a, using stepwise logistic regression. Compare the results.

12.13 Repeat Problem 11.14, using stepwise logistic regression. Compare the results.

12.14 Define low FEV1 to be an FEV1 measurement below the median FEV1 of the fathers in the family lung function data set given in Appendix B. What are the odds that a father in this data set has low FEV1? What are the odds that a father from Glendora has low FEV1? What are the odds that a father from Long Beach does? What is the odds ratio of having low FEV1 for these two groups?

12.15 Using the definition of low FEV1 given in Problem 12.14, perform a logistic regression of low FEV1 on area for the fathers. Include all four areas and use a dummy variable for each area. What is the intercept term? Is it what you expected (or could have expected)?

12.16 For the family lung function data set, define a new variable VALLEY (residence in San Gabriel or San Fernando Valley) to be one if the family lives in Burbank or Glendora and zero otherwise. Perform a stepwise logistic regression of VALLEY on mother's age and FEV1, father's age and FEV1, and number of children (1, 2, or 3). Are these useful predictor variables?

# Chapter Thirteen

# REGRESSION ANALYSIS USING SURVIVAL DATA

## 13.1 WHAT WILL YOU LEARN FROM THIS CHAPTER?

In Chapters 11 and 12, we presented methods for classifying individuals into one of two possible populations and for computing the probability of belonging to one of the populations. We examined the use of discriminant function analysis and logistic regression for identifying important variables and developing functions for describing the risk of occurrence of a given event. In this chapter, you will learn another method for further quantifying the probability of occurrence of an event, such as death or terminating employment. In particular you will learn:

- When survival analysis is used (13.2).
- How to understand survival data (13.3).
- How to describe a survival function (13.4).
- Which distributions are often used in survival analysis (13.5).
- How to use a log-linear regression model for the relationship between survival time and explanatory factors (13.6).

- How to perform and interpret the results of the Cox proportional hazards regression model (13.7).
- What the relationship is between the log-linear, Cox, and logistic regression models (13.8).
- What computer programs to use to obtain the output (13.9).
- What to watch out for in regression analysis using survival data (13.10).

## 13.2 WHEN IS SURVIVAL ANALYSIS USED?

Survival analysis can be used to analyze data on the length of time it takes a specific event to occur. This technique takes on different names, depending on the particular application on hand. For example, if the event under consideration is the death of a person, animal, or plant, then the name *survival analysis* is used. If the event is failure of a manufactured item, e.g., a light bulb, then one speaks of *failure time analysis* or reliability theory (see Nelson 1982). A more recent term, namely *event history analysis* (see Allison 1984) is used by social scientists to describe applications in their fields. For example, analysis of the length of time it takes an employee to retire or resign from a given job could be called event history analysis. In this chapter, we will use the term survival analysis to mean any of the analyses just mentioned.

Survival analysis is a way of describing the distribution of the length of time to a given event. Suppose the event is termination of employment. We could simply draw a histogram of the length of time individuals are employed. Alternatively, we could use log length of employment as a dependent variable and determine if it can be predicted by variables such as age, gender, education level, or type of position (this will be discussed in Section 13.6). Another possibility would be to use the Cox regression model as described in Section 13.7.

Readers interested in a comprehensive coverage of the subject of survival analysis are advised to study one of the texts given in the Bibliography. In this chapter, our objective is to describe regression-type techniques that allow the user to examine the relationship between length of survival and a set of explanatory variables. The explanatory variables, often called covariates, can be either continuous, such as age or income, or they can be dummy variables that denote treatment group. The material in Sections 13.3–13.5 is intended as a brief summary of the background necessary to understand the remainder of the chapter.

## 13.3 DATA EXAMPLES

In this section, we begin with a description of a hypothetical situation to illustrate the types of data typically collected. This will be followed by a real data example taken from a cancer study.

In a survival study we can begin at a particular point in time, say 1980, and start to enroll patients. For each patient we collect baseline data, that is data at the time of enrollment, and begin follow-up. We then record other information about the patient during follow-up and, especially, the time at which the patient dies if this occurs before termination of the study, say 1990. If some patients are still alive, we simply record the fact that they survived up to 10 years. However, we may enroll patients throughout a certain period, say the first eight years of the study. For each patient we similarly record the baseline and other data, the length of survival, or the length of time of study (if the patient does not die during the period of observation). Typical examples of five patients are shown in Figure 13.1 to illustrate the various possibilities. Patient 1 started in 1980 and died $1\frac{1}{2}$ years after. Patient 2 started in 1982 and also died during the study period. Patient 3 entered in 1988 and died $1\frac{1}{2}$ years later. These

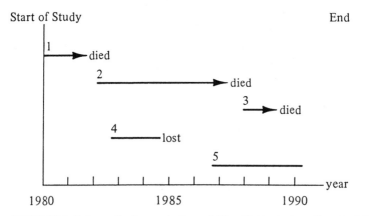

**FIGURE 13.1.** Schematic Presentation of Five Patients in a Survival Study

three patients were observed until death. Patient 4 was observed for two years, but was lost to follow-up. All we know about this patient is that survival was more than two years. This is an example of what is known as a *censored* observation. Another censored observation is represented by patient 5 who enrolled for $3\frac{1}{2}$ years and was still alive at the end of the study.

The data shown in Figure 13.1 can be represented as if all patients enrolled at the same time (see Figure 13.2). This makes it easier to compare the length of time the patients were followed. Note that this representation implicitly assumes that there have been no changes over time in the conditions of the study or in the types of patients enrolled.

One simple analysis of this data would be to compute the so-called $T$ year survival rate, for example $T = 3$. For all five patients we determine whether it is known that they survived three years:

Patient 1: did not survive three years

Patient 2: did survive three years and the time of later death is known

Patient 3: did not survive three years, but should not be counted in the analysis since the maximum period of observation would have been less than three years

Patient 4: was censored in less than three years and does not count for this analysis

Patient 5: was censored, but was known to live for more than three years and counts in this analysis

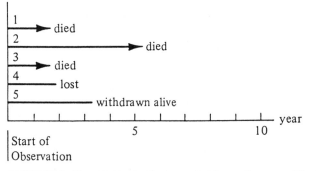

**FIGURE 13.2.** Five Patients Represented by a Common Starting Time

Note that three of the five patients can be used for the analysis and two cannot. The three-year survival rate is then calculated as:

$$\frac{\text{Number of patients known to survive 3 years}}{\text{Number of patients observed at risk for 3 years}}$$

In this expression, "patients at risk" means that they were enrolled more than three years before termination of the study. In the above

**TABLE 13.1.** Codebook for Lung Cancer Data*

| Variable Number | Variable Name | Description |
|---|---|---|
| 1 | ID | Identification Number from 1 to 401 |
| 2 | STAGET | Tumor Size<br>0 = small<br>1 = large |
| 3 | STAGEN | Stage of Nodes<br>0 = early<br>1 = late |
| 4 | HIST | Histology<br>1 = squamous cells<br>2 = other types of cells |
| 5 | TREAT | Treatment<br>0 = saline (control)<br>1 = BCG (experiment) |
| 6 | PERFBL | Performance Status at Baseline<br>0 = good<br>1 = poor |
| 7 | POINF | Post-Operative Infection<br>0 = no<br>1 = yes |
| 8 | SMOKFU | Smoking Status at Follow-up<br>1 = smokers<br>2 = ex-smokers<br>3 = never smokers |
| 9 | SMOKBL | Smoking Status at Baseline<br>1 = smokers<br>2 = ex-smokers<br>3 = never smokers |
| 10 | DAYS | Length of Observation Time in Days |
| 11 | DEATH | Status at End of Observation Time<br>0 = alive (censored)<br>1 = dead |

* Missing data are denoted as MIS in Appendix C.

example, two patients (patients 2 and 5) were known to survive three years, three patients (patients 1, 2, and 5) were observed at risk for three years, and therefore the three-year survival rate = 2/3.

In succeeding sections, we discuss methods that use more information from each observation. We end this section by describing a real data set used in various illustrations in this chapter. The data were taken from a multicenter clinical trial of patients with lung cancer who, in addition to the standard treatment, received either the experimental treatment known as BCG or the control treatment consisting of a saline solution injection (see Gail et al. 1984). The total length of the study was approximately 10 years. We selected a sample of 401 patients for analysis. The patients in the study were selected to be without distant metastasis or other life threatening conditions, and thus represented early stages of cancer. Data selected for presentation in this book are described in a code book shown as Table 13.1. In the binary variables, we used 0 to denote good or favorable values and 1 to denote unfavorable values. For HIST, subtract 1 to obtain 0, 1 data. The data are given in Appendix C.

## 13.4 SURVIVAL FUNCTIONS

The concept of a frequency histogram should be familiar to the reader. If not, the reader should consult an elementary statistics book. When the variable under consideration is the length of time to death, a histogram of the data from a sample can be constructed either by hand or by using a packaged program. If we imagine that the data were available on a very large sample of the population, then we can construct the frequency histogram using a large number of narrow frequency intervals. In constructing such a histogram, we use the relative frequency as the vertical axis so that the total area under the frequency curve is equal to one. This relative frequency curve is obtained by connecting the midpoints of the tops of the rectangles of the histogram (see Figure 13.3). This curve represents an approximation of what is known as the *death density function*. The actual death density function can be imagined as the result of letting the sample get larger and larger—the number of intervals gets larger and larger while the width of each interval gets smaller and smaller. In the limit, the death density is therefore a smooth curve.

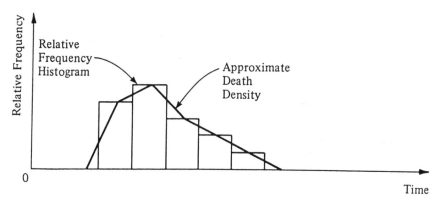

**FIGURE 13.3.** Relative Frequency Histogram and Death Density

Figure 13.4 presents an example of a typical death density function. Note that this curve is not symmetric, since most observed death density functions are not. In particular, it is not the familiar bell-shaped normal density function.

The death density is usually denoted by $f(t)$. For any given time $t$, the area under the curve to the left of $t$ is the proportion of individuals in the

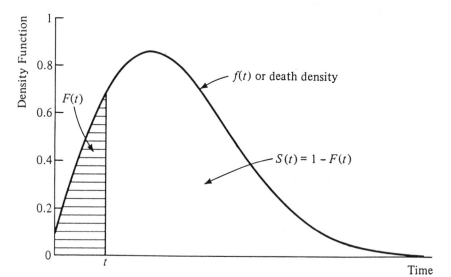

**FIGURE 13.4.** Graph of Death Density, Cumulative Death Distribution Function, and Survival Function

population who die up to time $t$. As a function of time $t$, this is known as the *cumulative death distribution function* and is denoted by $F(t)$. The area under the curve to the right of $t$ is $1 - F(t)$, since the total area under the curve is one. This latter proportion of individuals, denoted by $S(t)$, is the proportion of those surviving at least to time $t$ and is called the *survival function*. A graphic presentation of both $F(t)$ and $S(t)$ is given in Figure 13.5. In many statistical analyses, the cumulative distribution function $F(t)$ is presented as shown in Figure 13.5a. In survival analysis, it is more customary to plot the survival function $S(t)$ as shown in Figure 13.5b.

We will next define a quantity called the hazard function. Before doing so, we return to an interpretation of the death density function, $f(t)$. If time is measured in very small units, say seconds, then $f(t)$ is the likelihood, or probability, that a randomly selected individual from the population dies in the interval between $t$ and $t + 1$, where 1 represents one second. In this context, the population consists of all individuals regardless of their time of death. On the other hand, we may restrict our attention to only those members of the population who are known to be alive at time $t$. The probability that a randomly selected one of those members dies between $t$ and $t + 1$ is called the *hazard function*, and is denoted by $h(t)$. Mathematically, we can write $h(t) = f(t)/S(t)$. Other names of the hazard function are the *force of mortality, instantaneous death rate,* or *failure rate*.

To summarize, we described four functions: $f(t)$, $F(t)$, $S(t)$, and $h(t)$. Mathematically, any three of the four can be obtained from the fourth (see Lee 1980 or Miller 1981). In survival studies it is important to compute an estimated survival function. A theoretical distribution can be assumed, for example, one of those discussed in the next section. In that case, we need to estimate the parameters of the distribution and substitute in the formula for $S(t)$ to obtain its estimate for any time $t$. This is called a *parametric* estimate. If a theoretical distribution cannot be assumed based on our knowledge of the situation, we can obtain a *non-parametric* estimated survival function. Most packaged programs (see Section 13.9) use the Kaplan–Meier method for this purpose. Readers interested in the details of survival function estimation should consult one of the texts cited in the Bibliography. Many of these texts also discuss estimation of the hazard function and death density function.

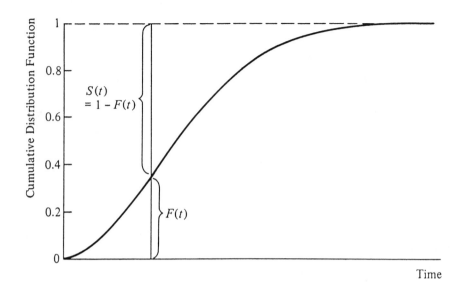

**a.** Cumulative Death Distribution Fuction

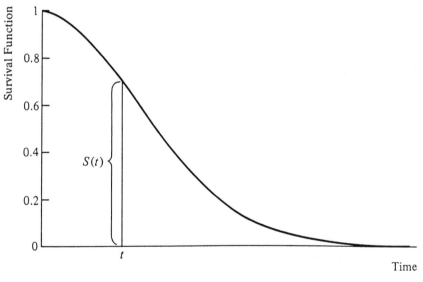

**b.** Survival Function

**FIGURE 13.5.** Cumulative Death Distribution Function and Survival Function

## 13.5 COMMON DISTRIBUTIONS USED IN SURVIVAL ANALYSIS

The simplest model for survival analysis assumes that the hazard function is constant across time, that is $h(t) = \lambda$, where $\lambda$ is any constant greater than zero. This results in the *exponential* death density $f(t) = \lambda \exp(-\lambda t)$, and the exponential survival function $S(t) = \exp(-\lambda t)$. Graphically, the hazard function and the death density function are displayed in Figures 13.6a and 13.7a. This model assumes that having survived up to a given point in time has no effect on the probability of dying in the next instant. Although simple, this model has been successful in describing many phenomena observed in real life. For example, it has been demonstrated that the exponential distribution closely describes the length of time from the first to the second myocardial infarction in humans and the time from diagnosis to death for some cancers. The exponential distribution can be easily recognized from a flat (constant) hazard function plot. Such plots are available in the output of many packaged programs, as we will discuss in Section 13.9.

If the hazard function is not constant, the *Weibull distribution* should be considered. For this distribution, the hazard function may be expressed as $h(t) = \alpha\lambda(\lambda t)^{\alpha-1}$. The expressions for the density function, the cumulative distribution function, and the survival function can be found in most of the references given at the end of the chapter. Figures 13.6b and 13.7b show plots of the hazard and density functions for $\lambda = 1$ and $\alpha = 0.5$, 1.5, and 2.5. The value of $\alpha$ determines the shape of the·distribution and for that reason it is called the *shape parameter* or *index*. Furthermore, as may be seen in Figure 13.6b, the value of $\alpha$ determines whether the hazard function increases or decreases over time. Namely, when $\alpha < 1$ the hazard function is decreasing and when $\alpha > 1$ it is increasing. When $\alpha = 1$ the hazard is constant, and the Weibull and exponential distributions are identical. In that case, the exponential distribution is used. The value of $\lambda$ determines how much the distribution is stretched, and therefore is called the *scale parameter*.

The Weibull distribution is used extensively in practice. Section 13.6 describes a model in which this distribution is assumed. However, the reader should note that other distributions are sometimes used. These include the lognormal, gamma, and others (see, for example, Cox and Oakes 1984).

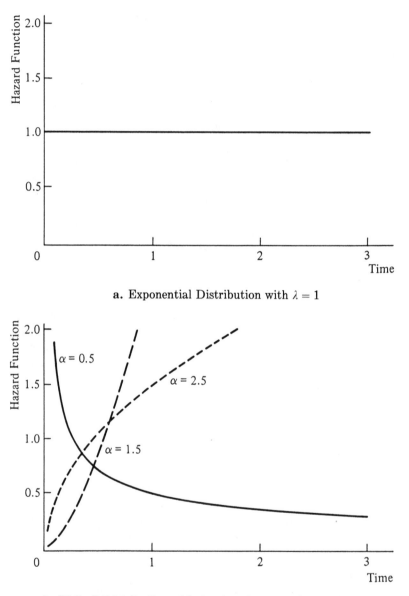

**a.** Exponential Distribution with $\lambda = 1$

**b.** Weibull Distribution with $\lambda = 1$ and $\alpha = 0.5$, 1.5, and 2.5

**FIGURE 13.6.** Hazard Functions for the Exponential and Weibull Distributions

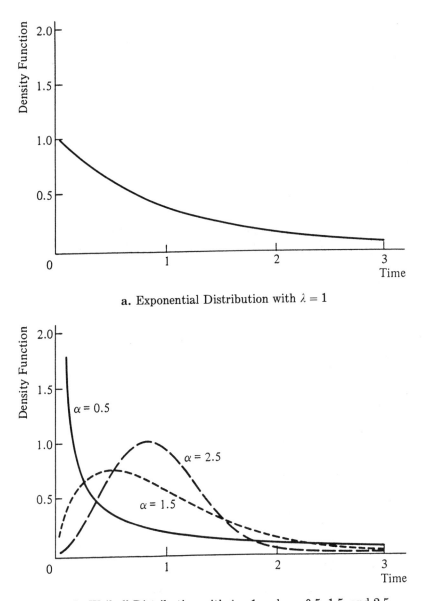

**a.** Exponential Distribution with $\lambda = 1$

**b.** Weibull Distribution with $\lambda = 1$ and $\alpha = 0.5$, 1.5, and 2.5

**FIGURE 13.7.** Death Density Functions for the Exponential and Weibull Distributions

A good way of deciding whether or not the Weibull distribution fits a set of data is to obtain a plot of $\log(-\log S(t))$ versus log time, and check whether the graph approximates a straight line (see Section 13.9). If it does, then the Weibull distribution is appropriate and the methods described in Section 13.6 can be used. If not, either another distribution may be assumed or the method described in Section 13.7 can be used.

## 13.6 THE LOG-LINEAR REGRESSION MODEL

In this section, we describe the use of multiple linear regression to study the relationship between survival time and a set of explanatory variables. Suppose that $t$ is survival time and $X_1, X_2, \ldots, X_p$ are the independent or explanatory variables. Let $Y = \log(t)$ be the dependent variable, where natural logarithms are used. Then the model assumes a linear relationship between $\log(t)$ and the $X$'s. The model equation is

$$\log(t) = \alpha + \beta_1 X_1 + \beta_2 X_2 + \cdots + \beta_p X_p + e$$

where $e$ is an error term. This model is known as the *log-linear regression model* since the log of survival time is a linear function of the $X$'s. If the distribution of $\log(t)$ were normal and if no censored observations exist in the data set, it would be possible to use the regression methods described in Chapter 7 to analyze the data. However, in most practical situations some of the observations are censored, as was described in Section 13.3. Furthermore, $\log(t)$ is usually not normally distributed ($t$ is often assumed to have a Weibull distribution). For those reasons, the method of maximum likelihood is used to obtain estimates of $\beta_i$'s and their standard errors. When the Weibull distribution is assumed, the log-linear model is sometimes known as the *accelerated life* or *accelerated failure time model* (Kalbfleisch and Prentice 1980).

For the data example presented in Section 13.3, it is of interest to study the relationship between length of survival time (since admission to the study) and the explanatory variables (variables 2 through 9) of Table 13.1. For the sake of illustration, we restrict our discussion to the variables Staget (tumor size: 0 = small, 1 = large), Perfbl (performance status at baseline: 0 = good, 1 = poor), Treat (treatment: 0 = control, 1 = experimental), and Poinf (post operative infection: 0 = no, 1 = yes).

A simple analysis that can shed some light on the effect of these variables is to determine the percent of those who died in the two categories of each of the variables. Table 13.2 gives the percent dying, and the results of a chi-square test of the hypothesis that the proportion of deaths is the same for each category. Based on this analysis, we may conclude that of the four variables, all except Treat may affect with the likelihood of death.

This simple analysis may be misleading, since it does not take into account the length of survival or the simultaneous effects of the explanatory variables. Previous analysis of this data (Gail et al. 1984) has demonstrated that the data fit the Weibull model well enough to justify this assumption. We estimated $S(t)$ from the data set and plotted $\log(-\log S(t))$ versus $\log(t)$ for small and large tumor sizes separately. The plots are approximately linear, further justifying the Weibull distribution assumption.

Assuming a Weibull distribution, the LIFEREG procedure of SAS was used to analyze the data. Table 13.3 displays some of the resulting output. Shown are the maximum-likelihood estimates of the intercept ($\alpha$) and regression ($\beta_i$) coefficients along with their estimated standard errors, and $P$ values for testing whether the corresponding parameters are zero.

Since all of the regression coefficients are negative except Treat, a value of one for any of the three status variables is associated with shorter survival time than is a value of zero. For example, for the variable Staget, a large tumor is associated with a shorter survival time. Furthermore, the $P$ values confirm the same results obtained from the previous simple analysis: three of the variables are significantly associated with survival

**TABLE 13.2.** Percentage of Deaths Versus Explanatory Variables

| Variable | | Percent Died | P Value |
|---|---|---|---|
| Staget | small | 42.7 | <0.01 |
| | large | 60.1 | |
| Perfbl | good | 48.4 | 0.02 |
| | poor | 64.5 | |
| Treat | control | 49.2 | 0.52 |
| | experiment | 52.4 | |
| Poinf | no | 49.7 | 0.03 |
| | yes | 75.0 | |

**TABLE 13.3.** Log-Linear Model for Lung Cancer Data: Results

| Variable | Estimate | Standard Error | Two-Sided P Value |
|----------|----------|----------------|-------------------|
| Intercept | 7.96 | 0.21 | <0.01 |
| Staget | −0.59 | 0.16 | <0.01 |
| Perfbl | −0.60 | 0.20 | <0.01 |
| Poinf | −0.71 | 0.31 | <0.02 |
| Treat | 0.08 | 0.15 | 0.59 |

(at the 5% significance level), whereas treatment is not. The information in Table 13.3 can also be used to obtain approximate confidence intervals for the parameters. Again, for the variable Staget, an approximate 95% confidence interval for $\beta_1$ is

$$-0.59 \pm (1.96)(0.16)$$

that is $-0.90 < \beta_1 < -0.28$.

The log-linear regression equation can be used to estimate typical survival times for selected cases. For example, for the least-serious case when each explanatory variable equals zero, we have $\log(t) = 7.96$. Since this is a natural logarithm, $t = \exp(7.96) = 2{,}864$ days $= 7.8$ years. This may be an unrealistic estimate of survival time caused by the extremely long tail of the distribution. On the other extreme, if every variable has a value of one, $t = \exp(6.14) = 464$ days, or somewhat greater than 1 year.

## 13.7 THE COX PROPORTIONAL HAZARDS REGRESSION MODEL

In this section, we describe the use of another method of modeling the relationship between survival time and a set of explanatory variables. In Section 13.4, we defined the hazard function and used the symbol $h(t)$ to indicate that it is a function of time $t$. Suppose that we use $X$, with no subscripts, as shorthand for all the $X_i$ variables. Since the hazard function may depend on $t$ and $X$, we now need to use the notation $h(t, X)$. The idea behind the *Cox model* is to express $h(t, X)$ as the product of two parts: one that depends on $t$ only and another that depends on the $X_i$ only. In symbols, the basic model is

$$h(t, X) = h_0(t)\exp(\beta_1 X_1 + \beta_2 X_2 + \cdots + \beta_p X_p)$$

where $h_0(t)$ does not depend on the $X_i$.

If all $X_i$'s are zero, then the second part of the equation would be equal to 1 and $h(t, X)$ reduces to $h_0(t)$. For this reason, $h_0(t)$ is sometimes called the *baseline hazard function*. In order to further understand the model, suppose that we have a single explanatory variable $X_1$, such that $X_1 = 1$ if the subject is from group 1 and $X_1 = 0$ if the subject is from group 2. For group 1, the hazard function is

$$h(t, 1) = h_0(t)\exp(\beta_1)$$

Similarly, for group 2, the hazard function is

$$h(t, 0) = h_0(t)\exp(0) = h_0(t)$$

The ratio of these two hazard functions is

$$h(t, 1)/h(t, 0) = \exp(\beta_1)$$

which is a constant that does not depend on time. In other words, the hazard function for group 1 is *proportional* to the hazard function for group 2. This property motivated D. R. Cox, the inventor of this model, to call it the *proportional hazards regression model*.

Another way to understand the model is to think in terms of two individuals in the study, each with a given set of values of the explanatory variables. The Cox model assumes that the ratio of the hazard functions for the two individuals, say $h_1(t)/h_2(t)$, is a constant that does not depend on time. The reader should note that it is difficult to check this proportionality assumption directly or visually in practice. An indirect method to check this assumption is to obtain a plot of $\log(-\log S(t))$ versus $t$ for different values of a given explanatory variable. This plot should exhibit an approximately constant difference between the curves corresponding to the levels of the explanatory variable.

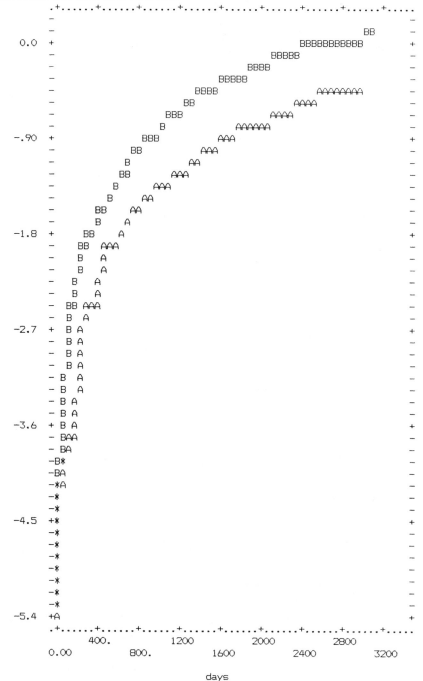

**FIGURE 13.8.** Computer-Generated Graph of $\log(-\log S(t))$ Versus $t$ for Lung Cancer Data (A = large tumor, B = small tumor)

**TABLE 13.4.** Cox's Model for Lung Cancer Data: Results

| Variable | Estimate | Standard Error | Two-Sided P Value |
|----------|----------|----------------|-------------------|
| Staget | 0.54 | 0.14 | $<0.01$ |
| Perfbl | 0.53 | 0.19 | $<0.01$ |
| Poinf | 0.67 | 0.28 | $<0.025$ |
| Treat | 0.07 | 0.14 | $>0.50$ |

Packaged programs use the maximum likelihood method to obtain estimates of the parameters and their standard errors. Some programs also allow variable selection by stepwise procedures similar to those we described for other models. One such program is BMDP2L, which we have used to analyze the same lung cancer survival data described earlier. In this analysis, we did not use the stepwise feature but rather obtained the output for the same variables as in Section 13.6. The results are given in Table 13.4.

Note that all except one of the signs of the coefficients in Table 13.4 are the opposites of those in Table 13.3. This is due to the fact that the Cox model describes the hazard function, whereas the log-linear model describes survival time. The reversal in sign indicates that a long survival time corresponds to low hazard and vice versa.

To check the assumption of proportionality, we obtained a plot of $\log(-\log S(t))$ versus $t$ for small and large tumors (variable Staget). The results, given in Figure 13.8, show that an approximately constant difference exists between the two curves, making the proportionality assumption a reasonable one. For this data set, it is reasonable to interpret the data in terms of the results of either this or the log-linear model. In the next section, we compare the two models with each other as well as with the logistic regression model presented in Chapter 12.

## 13.8 SOME COMPARISONS OF THE LOG-LINEAR, COX, AND LOGISTIC REGRESSION MODELS

### Log-Linear Versus Cox

In Tables 13.3 and 13.4, the estimates of the coefficients from the log-linear and Cox models are reported for the lung cancer data. As noted in the previous section, the coefficients in the two tables have opposite signs except for the variable Treat, which was nonsignificant. Also, in this particular example the magnitudes of the coefficients are not too different. For example, for the variable Staget, the *magnitude* of the coefficient was 0.59 for the log-linear model and 0.54 for the Cox model. One could ask, is there some general relationship between the coefficients obtained by using these two different techniques?

An answer to this question can be found when the Weibull distribution is assumed in fitting the log-linear model. For this distribution, the *population* coefficients have the following relationship

$$B(\text{Cox}) = -(\text{Shape})B(\text{log-linear})$$

where (Shape) is the value of the shape parameter $\alpha$. For a mathematical proof of this relationship, see Kalbfleisch and Prentice (1980), page 35. The sample coefficients will give an approximation to this relationship. For example, the sample estimate of shape is 0.918 for the lung cancer data and $-(0.918)(-0.59) = +0.54$, satisfying the relationship exactly for the Staget variable. But for Perfbl, $(-0.918)(-0.60) = +0.55$, close but not exactly equal to $+0.53$ given in Table 13.4.

If one is trying to decide which of these two models to use, then there are several points to consider. A first step would be to determine if either model fits the data, by obtaining the plots suggested in Section 13.5 and 13.7: (1) plot $\log(-\log S(t))$ versus $t$ for the Cox model separately for each category of the major dichotomous independent variables to see if the proportionality assumption holds; and (2) plot $\log(-\log S(t))$ versus $\log(t)$ to see if a Weibull fits, or plot $-\log S(t)$ versus $t$ to see if a straight line results, implying an exponential distribution fits. (The Weibull and exponential distributions are commonly used for the log-linear model.) In practice, these plots are sometimes helpful in discriminating between the two models, but often both models appear to fit equally well. This is the case with the lung cancer data. Neither plot looks perfect and neither fit is hopelessly poor.

## Cox Versus Logistic Regression

In Chapter 12, logistic regression analysis was presented as a method of classifying individuals into one of two categories. In analyzing data on survival, one possibility is to classify individuals as dead or alive after a *fixed* duration of follow-up. Thus we create two groups: those who survived, and those who did not survive beyond this fixed duration. Logistic regression can be used to analyze the data from those two groups with the objective of estimating the probability of surviving the given length of time. In the lung cancer data set, for example, the patients could be separated into two groups, those who died in less than one year and those who survived one year or more. In using this method, only patients whose results are known for at least one year should be considered for analysis. These are the same patients that could be used to compute a one-year survival rate as discussed in Section 13.3. Thus, excluded would be all patients who enrolled less than one year from the end of the study (whether they lived or died) and patients who had a censored survival time of less than one year. Logistic regression coefficients can be computed to describe the relationship between the independent variables and the log odds of death.

Several statisticians (Ingram and Kleinman 1989; Green and Symonds 1982; Mickey 1985) have examined the relationships between the results obtained from logistic regression analysis and those from the Cox model. Ingram and Kleinman (1989) assumed that the true population distribution followed a Weibull distribution with one independent variable that took on two values, and that the distribution could be thought of as group membership. Using a technique known as Monte Carlo simulation, the authors obtained the following results:

1. *Length of follow-up*: The estimated regression coefficients were similar (same sign and magnitude) for logistic regression and the Cox model when the patients were followed for a short time. They classified the cases as alive if they survived the follow-up period. Note that for a short time period, relatively few patients die. The range of survival times for those who die would be less than for a longer period. As the length of follow-up increased, the logistic regression coefficients increased in magnitude but those for the Cox model stayed the same. The magnitude of the standard error of the coefficients was somewhat larger for the logistic regression model. The estimates of the standard error for the Cox model decreased as the follow-up time increased.

2. *Censoring*: The logistic regression coefficients became very biased when there was greater censoring in one group than in the other, but

the regression coefficients from the Cox model remained unbiased (50% censoring was used).

3. When the proportion dying is small (10% and 19% in the two groups), the estimated regression coefficients were similar for the two methods. As the proportion dying increased, the regression coefficients for the Cox model stayed the same and their standard errors decreased. The logistic regression coefficients increased as the proportion dying increased.

4. Minor violations of the proportional hazards assumption had only a small effect on the estimates of the coefficients for the Cox model. More serious violations resulted in bias when the proportion dying is large or the effect of the independent variable is large.

Green and Symonds (1983) have concluded that the regression coefficients for the logistic regression and the Cox model will be similar when the outcome event is uncommon, the effect of the independent variables is weak, and the follow-up period is short. Mickey (1985) derived an approximate mathematical relationship between the formulas for the two sets of population coefficients and concluded that if less than 20% die, then the coefficients should agree within 12%, all else being equal.

The use of the Cox model is preferable to logistic regression if one wishes to compare results with other researchers who chose different follow-up periods. Logistic regression coefficients may vary with follow-up time or proportion dying, and therefore use of logistic regression requires care in describing the results and in making comparisons with other studies.

For a short follow-up period of fixed duration (when there is an obvious choice of the duration), some researchers may find logistic regression simpler to interpret and the options available in logistic regression programs useful. If it is thought that different independent variables best predict survival at different time periods (for example, less than two months versus greater than two months after an operation), then a separate logistic regression analysis on each of the time periods may be sensible.

## 13.9 DISCUSSION OF COMPUTER PROGRAMS

Of the three major packages, SPSS has only one program dealing with survival analysis, namely SURVIVAL. This program does not handle the regression techniques presented in this chapter and will therefore not be discussed further. SAS has LIFEREG for log-linear analysis (available both in the PC and mainframe versions) and two Cox regression programs, COXREGR and PHGLM (a stepwise procedure), that can be run under CMS or OS operating systems and thus are limited to larger computers. In addition, SAS has the SURVFIT and LIFETEST procedures (available for CMS and OS systems) that are useful in obtaining survival plots and information on when losses (censoring) occur. BMDP has 2L which fits the Cox model and 1L that can be used to obtain numerous types of survival and censoring plots. The 1990 version of 2L also includes the log-linear analysis. Both 1L and 2L are available for mainframe and personal computers. It should be noted that the computational techniques used in fitting the Cox and the log-linear models are iterative, so the number of independent variables used in a single run should be limited if a slower personal computer is used.

LIFEREG was used to obtain the log-linear coefficients in the cancer example. The Weibull distribution is the default option, so it was not necessary to state that we wished to assume that distribution. The survival time data were indicated by two variables: DAYS, the number of days the patient was in the study before death or censoring, and DEATH ($1 =$ died, $0 =$ censored), the patient's status.

The programs also offer the user the option of inputting the dates when the person entered and terminated the study. For example, for a study of employment, this could be the date of hire and the date of termination of employment or end of study. Dates could be given as month–day–year and be read by SAS using first the MMDDYYW. informat (see SAS Users' Guide: Informats). For BMDP, the use of calendar dates of entry and termination are standard options in the 1L and 2L programs. From the dates, the programs can compute the length of time from entry to termination.

To run the SAS LIFEREG procedure, we used the following MODEL statement:

MODEL DAY∗DEATH (0) = STAGET PERFBL POINF TREAT;

The independent variables are listed to the right of the equal sign. To the left, the dependent variable is listed first followed by the variable which indicates censored observations. Here the variable DEATH, when it takes on the value 0, indicates censored values of the dependent variable DAY. This model statement was preceded by a DATA statement that described the input data.

The default output from LIFEREG includes a count of the number of observations that were censored, noncensored, and missing. Estimates of the coefficients, their standard errors, chi-square statistics to test whether the population coefficients are zero, and $P$ values are printed. A variable labelled SCALE is also printed. This variable is equal to 1 divided by shape parameter for the Weibull distribution as defined in Section 13.5.

A plot of $\log(-\log S(t))$ versus $\log(t)$ was obtained by first using SURVFIT to calculate estimates of $S(t)$ for each patient. These values were saved into a data set along with the respective values of $t$. Both variables were appropriately transformed and the new variables, $\log(-\log(S(t)))$ and $\log(t)$, were plotted against each other using PLOT. Note that SURVFIT does not run on a PC, and currently there is no procedure available in PC-SAS to estimate $S(t)$ non-parametrically.

We used the 2L program of BMDP to obtain the coefficients for the Cox model. This program can be run with or without the stepwise option. The coefficients are printed at each step, thus the reader can examine the changes as new variables are added. The standard error, ratio of the coefficient to the standard error, and exponent of the coefficient are also

**TABLE 13.5.** Summary of Computer Output from SAS and BMDP Survival Regression Programs

| Options | SAS | BMDP |
|---|---|---|
| Data entry with dates | Informat statement | 2L |
| Number or time of censored and noncensored observations | LIFEREG, PHGLM | 2L |
| Graphical goodness of fit | SURVFIT + PLOT PHGLM + PLOT | 2L |
| Tests for goodness of fit | PHGLM | |
| Covariance matrix of coefficients | PHGLM | 2L |
| Coefficients and standard errors | LIFEREG, PHGLM | 2L |
| Test $\beta_i = 0$ | LIFEREG, PHGLM | 2L |
| $P$ value reported | LIFEREG, PHGLM | 2L |
| Stepwise available | PHGLM | 2L |
| Weights of observation | LIFEREG | |
| Quantiles | LIFEREG | |
| Time dependent covariates | | 2L |

provided together with a summary of the stepwise results. A plot of $\log(-\log S(t))$ versus $t$ is an option. More advanced features of this program not discussed in this book include the use of time-varying or time-dependent covariates. For example, a person's income or medical condition could change over time and the more recent measurement may be a better predictor of survival. For further discussion of this subject, see, e.g., Kalbfleisch and Prentice (1980).

The various features of BMDP2L, and SAS LIFEREG and PHGLM procedures are given in Table 13.5. SAS COXREGR is not included in the table because it offers fewer features than PHGLM. Note, however, that a special feature of COXREGR is that it allows the user to obtain tests of association between the independent variables.

## 13.10 WHAT TO WATCH OUT FOR

In addition to the remarks made in the regression analysis chapters, several potential problems exist for survival regression analysis:

1. If subjects are entered into the study over a long time period, the researcher needs to check that those entering early are like those entering later. If this is not the case, the sample may be a mixture of persons from different populations. For example, if the sample was taken from employment records 1960–80 and the employees hired in the 1970s were substantially different from those hired in the 1960s, putting everyone back to a common starting point does not take into account the change in the type of employees over time. It may be necessary to stratify by year and perform separate analyses, or to use a dummy independent variable to indicate the two time periods.

2. The analyses described in this chapter assume that censoring occurs independently of any possible treatment effect or subject characteristic (including survival time). If persons with a given characteristic that relates to the independent variables, including treatment, are censored early, then this assumption is violated and the above analyses are not valid. Whenever considerable censoring exists, it is recommended to check this assumption by comparing the patterns of censoring among subgroups of subjects

with different characteristics. Programs such as BMDP1L provide graphical output illustrating when the censoring occurs. A researcher can also reverse the roles of censored and non-censored values by temporarily switching the labels and calling the censored values "deaths." The researcher can then use the methods described in this chapter to study patterns in time to "death," in subgroups of subjects. If such examination reveals that the censoring pattern is not random, then other methods need to be used for analysis. Unfortunately, the little that is known about analyzing non-randomly censored observations is beyond the level of this book (see, e.g., Kalbfleisch and Prentice 1980). Also, to date such analyses have not been incorporated in general statistical packages.

**3.** In comparisons among studies it is critical to have the starting and end points defined in precisely the same way in each study. For example, death is a common end point for medical studies of life-threatening diseases such as cancer. But there is often variation in how the starting point is defined. It may be time of first diagnosis, time of entry into the study, just prior to an operation, post operation, etc. In medical studies, it is suggested that a starting point be used that is similar in the course of the disease for all patients.

**4.** If the subjects are followed until the end point occurs, it is important that careful and comprehensive follow-up procedures be used. One type of subject should not be followed more carefully than another, as differential censoring can occur. Note that although the survival analysis procedures allow for censoring, more reliable results are obtained if there is less censoring.

**5.** The investigator should be cautious concerning stepwise results with small sample sizes or in the presence of extensive censoring.

## SUMMARY

In this chapter, we presented two methods for performing regression analysis where the dependent variable was time until the occurrence of an event, such as death or termination of employment. These methods are called the log-linear or accelerated failure time model, and the Cox or proportional hazards model. One special feature of measuring time to an event is that some of the subjects may not be followed long enough for the event to occur, so they are classified as censored. The regression analyses allow for such censoring as long as it is independent of the characteristics of the subjects or the treatment they receive. In addition, we discussed the relationship among the results obtained when the log-linear, Cox, or logistic regression models are used.

Further information on survival analysis at a modest mathematical level can be found in either Allison (1984) or Lee (1980). Comprehensive references requiring a higher mathematical background are Kalbfleisch and Prentice (1980), Cox and Oakes (1984), Lawless (1982), and Miller (1981). For those interested in industrial applications, Nelson (1982) is highly recommended.

## BIBLIOGRAPHY

Abbot, R. D. 1985. Logistic regression in survival analysis. *American Journal of Epidemiology* 121:465–471.

Allison, P. D. 1984. *Event history analysis*, #46. Newbury Park, Calif.: Sage.

Cox, D. R., and Oakes, D. 1984. *Analysis of survival data*. London: Chapman and Hall.

Gail, M. H., Eagan, R. T., Feld, R., Ginsberg, R., Goodell, B., Hill, L., Holmes, E. C., Lukeman, J. M., Mountain, C. F., Oldham, R. K., Pearson, F. G., Wright, P. W., Lake, W. H., and the Lung Cancer Study Group. 1984. Prognostic factors in patients with resected stage 1 non-small lung cancer, *Cancer* 9:1802–1813.

Green, M. S., and Symonds, M. J. 1983. A comparison of the logistic risk function and the proportional hazards model in prospective epidemiologic studies. *Chronic Disease* 36:715–724.

Ingram, D. D., and Kleinman, J. C. 1989. Empirical comparison of proportional hazards and logistic regression models. *Statistics in Medicine* 8:525–538.

Kalbfleisch, J. D., and Prentice, R. L. 1980. *The statistical analysis of failure time data*. New York: Wiley.

Lawless, J. F. 1982. *Statistical models and methods for lifetime data*. New York: Wiley.

Lee, E. T. 1980. *Statistical methods for survival data analysis*. Belmont, Calif.: Lifetime Learning.

Mickey, M. R. 1985. Multivariable Analysis of One-Year Graft Survival. In: Terasaki, P. I., Editor, *Clinical Kidney Transplants*, Los Angeles: UCLA Tissue Typing Laboratory.

Miller, R. G. 1981. *Survival analysis.* New York: Wiley.

Nelson, W. 1982. *Applied life analysis.* New York: Wiley.

Slonim-Nevo, V., and Clark, V. A. 1989. An illustration of survival analysis: Factors affecting contraceptive discontinuation among American teenagers. *Social Work* 25: 7–19.

## PROBLEMS

The following problems all refer to the lung cancer data given in Appendix C and described in Table 13.1.

13.1   a. Find the effect of STAGEN and HIST upon survival by fitting a log-linear model. Check any assumptions and evaluate the fit using the graphical methods described in this chapter.

b. What happens in part a if you include STAGET in your model along with STAGEN and HIST as predictors of survival?

13.2   Repeat Problem 13.1, using a Cox proportional hazards model instead of a log-linear. Compare the results.

13.3   Do the patterns of censoring appear to be the same for smokers at baseline, ex-smokers at baseline, and non-smokers at baseline? What about for those who are smokers, ex-smokers, and non-smokers at follow-up?

13.4   Assuming a log-linear model for survival, does smoking status (i.e., the variables SMOKBL and SMOKFU) significantly affect survival?

13.5   Repeat Problem 13.4 assuming a proportional hazards model.

13.6   Assuming a log-linear model, do the effects of smoking status upon survival change depending on the tumor size at diagnosis?

13.7   Repeat Problem 13.6 assuming a proportional hazards model.

13.8   Define a variable SMOKCHNG that measures change in smoking status between baseline and follow-up, so that SMOKCHNG equals 0 if there is no change and 1 if there is a change.

a. Is there an association between STAGET and SMOKCHNG?

b. What effect does SMOKCHNG have upon survival? Is it significant? Is the effect the same if the smoking status variables are also included in the model as independent variables?

c. How else might you assign values to a variable that measures change in smoking status. Repeat part a using the new values.

# Chapter Fourteen

# PRINCIPAL COMPONENTS ANALYSIS

## 14.1 WHAT WILL YOU LEARN FROM THIS CHAPTER?

From this chapter you will learn:

- How to recognize when the use of principal components will make your data analysis simpler (14.2, 14.3).
- About the basic concepts of restructuring data by using principal components (14.4).
- How to decide if it was worthwhile doing the restructuring (14.5).
- How to use principal components to perform regression analysis of data with multicollinearity (14.6).
- How to decide on an appropriate computer program (14.7).
- What to watch out for in principal components analysis (14.8).

## 14.2 WHEN IS PRINCIPAL COMPONENTS ANALYSIS USED?

Principal components analysis is performed in order to simplify the description of a set of interrelated variables. In principal components analysis the variables are treated equally; i.e., they are not divided into dependent and independent variables, as in regression analysis.

The technique can be summarized as a method of transforming the original variables into new, uncorrelated variables. The new variables are called the *principal components*. Each principal component is a linear combination of the original variables. One measure of the amount of information conveyed by each principal component is its variance. For this reason the principal components are arranged in order of decreasing variance. Thus the most informative principal component is the first, and the least informative is the last (a variable with zero variance does not distinguish between the members of the population).

An investigator may wish to reduce the dimensionality of the problem, i.e., reduce the number of variables without losing much of the information. This objective can be achieved by choosing to analyze only the first few principal components. The principal components not analyzed convey only a small amount of information since their variances are small. This technique is attractive for another reason, namely, that the principal components are not intercorrelated. Thus instead of analyzing a large number of original variables with complex interrelationships, the investigator can analyze a small number of uncorrelated principal components.

The selected principal components may also be used to test for their normality. *If the principal components are not normally distributed, then neither are the original variables.* Another use of the principal components is to search for outliers. A histogram of each of the principal components can identify those individuals with very large or very small values; these values are candidates for outliers or blunders.

In regression analysis it is sometimes useful to obtain the first few principal components corresponding to the $X$ variables and then perform the regression on the selected components. This tactic is useful for overcoming the problem of multicollinearity since the principal components are uncorrelated (see Chatterjee and Price 1977). Principal components analysis can also be viewed as a step toward factor analysis (see Chapter 15).

Principal components analysis is considered to be an exploratory technique that may be useful in gaining a better understanding of the interrelationships among the variables. This idea will be discussed further in Section 14.5.

The original application of principal components analysis was in the field of education testing. Hotelling (1933) developed this technique and showed that there are two major components to responses on entry-examination tests: verbal and quantitative ability. Principal components and factor analysis are also used extensively in psychological applications in an attempt to discover underlying structure. In addition, principal components analysis has been used in biological and medical applications (see Seal 1964, Morrison 1976).

## 14.3 DATA EXAMPLE

The depression data set will be used later in the chapter to illustrate the technique. However, to simplify the exposition of the basic concepts, we generated a hypothetical data set using the BMDP package. These hypothetical data consist of 100 random pairs of observations, $X_1$ and $X_2$. The population distribution of $X_1$ is normal with mean 100 and variance 100. For $X_2$ the distribution is normal with mean 50 and variance 50. The population correlation between $X_1$ and $X_2$ is $1/\sqrt{2} = 0.707$. Figure 14.1 shows a scatter diagram of the 100 random pairs of points. The two variables are denoted in this graph by NORM1 and NORM2. The sample statistics are shown in Table 14.1. The sample correlation is $r = 0.757$.

The advantage of this data set is that it consists of only two variables, so it is easily graphed. In addition, it satisfies the usual normality assumption made in statistical theory.

**TABLE 14.1.** Sample Statistics for Hypothetical Data Set

|  | Variable | |
| --- | --- | --- |
| Statistic | NORM1 | NORM2 |
| N | 100 | 100 |
| Mean | 101.63 | 50.71 |
| Standard deviation | 10.47 | 7.44 |
| Variance | 109.63 | 55.44 |

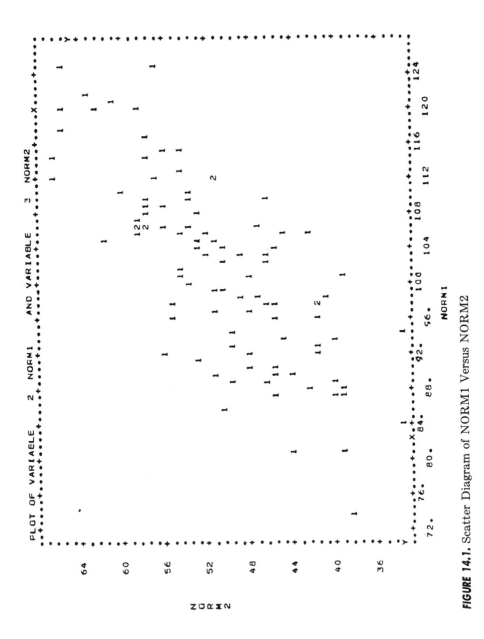

**FIGURE 14.1.** Scatter Diagram of NORM1 Versus NORM2

374

## 14.4 BASIC CONCEPTS OF PRINCIPAL COMPONENTS ANALYSIS

Again, to simplify the exposition of the basic concepts, we present first the case of two variables $X_1$ and $X_2$. Later we discuss the general case of $P$ variables.

Suppose that we have a random sample of $N$ observations on $X_1$ and $X_2$. For ease of interpretation we subtract the sample mean from each observation, thus obtaining

$$x_1 = X_1 - \bar{X}_1$$

and

$$x_2 = X_2 - \bar{X}_2$$

Note that this technique makes the means of $x_1$ and $x_2$ equal to zero but does not alter the sample variances $S_1^2$ and $S_2^2$ or the correlation $r$.

The basic idea is to create two new variables, $C_1$ and $C_2$, called the *principal components*. These new variables are linear functions of $x_1$ and $x_2$ and can therefore be written as

$$C_1 = a_{11}x_1 + a_{12}x_2$$

and

$$C_2 = a_{21}x_1 + a_{22}x_2$$

We note that for any set of values of the coefficients $a_{11}$, $a_{12}$, $a_{21}$, $a_{22}$, we can introduce the $N$ observed $x_1$ and $x_2$ and obtain $N$ values of $C_1$ and $C_2$. The means and variances of the $N$ values of $C_1$ and $C_2$ are

$$\text{mean } C_1 = \text{mean } C_2 = 0$$
$$\text{Var } C_1 = a_{11}^2 S_1^2 + a_{12}^2 S_2^2 + 2a_{11}a_{12}rS_1S_2$$

and

$$\text{Var } C_2 = a_{21}^2 S_1^2 + a_{22}^2 S_2^2 + 2a_{21}a_{22}rS_1S_2$$

where $S_i^2 = \text{Var } X_i$.

The *coefficients* are chosen to satisfy three requirements:

**1.** The Var $C_1$ is as large as possible.

**2.** The $N$ values of $C_1$ and $C_2$ are uncorrelated.

**3.** $a_{11}^2 + a_{12}^2 = a_{21}^2 + a_{22}^2 = 1$.

The mathematical solution for the coefficients was derived by Hotelling (1933). The solution is illustrated graphically in Figure 14.2. Principal components analysis amounts to rotating the original $x_1$ and $x_2$ axes to new $C_1$ and $C_2$ axes. The angle of rotation is determined uniquely by the requirements just stated. For a given point $x_1$, $x_2$ (see Figure 14.2) the values of $C_1$ and $C_2$ are found by drawing perpendicular lines to the new $C_1$ and $C_2$ axes. The $N$ values of $C_1$ thus obtained will have the largest variance according to requirement 1. The $N$ values of $C_1$ and $C_2$ will have a zero correlation.

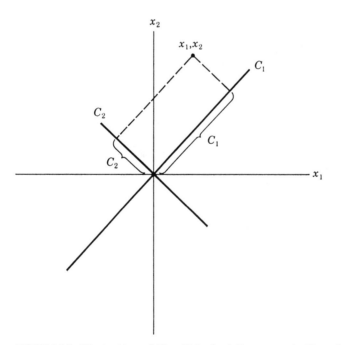

**FIGURE 14.2.** Illustration of Two Principal Components $C_1$ and $C_2$

In our hypothetical data example, the two principal components are

$$C_1 = 0.841x_1 + 0.539x_2$$
$$C_2 = -0.539x_1 + 0.841x_2$$

where

$$x_1 = \text{NORM1} - \text{mean(NORM1)}$$

and

$$x_2 = \text{NORM2} - \text{mean(NORM2)}$$

Note that

$$0.841^2 + 0.539^2 = 1$$

and

$$(-0.539)^2 + 0.841^2 = 1$$

as required by requirement 3 above. Also note that, for the *two*-variable case only, $a_{11} = a_{22}$ and $a_{12} = -a_{21}$.

Figure 14.3 illustrates the $x_1$ and $x_2$ axes (after subtracting the means) and the rotated $C_1$ and $C_2$ principal component axes. Also drawn in the graph is an ellipse of concentration of the original bivariate normal distribution.

The variance of $C_1$ is 147.44, and the variance of $C_2$ is 17.59. These two variances are commonly known as the first and second *eigenvalues*, respectively (synonyms used for eigenvalue are characteristic root, latent root, and proper value). Note that the sum of these two variances is 165.03. This quantity is equal to the sum of the original two variances (109.63 + 55.40 = 165.03). This result will always be the case; i.e., the total variance is preserved under rotation of the principal components. Note also that the lengths of the axes of the ellipse of concentration (Figure 14.3) are proportional to the sample standard deviations of $C_1$ and $C_2$, respectively. These standard deviations are the square roots of the eigenvalues. It is therefore easily seen from Figure 14.3 that $C_1$ has a larger

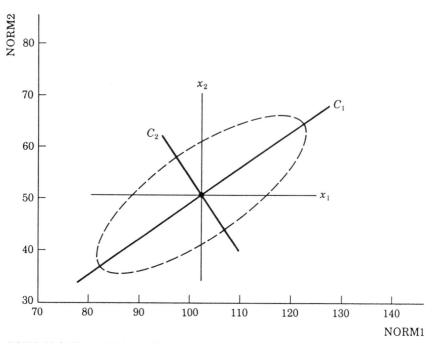

**FIGURE 14.3.** Plot of Principal Components for Bivariate Hypothetical Data

variance than $C_2$. In fact, $C_1$ has a larger variance than either of the original variables $x_1$ and $x_2$.

These basic ideas are easily extended to the case of $P$ variables $x_1$, $x_2, \ldots, x_P$. Each principal component is a linear combination of the $x$ variables. Coefficients of these linear combinations are chosen to satisfy the following three requirements:

**1.** Var $C_1 \geq$ Var $C_2 \geq \cdots \geq$ Var $C_P$.
**2.** The values of any two principal components are uncorrelated.
**3.** For any principal component the sum of the squares of the coefficients is one.

In other words, $C_1$ is the linear combination with the largest variance. Subject to the condition that it is uncorrelated with $C_1$, $C_2$ is the linear combination with the largest variance. Similarly, $C_3$ has the largest variance subject to the condition that it is uncorrelated with $C_1$ and $C_2$;

etc. The Var $C_i$ are the *eigenvalues*. These $P$ variances add up to the original total variance. In some packaged programs the set of coefficients of the linear combination for the $i$th principal component is called the $i$th *eigenvector* (also known as the characteristic or latent vector).

## 14.5 INTERPRETATION

In this section we discuss how many components should be retained for further analysis, and we present the analysis for standardized $x$ variables. Application to the depression data set is given as an example.

### Number of Components Retained

As mentioned earlier, one of the objectives of principal components analysis is *reduction of dimensionality*. Since the principal components are arranged in decreasing order of variance, we may select the first few as representatives of the original set of variables. The number of components selected may be determined by examining the proportion of total variance explained by each component. The cumulative proportion of total variance indicates to the investigator just how much information is retained by selecting a specified number of components.

In the hypothetical example the total variance is 165.03. The variance of the first component is 147.44, which is $147.44/165.03 = 0.893$, or 89.3%, of the total variance. It can be argued that this amount is a sufficient percentage of the total variation, and therefore the first principal component is a reasonable representative of the two original variables NORM1 and NORM2 (see Morrison 1976 or Eastment and Krzanowski 1982 for further discussion).

To interpret the meaning of the first principal component, we recall that it was expressed as

$$C_1 = 0.841x_1 + 0.539x_2$$

The coefficient 0.841 can be transformed into a correlation between $x_1$ and $C_1$. In general, the correlation between the $i^{\text{th}}$ principal component and the $j^{\text{th}}$ $x$ variable is

$$r_{ij} = \frac{a_{ij}\sqrt{\text{Var}\,C_i}}{\sqrt{\text{Var}\,x_j}}$$

where $a_{ij}$ is the coefficient of $x_j$ for the $i^{\text{th}}$ principal component. For example, the correlation between $C_1$ and $x_1$ is

$$r_{11} = \frac{0.84\sqrt{147.44}}{\sqrt{109.63}} = 0.975$$

and the correlation between $C_1$ and $x_2$ is

$$r_{12} = \frac{0.539\sqrt{147.44}}{\sqrt{55.40}} = 0.880$$

Note that both of these correlations are fairly high and positive. As can be seen from Figure 14.3, when either $x_1$ or $x_2$ increases, so will $C_1$. This result occurs often in principal components analysis whereby the first component is positively correlated with all of the original variables.

## Using Standardized Variables

Investigators frequently prefer to *standardize* the $x$ variables prior to performing the principal components analysis. Standardization is achieved by dividing each variable by its sample standard deviation. This analysis is then equivalent to analyzing the correlation matrix instead of the covariance matrix. When we derive the principal components from the correlation matrix, the interpretation becomes easier in two ways:

1. The total variance is simply the number of variables $P$, and the proportion explained by each principal component is the corresponding eigenvalue divided by $P$.
2. The correlation between the $i^{\text{th}}$ principal component $C_i$ and the $j^{\text{th}}$ variable $x_j$ is

$$r_{ij} = a_{ij}\sqrt{\operatorname{Var} C_i}$$

Therefore for a given $C_i$ we can compare the $a_{ij}$ to quantify the relative degree of dependence of $C_i$ on each of the standardized variables.

In our hypothetical example the correlation matrix is as follows:

| $x_1/S_1$ | $x_2/S_2$ |
|-----------|-----------|
| 1.000     | 0.757     |
| 0.757     | 1.000     |

Here $S_1$ and $S_2$ are the standard deviations of the first and second variables.

Analyzing this matrix results in the following two principal components:

$$C_1 = 0.707 \frac{\text{NORM1}}{S_1} + 0.707 \frac{\text{NORM2}}{S_2}$$

which explains $1.757/2 \times 100 = 87.8\%$ of the total variance, and

$$C_2 = 0.707 \frac{\text{NORM1}}{S_1} - 0.707 \frac{\text{NORM2}}{S_2}$$

explaining $0.243/2 \times 100 = 12.2\%$ of the total variance. So the first principal component is equally correlated with the two standardized variables. The second principal component is also equally correlated with the two standardized variables, but in the opposite direction.

The case of two standardized variables is illustrated in Figure 14.4. Forcing both standard deviations to be one causes the vast majority of the data to be contained in the square shown in the figure (e.g., 99.7% of normal data must fall within plus or minus three standard deviations of the mean). Because of the symmetry of this square, the first principal component will be in the direction of the 45° line for the case of two variates.

In terms of the original variables NORM1 and NORM2, the principal components based on the correlation matrix are

$$C_1 = 0.707 \frac{\text{NORM1}}{S_1} + 0.707 \frac{\text{NORM2}}{S_2} = 0.707 \frac{\text{NORM1}}{10.47} + 0.707 \frac{\text{NORM2}}{7.44}$$

$$= 0.0675\text{NORM1} + 0.0953\text{NORM2}$$

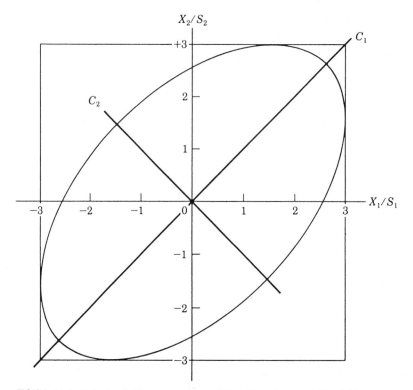

**FIGURE 14.4.** Principal Components for Two Standardized Variables

Similarly,

$$C_2 = +0.0675\text{NORM1} - 0.0953\text{NORM2}$$

Note that these results are very different from the principal components obtained from the covariance matrix (Section 14.4). This is the case in general. In fact, there is no easy way to convert the results based on the covariance matrix into those based on the correlation matrix, or vice versa. The majority of researchers prefer to use the correlation matrix because it compensates for the units of measurement of the different variables. But if it is used, then all interpretations must be made in terms of the standardized variables.

## Analysis of Depression Data Set

Next, we present a principal components analysis of a real data set, the depression data. We select for this example the 20 items that make up the CESD scale. Each item is a statement to which the response categories are ordinal. The answer rarely or none of the time (less than 1 day) is coded as 0, some or a little of the time (1–2 days) as 1, occasionally or a moderate amount of the time (3–4 days) as 2, and most or all of the time (5–7 days) as 3. The values of the response categories are reversed for the positive-affect items (see Table 14.2 for a listing of the items) so that a high score indicates likelihood of depression. The CESD score is simply a sum of the scores for these 20 items.

We emphasize that these variables do not satisfy the assumptions often made in statistics of a multivariate normal distribution. In fact, they cannot even be considered to be continuous variables. However, they are typical of what is found in real life applications.

In this example we used BMDP4R with the correlation matrix in order to be consistent with the factor analysis we present in Chapter 15. The eigenvalues (variances of the principal components) are plotted in Figure 14.5b. Since the correlation matrix is used, the total variance is the number of variables, 20. By dividing each eigenvalue by 20 and multiplying by 100, we obtain the percentage of total variance explained by each principal component. Adding these percentages successively produces the cumulative percentages plotted in 14.5a.

These eigenvalues and cumulative percentages are found in the output of standard packaged programs. They enable the user to determine whether and how many principal components should be used. If the variables were uncorrelated to begin with, then each principal component, based on the correlation matrix, would explain the *same* percentage of the total variance, namely $100/P$. If this were the case, a principal components analysis would be unnecessary. Typically, the first principal component explains a much larger percentage than the remaining components, as shown in Figure 14.5.

Ideally, we wish to obtain a small number of principal components, say two or three, which explain a large percentage of the total variance, say 80% or more. In this example, as is the case in many applications, this ideal is not achieved. We therefore must compromise by choosing as few principal components as possible to explain a reasonable percentage of the

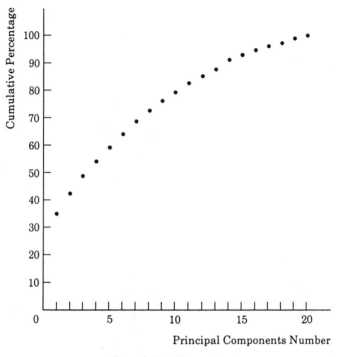

**a.** Cumulative Percentage

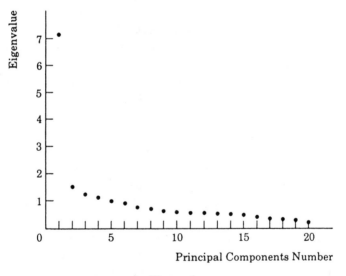

**b.** Eigenvalue

**FIGURE 14.5.** Eigenvalues and Cumulative Percentages of Total Variance for Depression Data

total variance. A rule of thumb adopted by many investigators is to select only the principal components explaining at least $100/P$ percent of the total variance (at least 5% in our example). This rule applies whether the covariance or the correlation matrix is used. In our example we would select the first five principal components if we followed this rule. These five components explain 59% of the total variance, as seen in Figure 14.5. Note, however, that the next two components each explain nearly 5%. Some investigators would therefore select the first seven components, which explain 69% of the total variance.

It should be explained that the eigenvalues are estimated variances of the principal components and are therefore subject to large sample variations. Arbitrary cutoff points should thus not be taken too seriously. Ideally, the investigator should make the selection of the number of principal components on the basis of some underlying theory.

Once the number of principal components is selected, the investigator should examine the coefficients defining each of them in order to assign an interpretation to the components. As was discussed earlier, a high coefficient of a principal component on a given variable is an indication of high correlation between that variable and the principal component (see the formulas given earlier). Principal components are interpreted in the context of the variables with high coefficients.

For the depression data example, Table 14.2 shows the coefficients for the first five components. For each principal component the variables with a correlation greater than 0.5 with that component are underlined. For the sake of illustration this value was taken as a cutoff point. Recall that the correlation $r_{ij}$ is $a_{ij}\sqrt{\text{Var } C_i}$, and therefore a coefficient $a_{ij}$ is underlined if it exceeds $0.5/\sqrt{\text{Var } C_i}$ (see Table 14.2).

As the table shows, many variables are highly correlated (greater than 0.5) with the first component. Note also that the correlations of all the variables with the first component are positive (recall that the scaling of the response for items 8–11 was reversed so that a high score indicates likelihood of depression). The first principal component can therefore be viewed as a weighted average of most of the items. A high value of $C_1$ is an indication that the respondent had many of the symptoms of depression.

On the other hand, the only item with more than 0.5 absolute correlation with $C_2$ is item 16, although items 17 and 14 have absolute correlations close to 0.5. The second principal component, therefore, can be interpreted as a measure of lethargy or energy. A low value of $C_2$ is an

**TABLE 14.2.** Principal Components Analysis for Standardized CESD Scale Items (Depression Data Set)

| Item | Principal Component 1 | 2 | 3 | 4 | 5 |
|---|---|---|---|---|---|
| **Negative Affect** | | | | | |
| 1. I felt that I could not shake off the blues even with the help of my family or friends. | 0.2774 | 0.1450 | 0.0577 | −0.0027 | 0.0883 |
| 2. I felt depressed. | 0.3132 | −0.0271 | 0.0316 | 0.2478 | 0.0244 |
| 3. I felt lonely. | 0.2678 | 0.1547 | 0.0346 | 0.2472 | −0.2183 |
| 4. I had crying spells. | 0.2436 | 0.3194 | 0.1769 | −0.0716 | −0.1729 |
| 5. I felt sad. | 0.2868 | 0.0497 | 0.1384 | 0.2794 | −0.0411 |
| 6. I felt fearful. | 0.2206 | −0.0534 | 0.2242 | 0.1823 | −0.3399 |
| 7. I thought my life had been a failure. | 0.2844 | 0.1644 | −0.0190 | −0.0761 | −0.0870 |
| **Positive Affect** | | | | | |
| 8. I felt that I was as good as other people. | 0.1081 | 0.3045 | 0.1103 | −0.5567 | −0.0976 |
| 9. I felt hopeful about the future. | 0.1758 | 0.1690 | −0.3962 | −0.0146 | 0.5355 |
| 10. I was happy. | 0.2766 | 0.0454 | −0.0835 | 0.0084 | 0.3651 |
| 11. I enjoyed life. | 0.2433 | 0.1048 | −0.1314 | 0.0414 | 0.2419 |
| **Somatic and Retarded Activity** | | | | | |
| 12. I was bothered by things that usually don't bother me. | 0.1790 | −0.2300 | 0.1634 | 0.1451 | 0.0368 |
| 13. I did not feel like eating; my appetite was poor. | 0.1259 | −0.2126 | 0.2645 | −0.5400 | 0.0953 |
| 14. I felt that everything was an effort. | 0.1803 | −0.4015 | −0.1014 | −0.2461 | −0.0847 |
| 15. My sleep was restless. | 0.2004 | −0.2098 | 0.2703 | 0.0312 | 0.0834 |
| 16. I could not "get going." | 0.1924 | −0.4174 | −0.1850 | −0.0467 | −0.0399 |
| 17. I had trouble keeping my mind on what I was doing. | 0.2097 | −0.3905 | −0.0860 | −0.0684 | −0.0499 |
| 18. I talked less than usual. | 0.1717 | −0.0153 | 0.2019 | −0.0629 | 0.2752 |
| **Interpersonal** | | | | | |
| 19. People were unfriendly. | 0.1315 | −0.0569 | −0.6326 | −0.0232 | −0.3349 |
| 20. I felt that people disliked me. | 0.2357 | 0.2283 | −0.1932 | −0.2404 | −0.2909 |
| Eigenvalues or Var $C_i$ | 7.055 | 1.486 | 1.231 | 1.066 | 1.013 |
| Cumulative proportion explained | 0.353 | 0.427 | 0.489 | 0.542 | 0.593 |
| $0.5/\sqrt{\mathrm{Var}C_i}$ | 0.188 | 0.410 | 0.451 | 0.484 | 0.497 |

indication of a lethargic state and a high value of $C_2$ is an indication of a high level of energy. By construction, $C_1$ and $C_2$ are uncorrelated.

Similarly, $C_3$ measures the respondent's feeling toward how others perceive him or her; a low value is an indication that the respondent believes that people are unfriendly. Similar interpretations can be made for the other two principal components in terms of the items corresponding to the underlined coefficients.

The numerical values of $C_i$ for $i = 1$ to 5, or 1 to 7, for each individual can be used in subsequent analyses. For example, we could use the value of $C_1$ instead of the CESD score as a dependent variable in a regression analysis such as that given in Problem 8.1. Had the first component explained a higher proportion of the variance, this procedure might have been better than simply using a sum of the scores.

The depression data example illustrates a situation in which the results are *not* clear-cut. This conclusion may be reached from observing Figure 14.5, where we saw that it is difficult to decide how many components to use. It is not possible to explain a very high proportion of the total variance with a small number of principal components. Also, the interpretation of the components in Table 14.2 is not straightforward. This is frequently the case in real life situations. Occasionally, situations do come up in which the results are clear-cut. For examples of such situations, see Seal (1964), Cooley and Lohnes (1971), or Harris (1975).

In biological examples the first component is often called the size component, and subsequent components are called the shape components. This terminology is especially appropriate when the data being analyzed are length and girth measurements of animals.

Test of hypotheses regarding principal components are discussed in Lawley (1956), Cooley and Lohnes (1971), Jackson and Hearne (1973), and Morrison (1976). These tests have not been implemented in the computer programs since the data are often not multivariate normal, and most users view principal components analysis as an exploratory technique.

## 14.6 USE OF PRINCIPAL COMPONENTS ANALYSIS IN REGRESSION

As mentioned in Section 9.5, when multicollinearity is present, principal components analysis may be used to alleviate the problem. The advantage of using principal components analysis is that it both helps in understanding what variables are causing the multicollinearity and provides a method for obtaining stable (though biased) estimates of the slope coefficients. If serious multicollinearity exists, the use of these coefficients will result in larger residuals in the data from which they were obtained and smaller multiple correlation than when least squares is used (the same holds for ridge regression). But the estimated standard error of the slopes could be smaller and the resulting equation could predict better in a fresh sample. Two programs exist to make the application of this method fairly straightforward for the user, RIDGEREG in SAS and 4R in BMDP. RIDGEREG can be used on mainframe computers that use the CMS or OS operating systems.

So far in this chapter we have concentrated our attention on the first few components that explain the highest proportion of the variance. To understand the use of principal components in regression analysis, it is useful to consider what information is available in, say, the last or $P$th component. The eigenvalue of that component will tell us how much of the total variance in the $X$'s it explains. Suppose we consider standardized data to simplify the magnitudes. If the eigenvalue of the last principal component (which must be smallest) is almost one, then the simple correlations among the $X$ variables must be close to zero. With low or zero simple correlations, the lengths of the principal component axes within the ellipse of concentration (see Figure 14.3) are all nearly the same and multicollinearity is not a problem.

At the other extreme, if the numerical value of the last eigenvalue is close to zero, then the length of its principal axis within the ellipse of concentration is very small. Since this eigenvalue can also be considered as the variance of the $P$th component, $C_P$, the variance of $C_P$ is close to zero. For standardized data with zero mean, the mean of $C_P$ is zero and its variance is close to zero. Approximately, we have

$$0 = a_{p1}x_1 + a_{p2}x_2 + \cdots + a_{pp}x_p$$

a linear relationship among the variables. The values of the coefficients of the principal components can give information on this interrelationship. For example, if two variables were almost equal to each other, then the value of $C_P$ might be $0 = 0.707x_1 - 0.707x_2$. For an example taken from actual data, see Chatterjee and Price (1977). In other words, when multi-collinearity exists, the examination of the last few principal components will provide information on which variables are causing the problem and on the nature of the interrelationship among them. It may be that two variables are almost a linear function of each other, or that two variables can almost be added to obtain a third variable.

In principal component regression analysis programs, first the principal components of the $X$ or independent variables are found. Then the regression equation is computed using the principal components instead of the original $X$ variables. From the regression equation using all the principal components, it is possible to obtain the original regression equation in the $X$'s, since there is a linear relationship between the $X$'s and the $C$'s (see Section 14.4). But what the special programs mentioned above allow the researcher to do is, to obtain a regression equation using the $X$ variables where one or more of the last principal components are not used in the computation. Based on the size of the eigenvalue (how small it is) and the possible relationship among the variables displayed in the later principal components, the user can decide how many components to discard. The default option for 4R uses 0.01 as the cutpoint for eigenvalues. A regression equation in the $X$'s can be obtained if only as few as one component is kept, but usually only one or two are discarded. The standard error of the slope coefficients when the components with very small eigenvalues are discarded tends to be smaller than when all are kept.

The use of principal component regression analysis is especially useful when the investigator understands the interrelationships of the $X$ variables well enough to interpret these last few components. Knowledge of the variables can help in deciding what to discard.

## 14.7 DISCUSSION OF COMPUTER PROGRAMS

Each of the three packages has at least one program to produce the principal component coefficients and the variances of the principal components (see Table 14.3). The coefficients are often called eigenvectors and the variances are called the eigenvalues.

The most straightforward program is the PRINCOMP procedure in SAS. In the depression example, we stored a SAS data file, called depress,

**TABLE 14.3.** Summary of Computer Output for Principal Components Analysis

| Output | BMDP | SAS | SPSS-X |
|---|---|---|---|
| Means, variances or standard deviations | 4R* | PRINCOMP | FACTOR |
| Covariance matrix | | PRINCOMP | |
| Correlation matrix | 4R | PRINCOMP | FACTOR |
| Coefficients from raw data | 4R | PRINCOMP | |
| Coefficients from standardized data | 4R | PRINCOMP | FACTOR |
| Eigenvalues | 4R | PRINCOMP | FACTOR |
| Cumulative proportion of variance explained | 4R | PRINCOMP | FACTOR |
| Plot of principal components | | PRINCOMP | PLOT** |

\* BMDP4M (factor analysis) may also be used for principal components; note discussion in this section.
\*\* After scores are saved in FACTOR.

in the computer and the analysis was done on a mainframe computer. It was only necessary to type

```
proc princomp data = in.depress;
var c1 -c20;
```

plus some job control language pertaining to the machine used. The output included means, standard deviations, the correlation matrix, the eigenvalues, and the coefficients for the 20 principal components. It is possible to obtain results for either the raw data or for standardized data.

For BMDP there are two possibilities. One is to use the 4R principal component regression program. In this case, the user takes any variable other than those needed for the principal component analysis, declares it the dependent variable and uses the desired $X$ variables as the principal components variables. To obtain all the principal components, use "limit = .000001,.000001." in the regression paragraph. This should be small enough for most applications. The part of the output relating to regression analysis is ignored, and the principal components and the eigenvalues are obtained directly from the program. Either raw (use the "no stand" option) or standardized coefficients can be obtained.

Alternatively, the 4M program can be used stating "constant = 0." and "form-cova." (if the raw data is preferred) in the factor paragraph. Also state "method = none." in the rotate paragraph. The 4M program was written to perform factor analysis (the subject of the next chapter). The desired output is called unrotated factor loadings. To obtain principal components coefficients from the loadings, multiply each vertical listing of the factor loadings by the square root of the corresponding eigenvalue (called "variance explained" in the output). For example, the first column of the factor loadings is multiplied by the square root of the first eigenvalue.

With the SPSS–X package the FACTOR program can be used to perform a principal components analysis. The correlation matrix is the only option available. The user should specify EXTRACTION = PA1 or PC and also CRITERIA = FACTORS(P), where P is the total number of variables.

## 14.8 WHAT TO WATCH OUT FOR

Principal components analysis is mainly used as an exploratory technique with little use of statistical tests or confidence intervals. Because of this, formal distributional assumptions are not necessary. Nevertheless, it is obviously simpler to interpret if certain conditions hold.

**1.** As with all statistical analyses, interpreting data from a random sample of a well-defined population is simpler than interpreting data from a sample that is taken in a haphazard fashion.

**2.** If the observations arise from or are transformed to a symmetric distribution, the results are easier to understand than if highly-skewed distributions are analyzed. Obvious outliers should be searched for and removed. If the data follow a multivariate normal distribution, then ellipses such as that pictured in Figure 14.4 are obtained and the principal components can be interpreted as axes of the ellipses. Statistical tests that can be found in some packaged programs usually require the assumption of multivariate normality. Note that the principal components offer a method for checking (in an informal sense) whether or not some data follow a multivariate normal distribution. For example, suppose that by using $C_1$ and $C_2$, 75% of the variance in the original variables can be explained. Their numerical values are obtained for each case. Then the $N$ values of $C_1$ are plotted using a cumulative normal probability plot, and checked to see if a straight line is obtained. A similar check is done for $C_2$. If both principal components are normally distributed, this lends some credence to the data set having a multivariate normal distribution. A scatter plot of $C_1$ versus $C_2$ can also be used to spot outliers. PRINCOMP allows the user to store the principal component scores of each case for future analysis.

**3.** If principal components analysis is to be used to check for redundancy in the data set (as discussed in Section 14.6), then it is important that the observations be measured accurately. Otherwise, it may be difficult to detect the interrelationships among the $X$ variables.

## SUMMARY

In the two previous chapters we presented methods of selecting variables for regression and discriminant function analyses. These methods include stepwise and subset procedures. In those analyses a dependent variable is present, implicitly or explicitly. In this chapter we presented another method for summarizing the data. It differs from variable selection procedures in two ways: (1) no dependent variable exists, and (2) variables are not eliminated but rather summary variables—i.e., principal components—are computed from all of the original variables.

The major ideas underlying the method of principal components analysis were presented in this chapter. We also discussed how to decide on the number of principal components retained and how to use them in subsequent analyses. Further, methods for attaching interpretations or "names" to the selected principal components were given.

A detailed discussion (at an introductory level) and further examples of the use of principal components analysis can be found in Dunteman (1989). At a higher mathematical level, Joliffe (1986) and Flury (1988) describe recent developments.

## BIBLIOGRAPHY

Chatterjee, S., and Price, B. 1977. *Regression analysis by example.* New York: Wiley.

*Cooley, W. W., and Lohnes, P. R. 1971. *Multivariate data analysis.* New York: Wiley.

Dunn, O. J., and Clark, V. A. 1987. *Applied statistics: Analysis of variance and regression*. 2nd ed. New York: Wiley.

Dunteman, G. H. 1989. *Principal components analysis*. Sage University Papers. Newbury Park, Calif.: Sage.

Eastment, H. T., and Krzanowski, W. J. 1982. Cross-validory choice of the number of components from a principal component analysis. *Technometrics* 24:73–77.

*Flury, B. 1988. *Common principal components and related multivariate models*. New York: Wiley.

Harris, R. J. 1985. *A primer of multivariate statistics*. 2nd ed. New York: Academic Press.

*Hotelling, H. 1933. Analysis of a complex of statistical variables into principal components. *Journal of Educational Psychology* 24:417–441.

Jackson, J. E., and Hearne, F. T. 1973. Relationships among coefficients of vectors in principal components. *Technometrics* 15:601–610.

*Joliffe, I. T. 1986. *Principal components analysis*. New York: Springer-Verlag.

*Lawley, D. N. 1956. Tests of significance for the latent roots of covariance and correlation matrices. *Biometrika* 43:128–136.

Morrison, D. F. 1976. *Multivariate statistical methods*. 2nd ed. New York: McGraw-Hill.

Seal, H. 1964. *Multivariate statistical analysis for biologists*. New York: Wiley.

*Tatsuoka, M. M. 1988. *Multivariate analysis: Techniques for educational and psychological research*. 2nd ed. New York: Wiley.

## PROBLEMS

14.1   For the depression data set, perform a principal components analysis on the last seven variables (DRINK through CHRONILL); see Table 3.2. Interpret the results.

14.2   Continuation of Problem 14.1: Perform a regression analysis of CASES on the last seven variables as well as on the principal components. What does the regression represent? Interpret the results.

14.3   For the data generated in Problem 7.7, perform a principal components analysis on $X_1, X_2, \ldots, X_9$. Compare the results with what is known about the population.

14.4   Continuation of Problem 14.3: Perform the regression of Y on the principal components. Compare the results with the multiple regression of Y on X1 to X9.

14.5   Perform a principal components analysis on the data in Table 8.1 (not including the variable P/E). Interpret the components. Then perform a regression analysis with P/E as the dependent variable, using the relevant principal components. Compare the results with those in Chapter 8.

14.6   Using the family lung function data in Appendix B, define a new variable RATIO = FEV1/FVC for the fathers. What is the correlation between RATIO and FEV1? Between RATIO and FVC? Perform a principal components analysis on FEV1 and FVC, plotting the results. Perform a principal components analysis on FEV1, FVC, and RATIO. Discuss the results.

14.7   Using the family lung function data, perform a principal components analysis on age, height, and weight for the oldest child.

14.8   Continuation of Problem 14.7: Perform a regression of FEV1 for the oldest child on the principal components found in Problem 14.7. Compare the results to those from Problem 7.15.

14.9   Using the family lung function data, perform a principal components analysis on mother's height, weight, age, FEV1, and FVC. Use the covariance matrix, then repeat using the correlation matrix. Compare the results and comment.

14.10  Perform a principal components analysis on AGE and INCOME using the depression data set. Include all the additional data points listed in Problem 6.9b. Plot the original variables and the principal components. Indicate the added points on the graph, and discuss the results.

# Chapter Fifteen

# FACTOR ANALYSIS

## 15.1 WHAT WILL YOU LEARN FROM THIS CHAPTER?

From this chapter you will learn a useful extension of principal components analysis called factor analysis that will enable you to obtain more distinct new variables. The methods given in this chapter for describing the interrelationships among variables and obtaining new variables have been deliberately limited to a small subset of the methods available in the literature in order to make the explanation more understandable. In particular, you will learn:

- When to use factor analysis (15.2, 15.3, 15.9).
- About the basic factor model (15.4).
- How to obtain the initial factors (15.5, 15.6).
- How to rotate the initial factors to obtain more distinct factors (15.7).
- How to obtain factor scores to use in future analyses (15.8).

- About which options are available on the various computer packages (15.10).
- What to watch out for in factor analysis (15.11).

## 15.2 WHEN IS FACTOR ANALYSIS USED?

*Factor analysis* is similar to principal components analysis in that it is a technique for examining the interrelationships among a set of variables. Both of these techniques differ from regression analysis in that we do not have a *dependent* variable to be explained by a set of *independent* variables. However, principal components analysis and factor analysis also differ from each other. In principal components analysis the major objective is to select a number of components that explain as much of the total variance as possible. The values of the principal components for a given individual are relatively simple to compute and interpret. On the other hand, the *factors* obtained in factor analysis are selected mainly to explain the interrelationships among the original variables. Ideally, the number of factors expected is known in advance. The major emphasis is placed on obtaining easily understandable factors that convey the essential information contained in the original set of variables.

Areas of application of factor analysis are similar to those mentioned in Section 14.2 for principal components analysis. Chiefly, applications have come from the social sciences, particularly psychometrics. It has been used mainly to explore the underlying structure of a set of variables. It has also been used in assessing what items to include in scales and to explore the interrelationships among the scores on different items. A certain degree of resistance to using factor analysis in other disciplines has been prevalent, perhaps because of the heuristic nature of the technique and the special jargon employed. Also, the multiplicity of methods available to perform factor analysis leaves some investigators uncertain of their results.

In this chapter no attempt will be made to present a comprehensive treatment of the subject. Rather, we adopt a simple geometric approach to explain only the most important concepts and options available in standard packaged programs. Interested readers should refer to the Bibliography for texts that give more comprehensive treatments of factor analysis. In particular, Dillon and Goldstein (1984), Everitt (1984),

Bartholomew (1987), and Long (1983) discuss confirmatory factor analysis and linear structural models, subjects not discussed in this book.

## 15.3 DATA EXAMPLE

As in Chapter 14, we generated a hypothetical data set to help us present the fundamental concepts. In addition, the same depression data set used in Chapter 14 will be subjected to a factor analysis here. This example should serve to illustrate the differences between principal components and factor analysis. It will also provide a real life application.

The hypothetical data set consists of 100 data points on five variables, $X_1, X_2, \ldots, X_5$. The data were generated from a multivariate normal distribution with zero means. (The statements used to generate these data are given in Section 15.10.) The sample means and standard deviations are shown in Table 15.1. The correlation matrix is presented in Table 15.2.

We note, on examining the correlation matrix, that there exists a high correlation between $X_4$ and $X_5$, a moderately high correlation between $X_1$

**TABLE 15.1.** Means and Standard Deviations of 100 Hypothetical Data Points

| Variable | Mean | Standard Deviation |
|---|---|---|
| $X_1$ | 0.163 | 1.047 |
| $X_2$ | 0.142 | 1.489 |
| $X_3$ | 0.098 | 0.966 |
| $X_4$ | −0.039 | 2.185 |
| $X_5$ | −0.013 | 2.319 |

**TABLE 15.2.** Correlation Matrix of 100 Hypothetical Data Points

|  | $X_1$ | $X_2$ | $X_3$ | $X_4$ | $X_5$ |
|---|---|---|---|---|---|
| $X_1$ | 1.000 | | | | |
| $X_2$ | 0.757 | 1.000 | | | |
| $X_3$ | 0.047 | 0.054 | 1.000 | | |
| $X_4$ | 0.115 | 0.176 | 0.531 | 1.000 | |
| $X_5$ | 0.279 | 0.322 | 0.521 | 0.942 | 1.000 |

and $X_2$, and a moderate correlation between $X_3$ and each of $X_4$ and $X_5$. The remaining correlations are fairly low.

## 15.4 BASIC CONCEPTS OF FACTOR ANALYSIS

In factor analysis we begin with a set of variables $X_1, X_2, \ldots, X_P$. These variables are usually standardized by the computer program so that their variances are each equal to one and their covariances are correlation coefficients. In the remainder of this chapter we therefore assume that each $x_i$ is a *standardized variable*, i.e., $x_i = (X_i - \bar{X}_i)/S_i$. In the jargon of factor analysis the $x_i$'s are called the original or *response variables*.

The object of factor analysis is to represent each of these variables as a linear combination of a smaller set of *common factors* plus a factor unique to each of the response variables. We express this representation as

$$x_1 = l_{11}F_1 + l_{12}F_2 + \cdots + l_{1m}F_m + e_1$$

$$x_2 = l_{21}F_1 + l_{22}F_2 + \cdots + l_{2m}F_m + e_2$$

$$\vdots$$

$$x_P = l_{P1}F_1 + l_{P2}F_2 + \cdots + l_{Pm}F_m + e_P$$

where the following assumptions are made:

**1.** $m$ is the number of common factors (typically this number is much smaller than $P$).

**2.** $F_1, F_2, \ldots, F_m$ are the *common factors*. These factors are assumed to have zero means and unit variances.

**3.** $l_{ij}$ is the coefficient of $F_j$ in the linear combination describing $x_i$. This term is called the *loading* of the $i$th variable on the $j$th common factor.

**4.** $e_1, e_2, \ldots, e_P$ are *unique factors*, each relating to one of the original variables.

The above equations and assumptions constitute the so-called *factor model*. Thus each of the response variables is composed of a part due to the common factors and a part due to its own unique factor. The part due to the common factors is assumed to be a linear combination of these factors.

As an example, suppose that $x_1, x_2, x_3, x_4, x_5$ are the standardized scores of an individual on five tests. If $m = 2$, we assume the following model:

$$x_1 = l_{11}F_1 + l_{12}F_2 + e_1$$

$$x_2 = l_{21}F_1 + l_{22}F_2 + e_2$$

$$\vdots$$

$$x_5 = l_{51}F_1 + l_{52}F_2 + e_5$$

Each of the five scores consists of two parts: a part due to the common factors $F_1$ and $F_2$ and a part due to the unique factor for that test. The common factors $F_1$ and $F_2$ might be considered the individual's verbal and quantitative abilities, respectively. The unique factors express the individual variation on each test score. The unique factor includes all other effects that keep the common factors from completely defining a particular $x_i$.

In a sample of $N$ individuals we can express the equations by adding a subscript to each $x_i$, $F_j$, and $e_i$ to represent the individual. For the sake of simplicity this subscript was omitted in the model presented here.

The factor model is, in a sense, the mirror image of the principal components model, where each principal component is expressed as a linear combination of the variables. Also, the number of principal components is equal to the number of original variables (although we may not use all of the principal components). On the other hand, in factor analysis we choose the number of factors to be smaller than the number of response variables. Ideally, the number of factors should be known in advance, although this is often not the case. However, as we will discuss later, it is possible to allow the data themselves to determine this number.

The factor model, by breaking each response variable $x_i$ into two parts, also breaks the variance of $x_i$ into two parts. Since $x_i$ is standardized, its variance is one. This variance of one is composed of the following two parts:

**1.** The *communality*, i.e., the part of the variance that is due to the common factors.

**2.** The *specificity*, i.e., the part of the variance that is due to the unique factor $e_i$.

Denoting the *communality* of $x_i$ by $h_i^2$ and the *specificity* by $u_i^2$, we can write the variance of $x_i$ as Var $x_i = 1 = h_i^2 + u_i^2$. In words, the variance of $x_i$ equals the communality plus the specificity.

The numerical aspects of factor analysis are concerned with finding estimates of the *factor loadings* $(l_{ij})$ and the *communalities* $(h_i^2)$. There are many ways available to numerically solve for these quantities. The solution process is called *initial factor extraction*. The next two sections discuss two such extraction methods. Once a set of initial factors is obtained, the next major step in the analysis is to obtain new factors, called the *rotated factors*, in order to improve the interpretation. Methods of rotation are discussed in Section 15.7.

In any factor analysis the number $m$ of common factors is required. As mentioned earlier, this number is, ideally, known prior to the analysis. If it is not known, most investigators use a default option available in standard computer programs whereby the number of factors is the number of eigenvalues greater than 1 (see Chapter 14 for the definition and discussion of eigenvalues). Also, since the numerical results are highly dependent on the chosen number $m$, many investigators run the analysis with several values in an effort to get further insights into their data.

## 15.5 INITIAL FACTOR EXTRACTION: PRINCIPAL COMPONENTS ANALYSIS

In this section and the next we discuss two methods for the initial extraction of common factors. We begin with the *principal components analysis method*, which can be found in most of the standard factor analysis programs. It is called PCA in BMDP, PA1 or PC in SPSS–X, and PRINCIPAL (PRIN or P) in SAS. The basic idea is to choose the first $m$ principal components and modify them to fit the factor model defined in the previous section. The reason for choosing the first $m$ principal components, rather than any others, is that they explain the greatest proportion of the variance and are therefore the most important. Note that the principal components are also uncorrelated and thus present an attractive choice as factors.

To satisfy the assumption of unit variances of the factors, we divide each principal component by its standard deviation. That is, we define the

$j$th common factor $F_j$ as $F_j = C_j/(\text{Var } C_j)^{1/2}$, where $C_j$ is the $j$th principal component.

To express each variable $x_i$ in terms of the $F_j$'s, we first recall the relationship between the variables $x_i$ and the principal components $C_j$. Specifically,

$$C_1 = a_{11}x_1 + a_{12}x_2 + \cdots + a_{1P}x_P$$
$$C_2 = a_{21}x_1 + a_{22}x_2 + \cdots + a_{2P}x_P$$
$$\vdots$$
$$C_P = a_{P1}x_1 + a_{P2}x_2 + \cdots + a_{PP}x_P$$

It may be shown mathematically that this set of equations can be inverted to express the $x_i$'s as functions of the $C_j$'s. The result is (see Harmon 1976 or Afifi and Azen 1979):

$$x_1 = a_{11}C_1 + a_{21}C_2 + \cdots + a_{P1}C_P$$
$$x_2 = a_{12}C_1 + a_{22}C_2 + \cdots + a_{P2}C_P$$
$$\vdots$$
$$x_P = a_{1P}C_1 + a_{2P}C_2 + \cdots + a_{PP}C_P$$

Note that the rows of the first set of equations become the columns of the second set of equations.

Now since $F_j = C_j/(\text{Var } C_j)^{1/2}$, it follows that $C_j = F_j(\text{Var } C_j)^{1/2}$, and we can then express the $i$th equation as

$$x_1 = a_{1i}F_1(\text{Var } C_1)^{1/2} + a_{2i}F_2(\text{Var } C_2)^{1/2} + \cdots + a_{Pi}F_P(\text{Var } C_P)^{1/2}$$

This last equation is now modified in two ways:

**1.** We use the notation $l_{ij} = a_{ji}(\text{Var } C_j)^{1/2}$ for the first $m$ components.
**2.** We combine the last $P - m$ terms and denote the result by $e_i$. That is,

$$e_i = a_{m+1,i}F_{m+1}(\text{Var } C_{m+1})^{1/2} + \cdots + a_{Pi}F_P(\text{Var } C_P)^{1/2}$$

With these manipulations we have now expressed each variable $x_i$ as

$$x_i = l_{i1}F_1 + l_{i2}F_2 + \cdots + l_{im}F_m + e_i$$

for $i = 1, 2, \ldots, P$. In other words, we have transformed the principal components model to produce the factor model. For later use, note also that when the original variables are standardized, the factor loading $l_{ij}$ turns out to be the *correlation* between $x_i$ and $F_j$ (see Section 14.5). The matrix of factor loadings is sometimes called the *pattern matrix*. When the factor loadings are correlations between the $x_i$'s and $F_j$'s as they are here, it is also called the *factor structure matrix* (see Dillon and Goldstein 1984). Furthermore, it can be shown mathematically that the communality of $x_i$ is $h_i^2 = l_{i1}^2 + l_{i2}^2 + \cdots + l_{im}^2$.

For example, in our hypothetical data set the eigenvalues of the correlation matrix (or Var $C_i$) are

$$\text{Var } C_1 = 2.578$$

$$\text{Var } C_2 = 1.567$$

$$\text{Var } C_3 = 0.571$$

$$\text{Var } C_4 = 0.241$$

$$\text{Var } C_5 = \underline{0.043}$$

$$\text{total} = 5.000$$

Note that the sum 5.0 is equal to $P$, the total number of variables. Based on the rule of thumb of selecting only those principal components corresponding to eigenvalues of one or more, we select $m = 2$. Using BMDP4M, we obtain the principal components analysis factor loadings, the $l_{ij}$'s, shown in Table 15.3. For example, the loading of $x_1$ on $F_1$ is $l_{11} = 0.511$ and on $F_2$ is $l_{12} = 0.782$. Thus the first equation of the factor model is

$$x_1 = 0.511F_1 + 0.782F_2 + e_1$$

Table 15.3 also shows the variance explained by each factor. For example, the variance explained by $F_1$ is 2.578, or 51.6%, of the total variance of 5.

**TABLE 15.3.** Initial Factor Analysis Summary for Hypothetical Data Set from Principal Components Extraction Method

| Variable | Factor Loadings | | Communality | Specificity |
|---|---|---|---|---|
| | $F_1$ | $F_2$ | $h_i^2$ | $u_i^2$ |
| $x_1$ | 0.511 | 0.782 | 0.873 | 0.127 |
| $x_2$ | 0.553 | 0.754 | 0.875 | 0.125 |
| $x_3$ | 0.631 | −0.433 | 0.586 | 0.414 |
| $x_4$ | 0.866 | −0.386 | 0.898 | 0.102 |
| $x_5$ | 0.929 | −0.225 | 0.913 | 0.087 |
| Variance explained | 2.578 | 1.567 | $\Sigma\, h_i^2 = 4.145$ | $\Sigma\, u_i^2 = 0.855$ |
| Percentage | 51.6% | 31.3% | 82.9% | 17.1% |

The communality column, $h_i^2$, in Table 15.3 shows the part of the variance of each variable explained by the common factors. For example,

$$h_1^2 = 0.511^2 + 0.782^2 = 0.873$$

Finally, the specificity $u_i^2$ is the part of the variance not explained by the common factors. In this example we use standardized variables $x_i$ whose variances are each equal to one. Therefore for each $x_i$, $u_i^2 = 1 - h_i^2$.

It should be noted that the sum of the communalities is equal to the cumulative part of the total variance explained by the common factors. In this example it is seen that

$$0.873 + 0.875 + 0.586 + 0.898 + 0.913 = 4.145$$

and

$$2.578 + 1.567 = 4.145$$

thus verifying the above statement. Similarly, the sum of the specificities is equal to the total variance minus the sum of the communalities.

A valuable graphical aid to interpreting the factors is the *factor diagram* shown in Figure 15.1. Unlike the usual scatter diagrams where each point

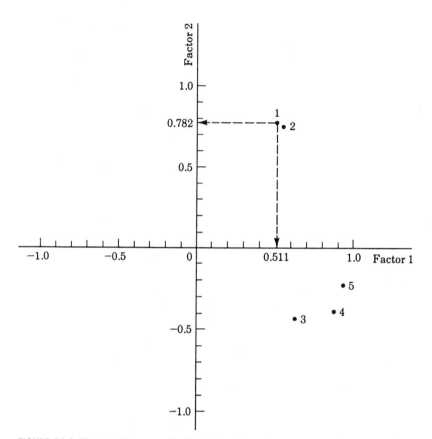

**FIGURE 15.1.** Factor Diagram for Principal Components Extraction Method

represents an individual case and each axis a variable, in this graph each point represents a response variable and each axis a common factor. For example, the point labeled 1 represents the response variable $x_1$. The coordinates of that point are the loadings of $x_1$ on $F_1$ and $F_2$, respectively. The other four points represent the remaining four variables.

Since the factor loadings are the correlations between the standardized variables and the factors, the range of values shown on the axes of the factor diagram is $-1$ to $+1$. It can be seen from Figure 15.1 that $x_1$ and $x_2$ load more on $F_2$ than they do on $F_1$. Conversely, $x_3$, $x_4$, and $x_5$ load more on $F_1$ than on $F_2$. However, these distinctions are not very clear-cut, and the technique of factor rotation will produce clearer results (see Section 15.7).

An examination of the correlation matrix shown in Table 15.2 confirms that $x_1$ and $x_2$ form a block of correlated variables. Similarly, $x_4$ and $x_5$ form another block, with $x_3$ also correlated to them. These two major blocks are only weakly correlated with each other. (Note that the correlation matrix for the standardized $x$'s is the same as that for the unstandardized $X$'s.)

## 15.6 INITIAL FACTOR EXTRACTION: ITERATED PRINCIPAL COMPONENTS

The second method of extracting initial factors is a modification of the principal components analysis method. It has different names in different packages. It is called the principal factor analysis or PFA method in BMDP, PA2 or PAF in SPSS–X, and PRINIT in SAS.

To understand this method, you should recall that the communality is the part of the variance of each variable associated with the common factors. The principle underlying the *iterated solution* states that we should perform the factor analysis by using the communalities in place of the original variance. This principle entails substituting communality estimates for the 1's representing the variances of the standardized variables along the diagonal of the correlation matrix. With 1's in the diagonal we are factoring the total variance of the variables; with communalities in the diagonal we are factoring the variance associated with the common factors. Thus with communalities along the diagonal we select those common factors that maximize the total communality.

Many factor analysts consider maximizing the total communality a more attractive objective than maximizing the total proportion of the explained variance, as is done in the principal components method. The problem is that communalities are not known before the factor analysis is performed. Some initial estimates of the communalities must be obtained prior to the analysis. Various procedures exist, and we recommend, in the absence of a priori estimates, that the investigator use the default option in the particular program since the resulting factor solution is usually little affected by the initial communality estimates.

The steps performed by a packaged program in carrying out the *iterated factor extraction* are summarized as follows:

**1.** Find the initial communality estimates.

**2.** Substitute the communalities for the diagonal elements (1's) in the correlation matrix.

**3.** Extract $m$ principal components from the modified matrix.

**4.** Multiply the principal components coefficients by the standard deviation of the respective principal components to obtain factor loadings.

**5.** Compute new communalities from the computed factor loadings.

**6.** Replace the communalities in step 2 with these new communalities and repeat steps 3, 4, and 5. This step constitutes an *iteration*.

**7.** Continue iterating, stopping when the communalities stay essentially the same in the last two iterations.

For our hypothetical data example the results of using this method are shown in Table 15.4. In comparing this table with Table 15.3, we note that the total communality is higher for the principal components method (82.9% versus 74.6%). This result is generally the case since in the iterative method we are factoring the total communality, which is by necessity smaller than the total variance. The individual loadings do not

**TABLE 15.4.** Initial Factor Analysis Summary for Hypothetical Data Set from Iterated Principal Factor Extraction Method

| Variable | Factor Loadings | | Communality | Specificity |
| | $F_1$ | $F_2$ | $h_i^2$ | $u_i^2$ |
|---|---|---|---|---|
| $x_1$ | 0.470 | 0.734 | 0.759 | 0.241 |
| $x_2$ | 0.510 | 0.704 | 0.756 | 0.244 |
| $x_3$ | 0.481 | −0.258 | 0.298 | 0.702 |
| $x_4$ | 0.888 | −0.402 | 0.949 | 0.051 |
| $x_5$ | 0.956 | −0.233 | 0.968 | 0.032 |
| Variance explained | 2.413 | 1.317 | $\Sigma h_i^2 = 3.730$ | $\Sigma u_i^2 = 1.270$ |
| Percentage | 48.3% | 26.3% | 74.6% | 25.4% |

seem to be very different in the two methods; compare Figures 15.1 and 15.2. The only apparent difference in loadings is in the case of the third variable. Note that Figure 15.2, the factor diagram for the iterated method, is constructed in the same way as Figure 15.1.

We point out that the factor loadings extracted by the iterated method depend on the number of factors extracted. For example, the loadings for the first factor would depend on whether we extract two or three common factors. This condition does not exist for the uniterated principal components method. It should also be pointed out that it is possible to obtain negative variances and eigenvalues. This occurs because the 1's in the

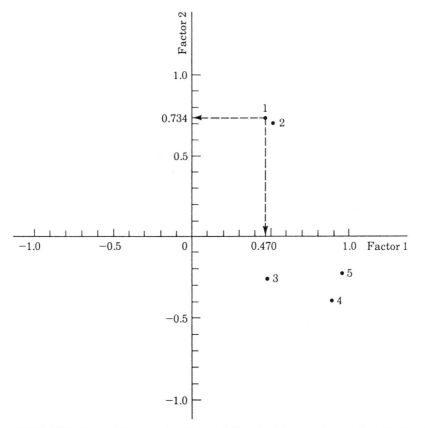

**FIGURE 15.2.** Factor Diagram for Iterated Principal Factor Extraction Method

diagonal of the correlation matrix have been replaced by the communality estimates which can be considerably less than one. This results in a matrix that does not necessarily have positive eigenvalues. These negative eigenvalues and the factors associated with them should not be used in the analysis.

Regardless of the choice of the extraction method, if the investigator does not have a preconceived number of factors ($m$) from knowledge of the subject matter, then several methods are available for choosing one numerically. Reviews of these methods are given in Harmon (1976), Gorsuch (1983), and Thorndike (1978). In Section 15.4 the commonly used technique of including any factor whose eigenvalue is greater than or equal to 1 was used. This criterion is based on theoretical rationales developed using true population correlation coefficients. It is commonly thought to yield about one factor to every three to five variables. It appears to correctly estimate the number of factors when the communalities are high and the number of variables is not too large. If the communalities are low, then the number of factors obtained with a cutoff equal to the average communality tends to be similar to that obtained by the principal components method with a cutoff of 1. As an alternative method, called the *scree method,* some investigators will simply plot the eigenvalues on the vertical axis versus the number of factors on the horizontal axis and look for the place where a change in the slope of the curve connecting successive points occurs. Examining the eigenvalues listed in Section 15.5, we see that compared with the first two eigenvalues (2.578 and 1.567), the values of the remaining eigenvalues are low (0.571, 0.241, and 0.043). This result indicates that $m = 2$ is a reasonable choice. The SAS FACTOR program will print a plot of the eigenvalues if the SCREE option is used.

In terms of initial factor extraction methods the iterated principal factor solution is the method employed most frequently by social scientists, the main users of factor analysis. For theoretical reasons many mathematical statisticians are more comfortable with the principal components method. Many other methods are available that may be preferred in particular situations (see Table 15.7 given later in the chapter). In particular, if the investigator is convinced that the factor model is valid and that the variables have a multivariate normal distribution, then the *maximum likelihood* (ML) *method* should be used. If these assumptions hold, then the ML procedure enables the investigator to perform certain tests of hypotheses or compute confidence intervals (see Lawley and Maxwell 1971).

In addition, the ML estimates of the factor loadings are invariant (do not change) with changes in scale of the original variables. Note that in Section 14.5, it was mentioned that for principal components analysis there was no easy way to go back and forth between the coefficients obtained from standardized and unstandardized data. The same is true for the principal components method of factor extraction. This invariance is therefore a major advantage of the ML extraction method. Finally, the ML procedure is the one used in confirmatory factor analysis (see Long 1983 or Dillon and Goldstein 1984). This procedure provides a method for testing causal hypotheses.

## *15.7 FACTOR ROTATIONS*

Recall that the main purpose of factor analysis is to derive from the data easily interpretable common factors. The initial factors, however, are often difficult to interpret. For the hypothetical data example we noted that the first factor is essentially an average of all variables. We also noted earlier that the factor interpretations in that example are not clear-cut. This situation is often the case in practice, regardless of the method used to extract the initial factors.

Fortunately, it is possible to find new factors whose loadings are easier to interpret. These new factors, called the *rotated factors*, are selected so that (ideally) some of the loadings are very large (near $\pm 1$) and the remaining loadings are very small (near zero). Conversely, we would ideally wish, for any given variable, that it have a high loading on only one factor. If this is the case, it is easy to give each factor an interpretation arising from the variables with which it is highly correlated (high loadings).

Theoretically, factor rotations can be done in an infinite number of ways. If you are interested in detailed descriptions of these methods, refer to some of the books listed in the Bibliography. In this section we highlight only two typical factor rotation techniques. The most commonly used technique is the *varimax rotation*. This method is the default option in most packaged programs and we present it first. The other technique, discussed later in this section, is called *oblique rotation*; we describe one commonly used oblique rotation, the *direct quartimin rotation*.

## Varimax Rotation

We have reproduced in Figure 15.3 the principal component factors that were shown in Figure 15.1 so that you can get an intuitive feeling for what factor rotation does. As shown in Figure 15.3, factor rotation consists of finding new axes to represent the factors. These new axes are selected so that they go through clusters or subgroups of the points representing the

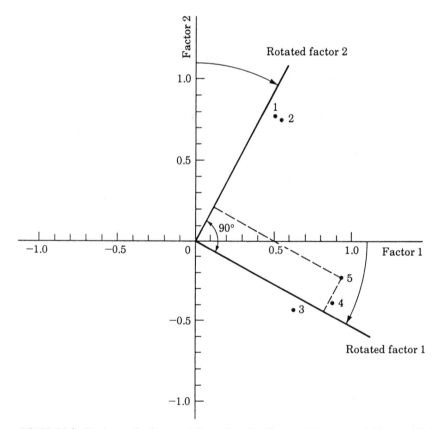

**FIGURE 15.3.** Varimax Orthogonal Rotation for Factor Diagram of Figure 15.1: Rotated Axes Go Through Clusters of Variables and Are Perpendicular to Each Other

response variables. The *varimax procedure* further restricts the new axes to being orthogonal (perpendicular) to each other. Figure 15.3 shows the results of the rotation. Note that the new axis representing rotated factor 1 goes through the cluster of variables $x_3$, $x_4$, and $x_5$. The axis for rotated factor 2 is close to the cluster of variables $x_1$ and $x_2$ but cannot go through them since it is restricted to being orthogonal to factor 1.

The result of this varimax rotation is that variables $x_1$ and $x_2$ have high loadings on rotated factor 2 and nearly zero loadings on rotated factor 1. Similarly, $x_3$, $x_4$, and $x_5$ load heavily on rotated factor 1 but not on rotated factor 2. Figure 15.3 clearly shows that the rotated factors are orthogonal since the angle between the axes representing the rotated factors is $90°$. In statistical terms the orthogonality of the rotated factors is equivalent to the fact that they are uncorrelated with each other.

Computationally, the *varimax rotation* is achieved by maximizing the sum of the variances of the squared factor loadings within each factor. Further, these factor loadings are adjusted by dividing each of them by the communality of the corresponding variable. This adjustment is known as the *Kaiser normalization* (see Harmon 1967 or Afifi and Azen 1979). This adjustment tends to equalize the impact of variables with varying communalities. If it were not used, the variables with higher communalities would highly influence the final solution.

Any method of rotation may be applied to any initially extracted factors. For example, in Figure 15.4 we show the varimax rotation of the factors initially extracted by the iterated principal factor method (Figure 15.2). Note the similarity of the graphs shown in Figures 15.3 and 15.4. This similarity is further reinforced by examination of the two sets of rotated factor loadings shown in Tables 15.5 and 15.6. In this example both methods of initial extraction produced a rotated factor 1 associated mainly with $x_3$, $x_4$, and $x_5$, and a rotated factor 2 associated with $x_1$ and $x_2$. The points in Figures 15.3 and 15.4 are plots of the rotated loadings (Tables 15.5 and 15.6) with respect to the rotated axes.

In comparing Tables 15.5 and 15.6 with Tables 15.3 and 15.4, we note that the *communalities are unchanged* after the varimax rotation. This result is always the case for any orthogonal rotation. Note also that the percentage of the variance explained by the rotated factor 1 is less than that explained by unrotated factor 1. However, the cumulative percentage of the variance explained by all common factors remains the same after orthogonal rotation. Furthermore, the loadings of any rotated factor

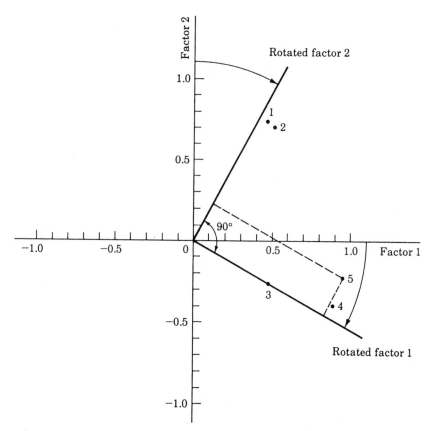

**FIGURE 15.4.** Varimax Orthogonal Rotation for Factor Diagram of Figure 15.2

**TABLE 15.5.** Varimax Rotated Factors: *Principal Components Extraction*

| Variable | Factor Loadings* | | Communality |
| | $F_1$ | $F_2$ | $h_i^2$ |
| --- | --- | --- | --- |
| $x_1$ | 0.055 | 0.933 | 0.873 |
| $x_2$ | 0.105 | 0.929 | 0.875 |
| $x_3$ | 0.763 | −0.062 | 0.586 |
| $x_4$ | 0.943 | 0.095 | 0.898 |
| $x_5$ | 0.918 | 0.266 | 0.913 |
| Variance explained | 2.328 | 1.817 | 4.145 |
| Percentage | 46.6% | 36.3% | 82.9% |

* Vertical lines indicate large loadings.

**TABLE 15.6.** Varimax Rotated Factors: *Iterated Principal Factors Extraction*

| Variable | Factor Loadings* $F_1$ | $F_2$ | Communality $h_i^2$ |
|---|---|---|---|
| $x_1$ | 0.063 | 0.869 ǀ | 0.759 |
| $x_2$ | 0.112 | 0.862 ǀ | 0.756 |
| $x_3$ | 0.546 ǀ | 0.003 | 0.298 |
| $x_4$ | 0.972 ǀ | 0.070 | 0.949 |
| $x_5$ | 0.951 ǀ | 0.251 | 0.968 |
| Variance explained | 2.164 | 1.566 | 3.730 |
| Percentage | 43.3% | 31.3% | 74.6% |

* Vertical lines indicate large loadings.

depend on how many other factors are selected, regardless of the method of initial extraction.

## Oblique Rotation

Some factor analysts are willing to relax the restriction of orthogonality of the rotated factors, thus permitting a further degree of flexibility. Nonorthogonal rotations are called *oblique rotations*. The origin of the term *oblique* lies in geometry, whereby two crossing lines are called oblique if they are not perpendicular to each other. Oblique rotated factors are correlated with each other, and in some applications it may be harmless or even desirable to have correlated common factors (see Mulaik 1972).

A commonly used oblique rotation is called the *direct quartimin procedure*. The results of applying this method to our hypothetical data example are shown in Figures 15.5 and 15.6. Note that the rotated factors go neatly through the centers of the two variable clusters. The angle between the rotated factors is not 90°. In fact, the cosine of the angle between the two factor axes is the sample correlation between them. For example, for the rotated principal components the correlation between rotated factor 1 and rotated factor 2 is 0.165, which is the cosine of 80.5°.

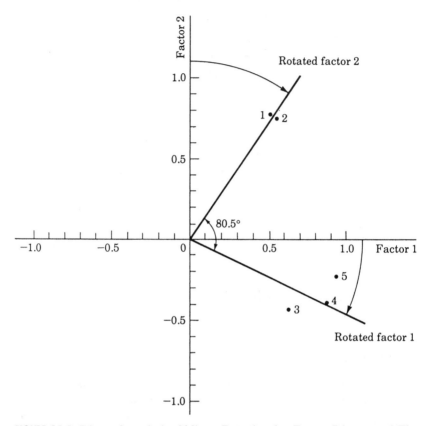

**FIGURE 15.5.** Direct Quartimin Oblique Rotation for Factor Diagram of Figure 15.1: Rotated Axes Go Through Clusters of Variables But Are Not Required to be Perpendicular to Each Other

For oblique rotations, the reported pattern and structure matrices are somewhat different. The structure matrix gives the correlations between $x_i$ and $F_j$. Also, the statements made concerning the proportion of the total variance explained by the communalities being unchanged by rotation do not hold for oblique rotations.

In general, a factor analysis computer program will print the correlations between rotated factors if an oblique rotation is performed. The rotated oblique factor loadings also depend on how many factors are selected. Finally, we note that packaged programs generally offer a choice of orthogonal and oblique rotations, as discussed in Section 15.10.

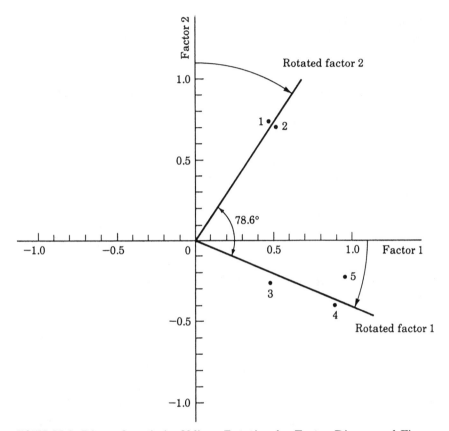

**FIGURE 15.6.** Direct Quartimin Oblique Rotation for Factor Diagram of Figure 15.2

## 15.8 ASSIGNING FACTOR SCORES TO INDIVIDUALS

Once the initial extraction of factors and the factor rotations are performed, it may be of interest to obtain the score an individual has for each factor. For example, if two factors, verbal and quantitative abilities, were derived from a set of test scores, it would be desirable to determine equations for computing an individual's scores on these two factors from a set of test scores. Such equations are linear functions of the original variables.

Theoretically, it is conceivable to construct factor score equations in an infinite number of ways (see Gorsuch 1983). But perhaps the simplest way is to add the values of the variables loading heavily on a given factor. For example, for the hypothetical data discussed earlier we would obtain the score of rotated factor 1 as $x_3 + x_4 + x_5$ for a given individual. Similarly, the score of rotated factor 2 in that example would be $x_1 + x_2$. In effect, the factor analysis identifies $x_3$, $x_4$, and $x_5$ as a subgroup of intercorrelated variables. One way of combining the information conveyed by the three variables is simply to add them up. A similar approach is used for $x_1$ and $x_2$. In some applications such a simple approach may be sufficient.

More sophisticated approaches do exist for computing factor scores. The three major computer packages include the so-called *regression procedure*. This method combines the intercorrelations among the $x_i$ variables and the factor loadings to produce quantities called *factor score coefficients*. These coefficients are used in a linear fashion to combine the values of the *standardized* $x_i$'s into factor scores. For example, in the hypothetical data example using principal component factors rotated by the varimax method (see Table 15.5), the scores for rotated factor 1 are obtained as

$$\text{factor score } 1 = -0.076x_1 - 0.053x_2 + 0.350x_3 + 0.414x_4 + 0.384x_5$$

The large factor score coefficients for $x_3$, $x_4$, and $x_5$ correspond to the large factor *loadings* shown in Table 15.5. Note also that the coefficients of $x_1$ and $x_2$ are close to zero and those of $x_3$, $x_4$, and $x_5$ are each approximately 0.4. Factor score 1 can therefore be approximated by $0.4x_3 + 0.4x_4 + 0.4x_5 = 0.4(x_3 + x_4 + x_5)$, which is proportional to the simple additive factor score given in the previous paragraph.

The factor scores can themselves be used as data for additional analyses. The major packages facilitate this technique by offering the user the option of storing the factor scores in a file to be used for subsequent analyses.

## 15.9 AN APPLICATION OF FACTOR ANALYSIS TO THE DEPRESSION DATA

In Chapter 14 we presented the results of a principal components analysis for the 20 CESD items in the depression data set. Table 14.2 listed the

**TABLE 15.7 Varimax Rotation, Principal Component Factors for Standardized CESD.** Scale Items (Depression Data Set)

| Item | Factor Loadings | | | | Communalities |
| --- | --- | --- | --- | --- | --- |
| | $F_1$ | $F_2$ | $F_3$ | $F_4$ | $h_i^2$ |
| **Negative Affect** | | | | | |
| 1. I felt that I could not shake off the blues even with the help of my family or friends. | 0.638 | 0.146 | 0.268 | 0.280 | 0.5784 |
| 2. I felt depressed. | 0.773 | 0.296 | 0.272 | -0.003 | 0.7598 |
| 3. I felt lonely. | 0.726 | 0.054 | 0.275 | 0.052 | 0.6082 |
| 4. I had crying spells. | 0.630 | -0.061 | 0.168 | 0.430 | 0.6141 |
| 5. I felt sad. | 0.797 | 0.172 | 0.160 | 0.016 | 0.6907 |
| 6. I felt fearful. | 0.624 | 0.234 | -0.018 | 0.031 | 0.4448 |
| 7. I thought my life had been a failure. | 0.592 | 0.157 | 0.359 | 0.337 | 0.6173 |
| **Positive Affect** | | | | | |
| 8. I felt that I was as good as other people. | 0.093 | -0.051 | 0.109 | 0.737 | 0.5655 |
| 9. I felt hopeful about the future. | 0.238 | 0.033 | 0.621 | 0.105 | 0.4540 |
| 10. I was happy. | 0.557 | 0.253 | 0.378 | 0.184 | 0.5516 |
| 11. I enjoyed life. | 0.498 | 0.147 | 0.407 | 0.146 | 0.4569 |
| **Somatic and Retarded activity** | | | | | |
| 12. I was bothered by things that usually don't bother me. | 0.449 | 0.389 | -0.049 | -0.065 | 0.3600 |
| 13. I did not feel like eating; my appetite was poor. | 0.070 | 0.504 | -0.173 | 0.535 | 0.5760 |
| 14. I felt that everything was an effort. | 0.117 | 0.695 | 0.180 | 0.127 | 0.5459 |
| 15. My sleep was restless. | 0.491 | 0.419 | -0.123 | -0.089 | 0.4396 |
| 16. I could not "get going." | 0.196 | 0.672 | 0.263 | -0.070 | 0.5646 |
| 17. I had trouble keeping my mind on what I was doing. | 0.270 | 0.664 | 0.192 | 0.000 | 0.5508 |
| 18. I talked less than usual. | 0.409 | 0.212 | -0.026 | 0.223 | 0.2628 |
| **Interpersonal** | | | | | |
| 19. People were unfriendly. | -0.015 | 0.237 | 0.746 | -0.088 | 0.6202 |
| 20. I felt that people disliked me. | 0.358 | 0.091 | 0.506 | 0.429 | 0.5770 |
| Variance explained | 4.795 | 2.381 | 2.111 | 1.551 | 10.838 |
| Percentage | 24.0% | 11.9% | 10.6% | 7.8% | 54.2% |

coefficients for the first five principal components. However, to be consistent with published literature on this subject, we now choose $m = 4$ factors (not 5) and proceed to perform an orthogonal varimax rotation using the principal components as the method of factor extraction. The results for the rotated factors are given in Table 15.7.

In comparing these results with those in Table 14.1, we note that, as expected, rotated factor 1 explains less of the total variance than does the first principal component. The four rotated factors together explain the same proportion of the total variance as do the first four principal components (54.2%). This result occurs because the unrotated factors were the principal components.

In interpreting the factors, we see that factor 1 loads heavily on variables 1 through 7. These items are known as negative-affect items. Factor 2 loads mainly on items 12 through 18, which measure somatic and retarded activity. Factor 3 loads on the two interpersonal items, 19 and 20, as well as some positive-affect items (9 and 11). Factor 4 does not represent a clear pattern. Its highest loading is associated with the positive-affect item 8. The factors can thus be loosely identified as negative affect, somatic and retarded activity, interpersonal relations, and positive affect, respectively.

Overall, these rotated factors are much easier to interpret than the original principal components. This example is an illustration of the usefulness of factor analysis in identifying important interrelations among measured variables. Further results and discussion regarding the application of factor analysis to the CESD scale may be found in Radloff (1977) and Clark et al. (1981).

## 15.10 DISCUSSION OF COMPUTER PROGRAMS

In this section we present the main features of the three packaged programs, and we illustrate their use with an example.

### Features of Packaged Programs

Each of the three packages has a major factor analysis program. The three programs offer a large variety of options, not all the same. Table 15.8

**TABLE 15.8.** Summary of Computer Output for Factor Analysis

| Output | BMDP | SAS | SPSS–X |
|---|---|---|---|
| **Matrix Analyzed** | | | |
| Correlation | 4M | FACTOR | FACTOR |
| Covariance | 4M | FACTOR | |
| Correlation about origin | 4M | FACTOR | |
| Covariance about origin | 4M | FACTOR | |
| **Method of Initial Extraction** | | | |
| Principal component | 4M | FACTOR | FACTOR |
| Iterative principal factor | 4M | FACTOR | FACTOR |
| Maximum likelihood | 4M | FACTOR | FACTOR |
| Alpha | | FACTOR | FACTOR |
| Image | | FACTOR | FACTOR |
| Harris | | FACTOR | |
| Least squares | | | FACTOR |
| Kaiser little jiffy | 4M | | |
| **Communality Estimates** | | | |
| Unaltered diagonal | 4M | FACTOR | FACTOR |
| Multiple correlation squared | 4M | FACTOR | FACTOR |
| Maximum correlation in row | 4M | FACTOR | |
| Estimates, user-supplied | 4M | FACTOR | FACTOR |
| **Number of Factors** | | | |
| Specify number | 4M | FACTOR | FACTOR |
| Specify minimum eigenvalue | 4M | FACTOR | FACTOR |
| Cumulative eigenvalues | | FACTOR | |
| **Orthogonal Rotations** | | | |
| Varimax | 4M | FACTOR | FACTOR |
| Equamax | 4M | FACTOR | FACTOR |
| Quartimax | 4M | FACTOR | FACTOR |
| Orthomax with gamma | | FACTOR | |
| Orthogonal with gamma | 4M | | |
| Promax | | FACTOR | |
| **Oblique Rotations** | | | |
| Direct quartimin | 4M | | |
| Promax | | FACTOR | |
| With gamma | 4M | | |
| Orthogonal oblique | 4M | | |
| Procrustes | | FACTOR | |
| Direct oblimin | | | FACTOR |
| Harris–Kaiser | | FACTOR | |
| **Factor Scores** | | | |
| Factor scores for individuals | 4M | FACTOR | FACTOR |
| Factor score coefficients | 4M | FACTOR | FACTOR |
| **Other Output** | | | |
| Inverse correlation matrix | 4M | FACTOR | FACTOR |
| Factor structure matrix | 4M | FACTOR | |
| Plots of factor loadings | 4M | FACTOR | |
| Scree | | FACTOR | FACTOR |
| **Tests of Hypotheses** | | | |
| Factors sufficient to explain Correlation (if initial factor method is ML) | | FACTOR | |
| Mahalanobis distances divided by df (to detect outliers) | 4M | | |

summarizes the most important options offered by each package, including some not discussed in the text here. If you are interested in those extra options, consult the corresponding package manual and the references listed in the Bibliography.

Each program can factor-analyze the correlation matrix, while BMDP4M and SAS FACTOR can analyze other matrices. The three packages agree on two methods of initial factor extraction, with each program offering other methods. In estimating the diagonal elements (i.e., communalities), each program offers the option of leaving the diagonal elements unaltered or accepting user-supplied estimates. The BMDP4M program and SAS FACTOR allow estimation of them by other methods. Each of the programs allows the user to specify the number of factors or specify that a factor should be selected only if its corresponding eigenvalue exceeds a certain value. In addition, SAS FACTOR allows selection of the number of factors to explain a certain cumulative proportion of the total variance if the PROPORTION option is used.

For orthogonal rotation each program allows the use of varimax, equimax, and quartimax. The latter two rotation methods are discussed in Gorsuch (1983). For oblique rotations each program has its own methods. In addition, BMDP4M offers two other rotation methods, one orthogonal and one that combines orthogonal and oblique ideas.

Finally, each of the programs can produce factor scores to be used in future analysis and can print, upon request, equations for computing these scores.

### Example

To illustrate some typical output for factor analysis, we used the BMDP4M factor analysis program to produce the results shown in Tables 15.3 and 15.5. For data we generated normal random deviates. The data were generated within the computer by the BMDP TRANSFORM paragraph. The following paragraphs were used:

```
/INPUT        VAR = 0. CASES = 100.
/VARIABLE     ADD = 5.
              NAMES = NORM1,NORM2,NORM3,NORM4,NORM5.
```

```
/TRANSFORM     NORM1 = RNDG(46791).
               NORM2 = RNDG(56705).
               NORM3 = RNDG(66717).
               NORM4 = RNDG(77729).
               NORM5 = RNDG(37255).
               NORM2 = NORM1 + NORM2.
               NORM4 = NORM3 + 2*NORM4 + 0.3*NORM2.
               NORM5 = NORM4 + 0.7*NORM5 + 0.3*NORM1.
/PRINT         FSFC.
/PLOT          INITIAL = 2.
/END
```

In the INPUT paragraph we are reading data on 100 cases but no variables. In effect, we are reserving space for 100 cases in the computer. In the VARIABLE paragraph we add five variables and give them names.

In the TRANSFORM paragraph we instruct the computer to create five new variables by using the built-in normal random generator. Each random number requires a five-digit number to start it off that must end in an odd digit. The other three statements modify these created variables to produce variables with the desired correlation structure. Otherwise, the five variables would be mutually independent and no further factor analysis would be necessary. No more new variables have to be added to the VARIABLE paragraph because the same names are used again. The variable is simply redefined.

If the investigator does not generate the data, the only paragraphs that are essential in order to run BMDP4M are the INPUT paragraph and the END paragraph. With the default options the initial factor extraction is done by the method of principal components and a varimax rotation is performed. The investigator is urged to use these default options for the first run in order to save time and to obtain some output that can be examined as an aid in planning other runs.

The output from the run that is necessary to interpret the factor analysis is described next. The program prints the first five cases so that the investigator can check that it is reading the proper data. The mean, standard deviation, smallest and largest values, and smallest and largest standardized scores are printed for each variable. If the data are normally distributed or symmetric, the absolute value of the smallest and largest standardized scores (each observation minus its mean and then divided by its standard deviation, done separately for each variable) should be approximately equal. Absolute value of standardized data that are greater than 4 are quite unusual if the data are normally distributed. Note that it is assumed that the data are normally distributed if tests of hypotheses are to be made or if the maximum likelihood method is to be used.

Next on the computer printout are the correlation matrix given in Table 15.2, the squared multiple correlation (SMC) of each variable with the remaining variables, and the estimated communalities listed in Table 15.3. If any variable has an extremely high squared multiple correlation with all other variables (greater than 0.9999), this variable can probably be expressed as an almost exact linear function of some of the other variables, and the investigator should consider removing it or some other variable. The variance explained is given as listed in Section 15.5, and the cumulative proportion of the total variance is also listed (variance divided by $P$ and cumulated). The unrotated factor loadings (Table 15.3 and Figure 15.1) are presented, followed by the rotated factor loadings as given in Figure 15.3.

In our example, we asked for factor score coefficients (FSFC) to be printed in the PRINT paragraph. These coefficients were presented for the first factor in Section 15.8 to illustrate that they were not equal to the factor loadings.

In the plot paragraph we requested a plot of the initial or unrotated factors, which was presented in Figure 15.1. The rotated factors, plotted as a default option, were displayed in Figure 15.3.

The two factor scores are printed out for each of the 100 cases. For example, for the first case and the first factor, the factor score is

$$-0.488 = -0.076x_1 - 0.053x_2 + 0.350x_3 + 0.414x_4 + 0.384x_5$$

For the first case and the first standardized variable,

$$x_1 = \frac{0.666 - 0.163}{1.047}$$

where 0.666 is the value of $x_1$ for the first case, 0.163 is the mean for the first variable, and 1.047 is the standard deviation for the first variable (see Table 15.1). The values of $x_2, x_3, x_4$, and $x_5$ are obtained in a similar fashion and then substituted to produce factor scores. These factor scores could be employed in subsequent analyses by using the SAVE paragraph with the default option.

The program also prints out three Mahalanobis distances divided by their degrees of freedom. These distances measure the following:

1. The distance from each case to the mean of all the cases.
2. The distance from each factor score to the mean of the factor scores.
3. The differences between these two distances adjusted for the appropriate degrees of freedom.

They are distributed approximately according to a chi-square distribution divided by its degrees of freedom if the original distribution is multivariate normal. The investigator may wish to examine values larger than 12.1 for 1 df, 7.6 for 2 df, 5.9 for 3 df, 5.0 for 4 df, 4.4 for 5 df, 4.0 for 6 df, 3.7 for 7 df, 3.5 for 8 df, and 3.1 for 10 df. These correspond to the 99.95 percentile of the chi-square distribution after dividing by the degrees of freedom. Doing so is useful in detecting multivariate outliers.

In the generated data used in this example, the largest distance from each case to the mean of all cases is 3.926 for case 85. The output indicates that this value should be compared with a chi-square divided by degrees of freedom with 5 degrees of freedom. It happens that case 85 has values of $x_1 = -1.304$, $x_2 = -0.949$, $x_3 = 0.689$, $x_4 = 2.457$, and $x_5 = 4.521$, none of which are the most extreme values when the variables are examined one at a time. The distance for the factor scores is only 1.911 for this case, so it does not result in an unusual factor score.

The simplest way of checking for outliers among the factor scores is to examine the plot of the factor scores. Figure 15.7 is taken directly from the BMDP4M output and illustrates such a plot. If an unusual point has been found, the listing of the factor scores can be used to determine which case it is. Note that the factor score plot in Figure 15.7 is typical of a bivariate normal plot with zero correlation; this data set was generated by using random normal deviates. Factor 1 scores range in value from $-1.98$ to $2.60$, and factor 2 scores from $-2.27$ to $2.21$.

To obtain Tables 15.4 and 15.6 where the iterated principal factor extraction method is used instead of the default option, the following FACTOR paragraph is added after the TRANSFORM paragraph:

/FACTOR      METHOD = PFA.

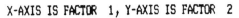

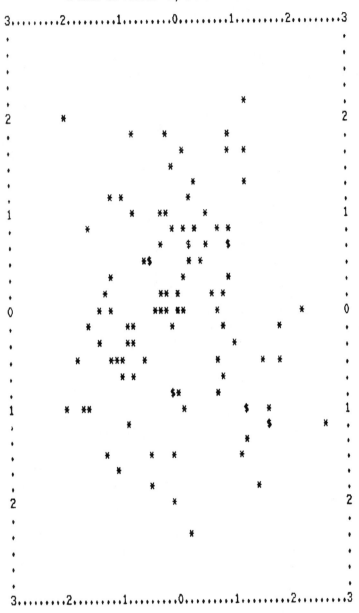

**FIGURE 15.7.** Plot of Factor Score 1 Versus Factor Score 2 for the Hypothetical Data Set

To obtain an oblique rotation such as the direct quartimin procedure, a ROTATE paragraph can be added, as follows:

```
/ROTATE      METHOD = DQUART.
```

There is one important difference in the default options of SAS and BMDP when the iterative principal factor method of factor extraction is used. If the user does not indicate how many factors to retain, the BMDP default cutoff point for the size of the variance explained (or eigenvalue) is one. The default cutoff in SAS is the total common variance, to be explained by the factors retained using the prior communality estimates, divided by the number of variables. This could be considerably less than one if the communality part of the total variance is not large. Hence, different numbers of factors may be retained by SAS and BMDP when the default options are used.

SAS has a large number of possible options to choose for factor analysis. One choice that is attractive, if an investigator is uncertain whether to use an orthogonal or oblique rotation, is to specify the promax rotation method. With this option a varimax rotation is obtained first, followed by an oblique rotation, so the two methods can be compared in a single run.

## 15.11 WHAT TO WATCH OUT FOR

In performing factor analysis, a linear model is assumed where each variable is seen as a linear function of common factors plus a unique factor. The method tends to yield more understandable results when at least moderate correlations exist among the variables. Also, since a simple correlation coefficient measures the full correlation between two variables when a linear relationship exists between them, factor analysis will fit the data better if only linear relationships exist among the variables. Specific points to look out for include:

1. The original sample should reflect the composition of the target population. Outliers should be screened, and linearity among the variables checked. It may be necessary to use transformations to linearize the relationships among the variables.

2. Do not accept the number of factors produced by the default options without checking that it makes sense. The results can change

drastically depending on the numbers of factors used. Note that the choice among different numbers of factors depends mainly on which is most interpretable to the investigator. The cutoff points for the choice of the number of factors do not take into account the variability of the results and can be considered to be rather arbitrary. It is advised that several different numbers of factors be tried, especially if the first run yields results that are unexpected.

**3.** Ideally, factor analyses should have at least two variables with non-zero weights per factor. If each factor has only a single variable, it can be considered a unique factor and one might as well be analyzing the original variables instead of the factors.

**4.** If the investigator expects the factors to be correlated (as might be the case in a factor analysis of items in a scale), then oblique factor analysis should be tried.

**5.** The ML method assumes multivariate normality. It also can take more computer time than many of the other methods. It is the method to use if statistical tests are desired.

**6.** Usually, the results of a factor analysis are evaluated by how much sense they make to the investigator rather than by use of formal statistical tests of hypotheses. Gorsuch (1983) and Dillon and Goldstein (1982) present comparisons among the various methods of extraction and rotation that can help the reader in evaluating the results.

**7.** As mentioned earlier, the factor analysis discussed in this chapter is an exploratory technique that can be an aid to analytical thinking. Statisticians have criticized factor analysis because it has sometimes been used by some investigators in place of sound theoretical arguments. However, when carefully used, for example to motivate or corroborate a theory rather than replace it, factor analysis can be a useful tool to be employed in conjunction with other statistical methods.

## SUMMARY

In this chapter we presented the most essential features of factor analysis, a technique heavily used in social science research. The major phases of a factor analysis are factor extraction, factor rotation, and factor score computation. We gave examples of its use for the depression data set.

A fact that should be remembered by investigators is that factor analysis is, in reality, an exploratory technique. Its main usefulness is to give the investigator an impression of the interrelationships present in the data and offer a preliminary method for summarizing them. Factor analysis often suggests certain hypotheses to be further examined in future research. On the one hand, the investigator should not hesitate to replace the results of any factor analysis by scientifically based theories derived at a later time. On the other hand, in the absence of such theory, factor analysis is a convenient and useful tool for searching for relationships among variables.

Since factor analysis is an exploratory technique, there does not exist an optimum way of performing it. We advise the investigator to try various combinations of extraction and rotation of main factors. Subjective judgment is then needed to select those results that seem most appealing, on the basis of the investigator's own knowledge of the underlying subject matter.

## BIBLIOGRAPHY

Afifi, A. A., and Azen, S. P. 1979. *Statistical analysis: A computer oriented approach.* 2nd ed. New York: Academic Press.

*Bartholomew, D. J. 1987. *Latent variable models and factor analysis.* London: Charles Griffin and Company.

Clark, V. A., Aneshensel, C. S., Frerichs, R. R., and Morgan, T. M. 1981. Analysis of effects of sex and age in response to items on the CES-D scale. *Psychiatry Research* 5:171–181.

Dillon, W. R., and Goldstein, M. 1984. *Multivariate analysis: Methods and applications.* New York: Wiley.

Everitt, B. S. 1984. *An introduction to latent variable models.* London: Chapman and Hall.

*Gorsuch, R. L. 1983. *Factor analysis.* 2nd ed. Hillsdale, N.J.: L. Erlbaum Associates.

*Harmon, H. H. 1976. *Modern factor analysis.* Chicago: University of Chicago Press.

Kim, J. O., and Mueller, C. W. 1978a. *Introduction to factor analysis.* Beverly Hills: Sage.

———. 1978b. *Factor analysis.* Beverly Hills: Sage.

*Lawley, D. N., and Maxwell, A. E. 1971. *Factor analysis as a statistical method.* 2nd ed. New York: Elsevier.

Long, J. S. 1983. *Confirmatory factor analysis.* Newbury Park, Calif.: Sage.

Long, J. S. 1983. *Covariance structural models: An introduction to LISREL.* Newbury Park, Calif.: Sage.

Mulaik, S. A. 1972. *The foundations of factor analysis.* New York: McGraw-Hill.

Radloff, L. S. 1977. The CES-D scale: A self-report depression scale for research in the general population. *Applied Psychological Measurement* 1:385–401.

Thorndike, R. M. 1978. *Correlational procedures for research.* New York: Gardner Press.

## PROBLEMS

15.1 The CESD scale items (C1–C20) from the depression data set in Chapter 3 were used to obtain the factor loadings listed in Table 15.7. The initial factor solution was obtained from the principal components method, and a varimax rotation was performed. Analyze this same data set by using an oblique rotation such as the direct quartimin procedure. Compare the results.

15.2 Repeat the analysis of Problem 15.1 and Table 15.7, but use an iterated principal factor solution instead of the principal components method. Compare the results.

15.3 Using the sentences given in the TRANSFORM paragraph in Section 15.10, which were used to generate the hypothetical data set, explain why the correlations given in Table 15.2 were obtained.

15.4 Use a BMDP program with the same TRANSFORM paragraph (including the same numbers for each random normal deviate generated) given in Section 15.10. Did the same data result. Run BMDP4M with these data but use the maximum likelihood method of extracting factors. Examine the output to determine whether this method leads to different results.

15.5 For the data generated in Problem 7.7, perform four factor analyses, using two different initial extraction methods and both orthogonal and oblique rotations. Interpret the results.

15.6 Separate the depression data set into two subgroups, men and women. Using four factors, repeat the factor analysis in Table 15.7. Compare the results of your two factor analyses to each other and to the results in Table 15.7.

15.7 For the depression data set, perform four factor analyses on the last seven variables (DRINK through CHRONILL); see Table 3.2. Use two different initial extraction methods, and both orthogonal and oblique rotations. Interpret the results. Compare the results to those from Problem 14.1.

# Chapter Sixteen

# CLUSTER ANALYSIS

## 16.1 WHAT WILL YOU LEARN FROM THIS CHAPTER?

From this chapter you will learn:

- What cluster analysis is (16.2).
- About distance measures used in cluster analysis (16.4).
- When hierarchical clustering is appropriate (16.5).
- When $K$-means clustering is appropriate (16.5).
- About simple graphical descriptions of data patterns that are useful in interpreting clusters (16.3, 16.4, 16.6).
- How to choose the appropriate computer program (16.7).
- What to watch out for in cluster analysis (16.8).

## 16.2 WHEN IS CLUSTER ANALYSIS USED?

*Cluster analysis* is a technique for grouping individuals or objects into *unknown* groups. It differs from other methods of classification, such as discriminant analysis, in that in cluster analysis the number and characteristics of the groups are to be derived from the data and are not usually known prior to the analysis.

In biology, cluster analysis has been used for decades in the area of taxonomy. In taxonomy, living things are classified into arbitrary groups on the basis of their characteristics. The classification proceeds from the most general to the most specific, in steps. For example, classifications for domestic roses and for modern humans are illustrated in Figure 16.1 (see Wilson et al. 1973). The most general classification is kingdom, followed by phylum, subphylum, etc. The use of cluster analysis in taxonomy is explained by Sneath and Sokal (1973).

Cluster analysis has been used in medicine to assign patients to specific diagnostic categories on the basis of their presenting symptoms and signs. In particular, cluster analysis has been used in classifying types of

A.  Modern Humans

      KINGDOM:      Animalia (animals)
        PHYLUM:      Chordata (chordates)
          SUBPHYLUM:      Vertebrata (vertebrates)
            CLASS:      Mammalia (mammals)
            ORDER:      Primates (primates)
              FAMILY:      Hominidae (humans and close relatives)
                GENUS:      Homo (modern humans and precursors)
              SPECIES:      sapiens (modern humans)

B.  Domestic Rose

      KINGDOM:      Plantae (plants)
        PHYLUM:      Tracheophyta (vascular plants)
          SUBPHYLUM:      Pteropsida (ferns and seed plants)
            CLASS:      Dicotyledoneae (dicots)
            ORDER:      Rosales (saxifrages, psittosporums, sweet gum, plane trees, roses, and relatives)
              FAMILY:      Rosaceae (cherry, plum, hawthorn, roses, and relatives)
                GENUS:      Rosa (roses)
              SPECIES:      galliea (domestic roses)

**FIGURE 16.1.** Example of Taxonomic Classification

depression (see, e.g., Andreason and Grove 1982). It has also been used in anthropology to classify stone tools, shards, or fossil remains by the civilization that produced them. Consumers can be clustered on the basis of their choice of purchases in marketing research. In short, it is possible to find applications of cluster analysis in virtually any field of research.

We point out that cluster analysis is highly empirical. Different methods can lead to very different groupings, both in number and in content. Furthermore, since the groups are not known a priori, it is usually difficult to judge whether the results make sense in the context of the problem being studied. We note also that programs exist for clustering variables. However, we will discuss in detail only clustering cases or observations.

## 16.3 DATA EXAMPLES

A hypothetical data set was created to illustrate several of the concepts discussed in this chapter. Figure 16.2 shows a plot of five observations for

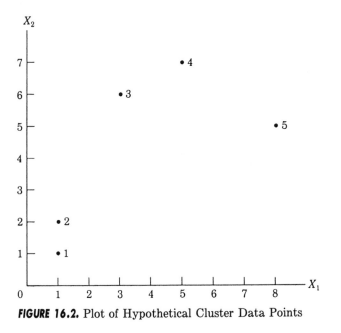

**FIGURE 16.2.** Plot of Hypothetical Cluster Data Points

**TABLE 16.1.** Financial Performance Data for Diversified Chemical, Health, and Supermarket Companies

| Type | Symbol | Observation Number | ROR5 | D/E | SALESGR5 | EPS5 | NPM1 | P/E | PAYOUTR1 |
|---|---|---|---|---|---|---|---|---|---|
| Chem | dia | 1 | 13.0 | 0.7 | 20.2 | 15.5 | 7.2 | 9 | 0.426398 |
| Chem | dow | 2 | 13.0 | 0.7 | 17.2 | 12.7 | 7.3 | 8 | 0.380693 |
| Chem | stf | 3 | 13.0 | 0.4 | 14.5 | 15.1 | 7.9 | 8 | 0.406780 |
| Chem | dd | 4 | 12.2 | 0.2 | 12.9 | 11.1 | 5.4 | 9 | 0.568182 |
| Chem | uk | 5 | 10.0 | 0.4 | 13.6 | 8.0 | 6.7 | 5 | 0.324544 |
| Chem | psm | 6 | 9.8 | 0.5 | 12.1 | 14.5 | 3.8 | 6 | 0.508083 |
| Chem | gra | 7 | 9.9 | 0.5 | 10.2 | 7.0 | 4.8 | 10 | 0.378913 |
| Chem | hpc | 8 | 10.3 | 0.3 | 11.4 | 8.7 | 4.5 | 9 | 0.481928 |
| Chem | mtc | 9 | 9.5 | 0.4 | 13.5 | 5.9 | 3.5 | 11 | 0.573248 |
| Chem | acy | 10 | 9.9 | 0.4 | 12.1 | 4.2 | 4.6 | 9 | 0.490798 |
| Chem | cz | 11 | 7.9 | 0.4 | 10.8 | 16.0 | 3.4 | 7 | 0.489130 |
| Chem | ald | 12 | 7.3 | 0.6 | 15.4 | 4.9 | 5.1 | 7 | 0.272277 |
| Chem | rom | 13 | 7.8 | 0.4 | 11.0 | 3.0 | 5.6 | 7 | 0.315646 |
| Chem | rci | 14 | 6.5 | 0.4 | 18.7 | -3.1 | 1.3 | 10 | 0.384000 |
| Heal | hum | 15 | 9.2 | 2.7 | 39.8 | 34.4 | 5.8 | 21 | 0.390879 |
| Heal | hca | 16 | 8.9 | 0.9 | 27.8 | 23.5 | 6.7 | 22 | 0.161290 |
| Heal | nme | 17 | 8.4 | 1.2 | 38.7 | 24.6 | 4.9 | 19 | 0.303030 |
| Heal | ami | 18 | 9.0 | 1.1 | 22.1 | 21.9 | 6.0 | 19 | 0.303318 |
| Heal | ahs | 19 | 12.9 | 0.3 | 16.0 | 16.2 | 5.7 | 14 | 0.287500 |
| Groc | lks | 20 | 15.2 | 0.7 | 15.3 | 11.6 | 1.5 | 8 | 0.598930 |
| Groc | win | 21 | 18.4 | 0.2 | 15.0 | 11.6 | 1.6 | 9 | 0.578313 |
| Groc | sgl | 22 | 9.9 | 1.6 | 9.6 | 24.3 | 1.0 | 6 | 0.194946 |
| Groc | slc | 23 | 9.9 | 1.1 | 17.9 | 15.3 | 1.6 | 8 | 0.321070 |
| Groc | kr | 24 | 10.2 | 0.5 | 12.6 | 18.0 | 0.9 | 6 | 0.453731 |
| Groc | sa | 25 | 9.2 | 1.0 | 11.6 | 4.5 | 0.8 | 7 | 0.594966 |
| Means | | | 10.45 | 0.70 | 16.79 | 13.18 | 4.30 | 10.16 | 0.408 |
| Standard deviation | | | 2.65 | 0.54 | 7.91 | 8.37 | 2.25 | 4.89 | 0.124 |

Data abstracted from *Forbes* 127, no. 1 (January 5, 1981).

the two variables $X_1$ and $X_2$. This small data set will simplify the presentation since the analysis can be performed by hand.

Another data set we will use includes financial performance data from the January 1981 issue of *Forbes*. The variables used are those defined in Section 8.3. Table 16.1 shows the data for 25 companies from three industries: chemical companies (the first 14 of the 31 discussed in Section 8.3), health care companies, and supermarket companies. The column labeled "Type" in Table 16.1 lists the abbreviations Chem, Heal, and Groc for these three industries. In Section 16.6, we will use two clustering techniques to group these companies and then check the agreement with their industrial type. These three industries were selected because they represent different stages of growth, different product lines, different management philosophies, different labor and capital requirements, etc. Among the chemical companies all of the large diversified firms were selected. From the major supermarket chains, the top six rated for return on equity were included. In the health care industry four of the five companies included were those connected with hospital management; the remaining company involves hospital supplies and equipment.

## 16.4 BASIC CONCEPTS: INITIAL ANALYSES AND DISTANCE MEASURES

In this section we present some preliminary graphical techniques for clustering. Then we discuss distance measures that will be useful in later sections.

### Scatter Diagrams

Prior to using any of the analytical clustering procedures (see Section 16.5), most investigators begin with simple graphical displays of their data. In the case of two variables a scatter diagram can be very helpful in displaying some of the main characteristics of the underlying clusters. In the hypothetical data example shown in Figure 16.2, the points closest to each other are points 1 and 2. This observation may lead us to consider these two points as one cluster. Another cluster might contain points 3 and 4, with point 5 perhaps constituting a third cluster. On the other hand, some investigators may consider points 3, 4, and 5 as the second cluster. This example illustrates the indeterminacy of cluster analysis, since even the number of clusters is usually unknown. Note that the concept of closeness was implicitly used in defining the clusters. Later in this section we expand on this concept by presenting several definitions of distance.

If the number of variables is small, it is possible to examine scatter diagrams of each pair of variables and search for possible clusters. But this technique may become unwieldy if the number of variables exceeds four, particularly if the number of points is large.

## Profile Diagram

A helpful technique for a moderate number of variables is a *profile diagram*. To plot a profile of an individual case in the sample, the investigator customarily first standardizes the data by subtracting the mean and dividing by the standard deviation for each variable. However, this step is omitted by some researchers, especially if the units of measurement of the variables are comparable. In the financial data example the units are not the same, so standardization seems helpful. The standardized financial data for the 25 companies are shown in Table 16.2. A profile diagram, as shown in Figure 16.3, lists the variables along the horizontal axis and the standardized value scale along the vertical axis. Each point on the graph indicates the value of the corresponding variable. The profile for the first company in the sample has been graphed in Figure 16.3. The points are connected in order to facilitate the visual interpretation. We see that this company hovers around the mean, being at most 1.3 standard deviations away on any variable.

A preliminary clustering procedure is to graph the profiles of all cases on the same diagram. To illustrate this procedure, we plotted the profiles of seven companies (15 through 21) in Figure 16.4. To avoid unnecessary clutter, we reversed the sign of the values of ROR5 and PAYOUTR1. Using the original data on these two variables would have caused the lines connecting the points to cross each other excessively, thus making it difficult to identify single company profiles.

Examining Figure 16.4, we note the following:

**1.** Companies 20 and 21 are very similar.

**2.** Companies 16 and 18 are similar.

**3.** Companies 15 and 17 are similar.

**4.** Company 19 stands out alone.

**TABLE 16.2.** Standardized Financial Performance Data for Diversified Chemical, Health, and Supermarket Companies

| | | | Standardized Input Data | | | | | | |
|---|---|---|---|---|---|---|---|---|---|
| Type | Symbol | Observation Number | ROR5 | D/E | SALESGR5 | EPS5 | NPM1 | P/E | PAYOUTR1 |
| Chem | dia | 1 | 0.963 | -0.007 | 0.431 | 0.277 | 1.289 | -0.237 | 0.151 |
| Chem | dow | 2 | 0.963 | -0.007 | 0.052 | -0.057 | 1.334 | -0.442 | -0.193 |
| Chem | stf | 3 | 0.963 | -0.559 | -0.290 | 0.230 | 1.601 | -0.442 | -0.007 |
| Chem | dd | 4 | 0.661 | -0.927 | -0.492 | -0.248 | 0.488 | -0.237 | 1.291 |
| Chem | uk | 5 | -0.171 | -0.559 | -0.403 | -0.618 | 1.067 | -1.056 | -0.668 |
| Chem | psm | 6 | -0.246 | -0.375 | -0.593 | 0.158 | -0.224 | -0.851 | 0.807 |
| Chem | gra | 7 | -0.209 | -0.375 | -0.833 | -0.737 | 0.221 | -0.033 | -0.231 |
| Chem | hpc | 8 | -0.057 | -0.743 | -0.681 | -0.534 | 0.087 | -0.237 | 0.597 |
| Chem | mtc | 9 | -0.360 | -0.559 | -0.416 | -0.869 | -0.358 | 0.172 | 1.331 |
| Chem | acy | 10 | -0.209 | -0.559 | -0.593 | -1.072 | 0.132 | -0.237 | 0.668 |
| Chem | cz | 11 | -0.964 | -0.589 | -0.757 | 0.337 | -0.402 | -0.647 | 0.655 |
| Chem | ald | 12 | -1.191 | -0.191 | -0.176 | -0.988 | 0.354 | -0.647 | -1.089 |
| Chem | rom | 13 | -1.002 | -0.559 | -0.732 | -1.215 | 0.577 | -0.647 | -0.740 |
| Chem | rei | 14 | -1.494 | -0.559 | 0.241 | -1.943 | -1.337 | -0.033 | -0.190 |
| Heal | hum | 15 | -0.473 | 3.672 | 2.908 | 2.534 | 0.666 | 2.218 | -0.135 |
| Heal | hca | 16 | -0.587 | 0.361 | 1.366 | 1.233 | 1.067 | 2.422 | -1.981 |
| Heal | nme | 17 | -0.775 | 0.913 | 2.769 | 1.364 | 0.265 | 1.809 | -0.841 |
| Heal | ami | 18 | -0.549 | 0.729 | 0.671 | 1.042 | 0.755 | 1.809 | -0.839 |
| Heal | ahs | 19 | 0.925 | -0.743 | -0.100 | 0.361 | 0.621 | 0.786 | -0.966 |
| Groc | lks | 20 | 1.794 | -0.007 | -0.189 | -0.188 | -1.248 | -0.442 | 1.538 |
| Groc | win | 21 | 3.004 | -0.927 | -0.226 | -0.188 | -1.204 | -0.237 | 1.372 |
| Groc | sgl | 22 | -0.209 | 1.649 | -0.909 | 1.328 | -1.471 | -0.851 | -1.710 |
| Groc | slc | 23 | -0.209 | 0.729 | 0.140 | 0.254 | -1.204 | -0.442 | -0.696 |
| Groc | kr | 24 | -0.095 | -0.375 | -0.530 | 0.576 | -1.515 | -0.851 | 0.370 |
| Groc | sa | 25 | -0.473 | 0.545 | -0.656 | -1.036 | -1.560 | -0.647 | 1.506 |

*Note:* * See Table 16.1 for the original data source.

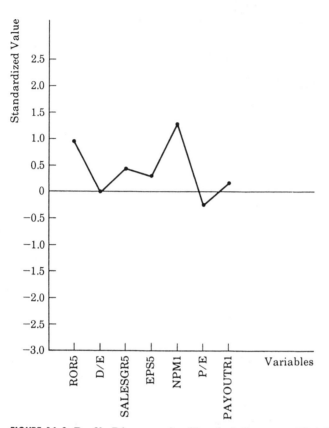

**FIGURE 16.3.** Profile Diagram of a Chemical Company (dia) Using Standardized Financial Performance Data

Thus it is possible to identify the above four clusters. It is also conceivable to identify three clusters: (15,16,17,18), (20,21), and (19). These clusters are consistent with the types of the companies, especially noting that company 19 deals with hospital supplies.

Although this technique's effectiveness is not affected by the number of variables, it fails when the number of observations is large. In Figure 16.4 the impression is clear, because we plotted only seven companies. Plotting all 25 companies would have produced too cluttered a picture.

## Distance Measures

For a large data set analytical methods such as those described in the next section are necessary. All of these methods require defining some measure

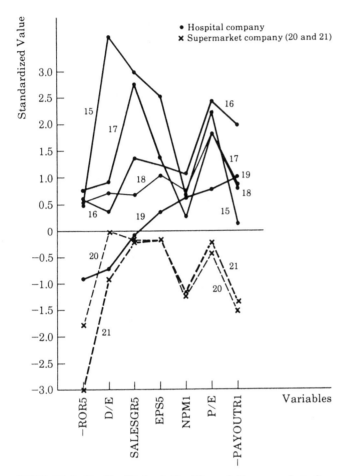

**FIGURE 16.4.** Profile Plot of Health and Supermarket Companies with Standardized Financial Performance Data

of closeness or *similarity* of two observations. The converse of similarity is *distance*. Before defining distance measures, though, we warn the investigator that many of the analytical techniques are particularly sensitive to outliers. Some preliminary checking for outliers and blunders is therefore advisable. This check may be facilitated by the graphical methods just described.

The most commonly used distance is the *Euclidian distance*. In two dimensions, suppose that two points have coordinates $(X_{11}, X_{21})$ and

$(X_{12}, X_{22})$, respectively. Then the Euclidian distance between the two points is defined as

$$distance = \sqrt{(X_{11} - X_{12})^2 + (X_{21} - X_{22})^2}$$

For example, the distance between points 4 and 5 in Figure 16.2 is

$$distance \; (points \; 4,5) = \sqrt{(5 - 7)^2 + (8 - 5)^2} = \sqrt{13} = 3.61$$

For $P$ variables the Euclidian distance is the square root of the sum of the squared differences between the coordinates of each variable for the two observations. Unfortunately, the Euclidian distance is not invariant to changes in scale and the results can change appreciably by simply changing the units of measurement.

In computer program output the distances between all possible pairs of points are usually summarized in the form of a matrix. For example, for our hypothetical data, the Euclidian distances between the five points are as given in Table 16.3. Since the distance between a point and itself is zero, the diagonal elements of this matrix are always zero. Also, since the distances are symmetric, many programs print only the distances above or below the diagonal.

Since the square root operation does not change the order of how close the points are to each other, some programs use the sum of the *squared differences* instead of the Euclidian distance (i.e., they don't take the square root). Another option available in some programs is to replace the squared differences by another *power* of the absolute differences. For example, if the power of 1 is chosen, the distance is the sum of the absolute differences of the coordinates. The distance is the so-called city-block

**TABLE 16.3.** Euclidian Distance Between Five Hypothetical Points

|   | 1 | 2 | 3 | 4 | 5 |
|---|---|---|---|---|---|
| 1 | 0 | 1.00 | 5.39 | 7.21 | 8.06 |
| 2 | 1.00 | 0 | 4.47 | 6.40 | 7.62 |
| 3 | 5.39 | 4.47 | 0 | 2.24 | 5.10 |
| 4 | 7.21 | 6.40 | 2.24 | 0 | 3.61 |
| 5 | 8.06 | 7.62 | 5.10 | 3.61 | 0 |

distance. In two dimensions it is the distance you must walk to get from one point to another in a city divided into rectangular blocks.

Several other definitions of distance exist (see Gower 1971). Here we will give only one more commonly used definition, the Mahalanobis distance discussed earlier in Chapter 11. In effect, the *Mahalanobis distance* is a generalization of the idea of standardization. The squared Euclidian distance based on standardized variables is the sum of the squared differences, each divided by the appropriate variance. When the variables are correlated, a distance can be defined to take this correlation into account. The Mahalanobis distance does just that. For two variables $X_1$ and $X_2$, suppose that the sample variances are $S_1^2$ and $S_2^2$, respectively, and that the correlation is $r$. The squared Euclidian distance based on the original values is

$$(\text{Euclidian distance})^2 = (X_{11} - X_{22})^2 + (X_{21} - X_{22})^2$$

The same quantity based on standardized variables is

$$(\text{standardized Euclidian distance})^2 = \frac{(X_{11} - X_{12})^2}{S_1^2} + \frac{(X_{21} - X_{22})^2}{S_2^2}$$

If $r = 0$, then the last quantity is also the Mahalanobis distance. If $r \neq 0$, the Mahalanobis distance is

$$D^2 = \frac{1}{1 - r^2} \left[ \frac{(X_{11} - X_{12})^2}{S_1^2} + \frac{(X_{21} - X_{22})^2}{S_2^2} - \frac{2r(X_{11} - X_{12})(X_{21} - X_{22})}{S_1 S_2} \right]$$

For more than two variables the Mahalanobis distance is easily defined in terms of vectors and matrices (see, e.g., Afifi and Azen 1979). Some computer programs, such as BMDPKM, offer the Mahalanobis distance as an option, with estimates of the sample variances and correlations obtained from the data. It is noted that, strictly speaking, the within-group sample covariance matrix should be used. However, before the investigator finds clusters, no such estimates exist, and the total group covariance matrix is the only one available. When the latter is used, the computed Mahalanobis distance is then somewhat different from that presented in Chapter 11.

One method for obtaining the within-cluster sample covariance matrix is to recompute it after each assignment of observations to each affected cluster. Thus, the within-cluster covariance matrix changes as the computation proceeds. These within-cluster covariance matrices are then pooled across clusters. This is an available option in the BMDPKM program. In the ACECLUS procedure in SAS, a different approach is used. Using a special algorithm, they estimate the within-cluster sample covariance matrix without knowledge of which points are in which clusters (see the discussion of the AGK method in the manual). In using this method it is assumed that the clusters are multivariately normally distributed with equal covariance matrices.

In most situations different distance measures yield different distance matrices, in turn leading to different clusters. When variables have different units, it is advisable to standardize the data before computing distances. Further, when high positive or negative correlations exist among the variables, it may be helpful to consider techniques of standardization that take the covariances as well as the variances into account. An example is given in the ACECLUS write-up using real data where actual cluster membership is known. It is therefore possible to say which observations are misclassified, and the results are striking.

In the next section we discuss two of the more commonly used analytical cluster techniques. These techniques make use of the distance functions just defined.

## *16.5 ANALYTICAL CLUSTERING TECHNIQUES*

The commonly used methods of clustering fall into two general categories: hierarchical and nonhierarchical. First, we discuss the hierarchical techniques.

### *Hierarchical Clustering*

Hierarchical methods can be either agglomerative or divisive. In the *agglomerative methods* we begin with $N$ clusters; i.e., each observation constitutes its own cluster. In successive steps we combine the two closest clusters, thus reducing the number of clusters by one in each step. In the final step all observations are grouped into one cluster. In *divisive methods* we begin with one cluster containing all of the observations. In successive steps we split off the cases that are most dissimilar to the remaining ones. Most of the commonly used programs are of the agglomerative type, and we therefore do not discuss divisive methods further.

The centroid procedure is a widely-used example of agglomerative methods. In the centroid method the distance between two clusters is defined as the distance between the group centroids (the centroid is the point whose coordinates are the means of all the observations in the cluster). If a cluster has one observation, then the centroid is the observation itself. The process proceeds by combining groups according to the distance between their centroids, the groups with the shortest distance being combined first. The centroid method is available in both BMDP2M, SAS CLUSTER, and SPSS CLUSTER.

The centroid method is illustrated in Figure 16.5 for our hypothetical data. Initially, the closest two centroids (points) of the five hypothetical observations plotted in Figure 16.2 are points 1 and 2, so they are

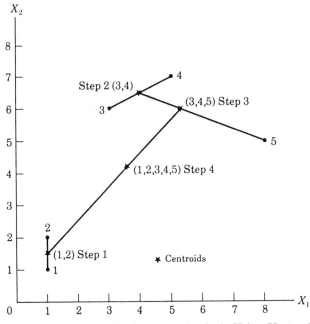

**FIGURE 16.5.** Hierarchical Cluster Analysis Using Unstandardized Hypothetical Data Set

combined first and their centroid is obtained in step 1. In step 2, centroids (points) 3 and 4 are combined (and their centroid is obtained), since they are the closest now that points 1 and 2 have been replaced by their centroid. At step 3 the centroid of points 3 and 4 and centroid (point) 5 are combined, and the centroid is obtained. Finally, at the last step the centroid of points 1 and 2 and the centroid of points 3, 4, and 5 are combined to form a single group.

Both BMDP and SAS offer optional distances based on standardized variables. Figure 16.6 illustrates the clustering steps based on the stand-

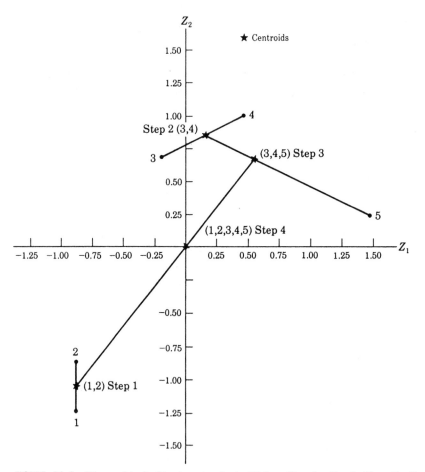

**FIGURE 16.6.** Hierarchical Cluster Analysis Using Standardized Hypothetical Data Set

ardized hypothetical data, using BMDP2M. The results are identical to the previous results, although this is not the case in general. In addition, BMDP2M and SPSS CLUSTER allow the optional distance of the sum of powers of differences (note that this option includes a power of 1, which would represent the city-block distance). When using the SPSS CLUSTER or SAS CLUSTER procedures, the input can be either the numerical values of the various variables or the distances themselves.

With two variables and a large number of data points, the representation of the steps in a graph similar to Figure 16.5 can get too cluttered to interpret. Also, if the number of variables is more than two, such a graph is not feasible. A clever device called the *dendrogram* or *tree graph* has therefore been incorporated into packaged computer programs to summarize the clustering at successive steps. The dendrogram for the hypothetical data set is illustrated in Figure 16.7. The horizontal axis lists the observations in a particular order. In this example the natural order is convenient. The vertical axis shows the successive steps. At step 1 points 1 and 2 are combined. Similarly, points 3 and 4 are combined in step 2; point 5 is combined with cluster (3,4) in step 3; and finally clusters (1,2) and (3,4,5) are combined. In each step two clusters are combined.

A tree graph is a default output of BMDP2M and an optional output of SPSS CLUSTER. It can be obtained from SAS by calling the TREE procedure, using the output from the CLUSTER program. In TREE the

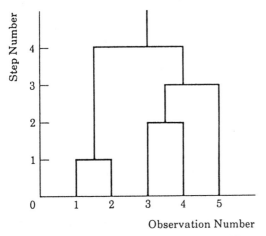

**FIGURE 16.7.** Dendrogram for Hierarchical Cluster Analysis of Hypothetical Data Set

number of clusters is printed on the vertical axis instead of the step number shown in Figure 16.7, while in BMDP2M the distance between the clusters just combined is printed. The order of the observations on the horizontal axis is helpful in indicating which observations are sufficiently similar to be combined in the early steps. The investigator should be aware that it is difficult to produce connected straight lines on a dot matrix computer printer. For this reason the dendrograms produced by standard packaged programs are not as easy to read as the one shown in Figure 16.7.

BMDP offers two additional options for combining clusters: single linkage and $k$th-nearest-neighbor linkage. SAS offers ten additional methods, while SPSS CLUSTER offers six. The manuals should be read for descriptions of the various options.

Hierarchical procedures are appealing in a taxonomic application. Such procedures can be misleading, however, in certain situations. For example, an undesirable early combination persists throughout the analysis and may lead to artificial results. The investigator may wish to perform the analysis several times after deleting certain suspect observations.

For large sample sizes the printed dendrograms become very large and unwieldy to read. One statistician noted that they were more like wallpaper than comprehensible results for large $N$.

An important problem is how to select the number of clusters. No standard objective procedure exists for making the selection. The distances between clusters at successive steps may serve as a guide. The investigator can stop when this distance exceeds a specified value or when the successive differences in distances between steps make a sudden jump. Also, the underlying situation may suggest a natural number of clusters. If such a number is known, a particularly appropriate technique is the $K$-means clustering techniques of MacQueen (1967).

## K-Means Clustering

The $K$-means clustering is a popular nonhierarchical clustering technique. For a specified number of clusters $K$ the basic algorithm proceeds in the following steps:

1. Divide the data into $K$ initial clusters. The members of these clusters may be specified by the user or may be selected by the program, according to an arbitrary procedure.

**2.** Calculate the means or centroids of each of the $K$ clusters.

**3.** For a given case, calculate its distance to each centroid. If the case is closest to the centroid of its own cluster, leave it in that cluster; otherwise, reassign it to the cluster whose centroid is closest to it.

**4.** Repeat step 3 for each case.

**5.** Repeat steps 2, 3, and 4 until no cases are reassigned.

Individual programs implement the basic algorithm in different ways. The default option of BMDPKM begins by considering all of the data as one cluster. For the hypothetical data set this step is illustrated in Figure 16.8a. The program then searches for the variable with the highest variance, in this case $X_1$. The original cluster is now split into two clusters, using the midrange of $X_1$ as the dividing point, as shown in Figure 16.8b. If the data are standardized, then each variable has a variance of one. In that case the variable with the smallest range is selected to make the split. The program, in general, proceeds in this manner by further splitting the clusters until the specified number $K$ is achieved. That is, it successively finds that particular variable and the cluster producing the largest variance and splits that cluster accordingly, until $K$ clusters are obtained. At this stage step 1 of the basic algorithm is completed and the program proceeds with the other steps.

For the hypothetical data example with $K = 2$, it is seen that every case already belongs to the cluster whose centroid is closest to it (Figure 16.8c). For example, point 3 is closer to the centroid of cluster $(1,2,3)$ than it is to the centroid of cluster $(4,5)$. Therefore it is not reassigned. Similar results hold for the other cases. Thus the algorithm stops, with the two clusters selected being cluster $(1,2,3)$ and cluster $(4,5)$.

The SAS procedure FASTCLUS is recommended especially for large data sets. The user specifies the maximum number of clusters allowed, and the program starts by first selecting cluster "seeds," which are used as initial guesses of the means of the clusters. The first observation with no missing values in the data set is selected as the first seed. The next complete observation that is separated from the first seed by an optional, specified minimum distance becomes the second seed (the default minimum distance is zero). The program then proceeds to assign each observation to the cluster with the nearest seed. The user can decide whether to update the cluster seeds by cluster means each time an

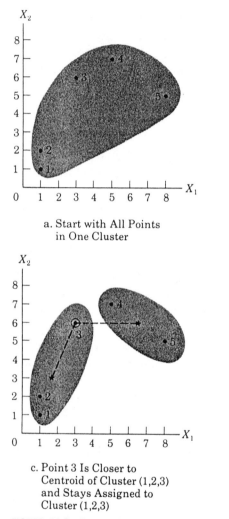

a. Start with All Points
   in One Cluster

b. Cluster Is Split into
   Two Clusters at Midrange
   of $X_1$ (Variable with
   Largest Variance)

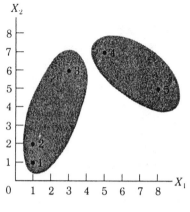

c. Point 3 Is Closer to
   Centroid of Cluster (1,2,3)
   and Stays Assigned to
   Cluster (1,2,3)

d. Every Point Is Now Closest
   to Centroid of Its Own
   Cluster; Stop

**FIGURE 16.8.** Successive Steps in $K$-Means Cluster Analysis for $K = 2$, Using BMDPKM for Hypothetical Data Set with Default Options

observation is assigned, using the DRIFT option, or only after all observations are assigned. Limiting seed replacement results in the program using less computer time. Note that the initial, and possibly the final, results depend on the order of the observations in the data set. Specifying the initial cluster seeds can lessen this dependence. Finally, we note that BMDPKM also permits the user to specify seed points. For either program judicious choice of the seeds or leaders can improve the results.

As mentioned earlier, the selection of the number of clusters is a troublesome problem. If no natural choice of $K$ is available, it is useful to try various values and compare the results. One aid to this examination is a plot of the points in the clusters, labeled by the cluster they belong to. If the number of variables is more than two, some two-dimensional plots are available from certain programs, such as BMDPKM. Another possibility is to plot the first two principal components, labeled by cluster membership. A second aid is to examine the $F$ ratio for testing the hypothesis that the cluster means are equal. This test may be done for each variable separately, as in BMDPKM. An alternative possibility is to supply the data with each cluster as a group as input to a $K$ group discriminant analysis program. Such a program would then produce a test statistic for the equality of means of all variables simultaneously, as was discussed in Chapter 11. Then for each variable, separately or as a group, comparison of the $P$ values for various values of $K$ may be a helpful indication of the value of $K$ to be selected. Note, however, that these $P$ values are valid only for comparative purposes and not as significance levels in a hypothesis-testing sense, even if the normality assumption is justified. The individual $F$ statistics for each variable indicate the relative importance of the variables in determining cluster membership (again, they are not valid for hypothesis testing).

Since cluster analysis is an empirical technique, it may be advisable to try several approaches in a given situation. In addition to the hierarchical and $K$-means approaches discussed above, several other methods are available in the literature cited in the Bibliography. It should be noted that the $K$-means approach is gaining acceptability in the literature over the hierarchical approach. In any case, unless the underlying clusters are clearly separated, different methods can produce widely different results. Even with the same program, different options can produce quite different results.

## 16.6 CLUSTER ANALYSIS FOR FINANCIAL DATA SET

In this section we apply some of the standard procedures to the financial performance data set shown in Table 16.1. In all of our runs the data are first standardized as shown in Table 16.2. Recall that in cluster analysis the total sample is considered as a single sample. Thus the information on type of company is not used to derive the clusters. However, this information will be used to interpret the results of the various analyses.

## Hierarchical Clustering

The dendrogram resulting from the hierarchical BMDP2M program is shown in Figure 16.9. Default options including the centroid method with Euclidian distance were used with the standardized data. The horizontal axis lists the observation numbers in a particular order, which prevents the lines in the dendrogram from crossing each other. One result of this arrangement is that certain subgroups appearing near each other on the

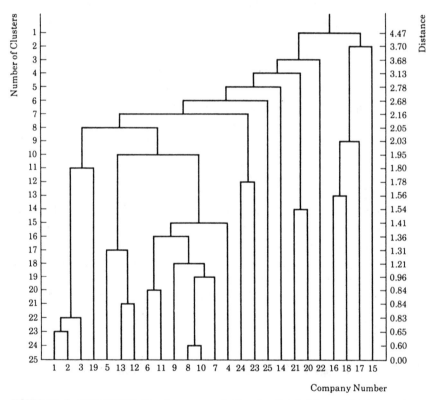

**FIGURE 16.9.** BMDP2M Dendrogram of Standardized Financial Performance Data Set

horizontal axis constitute clusters at various steps. Note that the distance is shown on the right vertical axis. These distances are measured between the *centers* of the two clusters just joined. On the left vertical axis the number of clusters is listed.

In the figure, companies 1, 2, and 3 form a single cluster, with the grouping being completed when there are 22 clusters. Similarly, at the opposite end 15, 17, 18, and 16 (all health care companies) form a single cluster at the step in which there are two clusters. Company 22 stays by itself until there are only three clusters.

The distance axis indicates how disparate the two clusters just joined are. The distances are progressively increasing. As a general rule, large increases in the sequence should be a signal for examining those particular steps. A large distance indicates that at least two very different observations exist, one from each of the two clusters just combined. Note, for example, the large increase in distance occurring when we combine the last two clusters into a single cluster.

It is instructive to look at the industry groups and ask why the cluster analysis, as in Figure 16.9, for example, did not completely differentiate them. It is clear that the clustering is quite effective. First, 13 of the chemical companies, all except no. 14 (rci), are clustered together with only one nonchemical firm, no. 19 (ahs), when the number of clusters is eight. This result is impressive when one considers that these are large diversified companies with varied emphasis, ranging from industrial chemicals to textiles to oil and gas production.

At the level of nine clusters three of the four hospital management firms, nos. 16, 17, and 18 (hca, nme, and ami), have also been clustered together, and the other, no. 15 (hum), is added to that cluster before it is aggregated with any nonhospital management firms. A look at the data in Tables 16.1 and 16.2 shows that no. 15 (hum) has a clearly different D/E value than the others, suggesting a more highly leveraged operation and probably a different management style. The misfit in this health group is no. 19 (ahs), clustered with the chemical firms instead of with the hospital management firms. Further examination shows that, in fact, no. 19 is a large, established supplies and equipment firm, probably more akin to drug firms than to the fast-growing hospital management business, and so it could be expected to share some financial characteristics with the chemical firms.

The grocery firms do not cluster tightly. In scanning the data of Tables 16.1 and 16.2 we note that they vary substantially on most variables. In particular, no. 22 (sgl) is highly leveraged (high D/E), and no. 21 (win) has low leverage (low D/E) relative to the others. Further, if we examine other characteristics not included in this data set, important disparities show up. Three of the six, nos. 21, 24, and 25 (win, kr, sa), are three of the four largest United States grocery supermarket chains. Two others, nos. 20 and 22 (kls, sgl), have a diversified mix of grocery, drug, department, and other stores. The remaining firm, no. 24 (slc), concentrates on convenience stores (7-Eleven) and has a substantial franchising operation. Thus the six, while all grocery-related, are quite different from each other.

### Various K-Means Clusters

Since these companies do not present a natural application of hierarchical clustering such as taxonomy, the $K$-means procedure may be more appropriate. The natural value $K$ is three since there are three types of companies. The first four columns of Table 16.4 show the results of one run of the SAS FASTCLUS procedure and three runs of BMDPKM. The numbers in each column indicate the cluster to which each company is assigned. In columns (1) and (2) the default options of the two programs were used with $K = 3$. In column (3) two values of $K$, 2 and 3, were specified in the same run. Note that the results are different from those in column (2) although the same program was used. The difference is due to the fact that in column (3) the initial three groups are selected by splitting one of the two clusters obtained for $K = 2$, whereas in column (2) the initial splitting is done before any reassignment is carried out (see Section 16.5). Different results still were obtained when the Mahalanobis distance option was used, as shown in column (4).

One method of summarizing the first four columns is shown in column (5). For example, all four runs agreed on assigning companies 1 through 4 to cluster 1. Similarly, every company assigned unanimously to a cluster is so indicated. Other companies were assigned to a cluster if three out of four runs agreed on it. For example, company 8 was assigned to cluster 1 and company 18 was assigned to cluster 2. Seven of the companies could not be assigned to a unique cluster. Each received two votes for cluster 1 and two votes for cluster 3.

It seems, therefore, that the value of $K$ should perhaps be 4, not 3. With $K = 4$ and the BMDPKM program the results are as shown in column (6). This run seems basically to group the companies labeled 1 and 3 in column

**TABLE 16.4.** Companies Clustered Together from K-Means Standardized Cluster Analysis (Financial Performance Data Set)

| Type of Company | | (1) Fast Clus* K = 3 Default | (2) BMDPKM K = 3 Default | (3) BMDPKM K = 2,3 Default | (4) BMDPKM K = 3 Mahalanobis | (5) Summary Of 4 Runs, K = 3 | (6) BMDPKM K = 4 Mahalanobis |
|---|---|---|---|---|---|---|---|
| 1 | Chem | 1 | 1 | 1 | 1 | 1 | 1 |
| 2 | Chem | 1 | 1 | 1 | 1 | 1 | 1 |
| 3 | Chem | 1 | 1 | 1 | 1 | 1 | 1 |
| 4 | Chem | 1 | 1 | 1 | 1 | 1 | 1 |
| 5 | Chem | 3 | 1 | 3 | 1 | 1,3 | 4 |
| 6 | Chem | 1 | 3 | 3 | 1 | 1,3 | 3 |
| 7 | Chem | 3 | 1 | 3 | 1 | 1,3 | 4 |
| 8 | Chem | 1 | 1 | 3 | 1 | 1 | 4 |
| 9 | Chem | 1 | 3 | 3 | 1 | 1,3 | 4 |
| 10 | Chem | 3 | 1 | 3 | 1 | 1 | 4 |
| 11 | Chem | 3 | 3 | 3 | 1 | 3 | 3 |
| 12 | Chem | 3 | 1 | 3 | 1 | 1,3 | 4 |
| 13 | Chem | 3 | 1 | 3 | 1 | 1,3 | 4 |
| 14 | Chem | 3 | 3 | 3 | 1 | 3 | 4 |
| 15 | Heal | 2 | 2 | 2 | 2 | 2 | 2 |
| 16 | Heal | 2 | 2 | 2 | 2 | 2 | 2 |
| 17 | Heal | 2 | 2 | 2 | 2 | 2 | 2 |
| 18 | Heal | 2 | 2 | 2 | 2 | 2 | 2 |
| 19 | Heal | 1 | 1 | 1 | 3 | 1 | 1 |
| 20 | Groc | 1 | 3 | 1 | 1 | 1 | 1 |
| 21 | Groc | 1 | 3 | 1 | 1 | 1 | 1 |
| 22 | Groc | 3 | 3 | 3 | 3 | 3 | 3 |
| 23 | Groc | 3 | 3 | 3 | 3 | 3 | 3 |
| 24 | Groc | 3 | 3 | 3 | 3 | 3 | 3 |
| 25 | Groc | 1 | 3 | 3 | 1 | 1,3 | 4 |

* Changing the order of companies 1 and 11 did not change the results.

(5) into a new cluster, 4. Otherwise, with the exception of companies 8, 10, and 14, all other companies remain in their original clusters. Cluster 1 now consists of companies 1, 2, 3, 4 (chemicals), 19 (health), and 20, 21 (grocery). Cluster 2 consists exclusively of health companies. Cluster 3 consists of three grocery companies and two chemicals. Finally, cluster 4 consists exclusively of chemical companies. It thus seems, with some exceptions, that the companies within an industry have similar financial data. For example, an exception is company 19, a company that deals with hospital supplies. This exception was evident in the hierarchical results as well as in $K$-means, where company 19 joined the chemical companies and not the other health companies.

## Profile Plots of Means

Some of the output of BMDPKM may be useful in further interpretations of the clusters. A plot of each observation labeled by its cluster is given as part of the default output. The variable means for each cluster are printed and graphed to provide cluster profiles, which aids in understanding the characteristics of each cluster. Figure 16.10 shows the cluster profiles based on the means of the standardized variables ($K = 4$). Note that the values of ROR5 and PAYOUTR1 are plotted with the signs reversed in order to keep the lines from crossing too much. It is immediately apparent that cluster 2 is very different from the remaining three clusters. Cluster 1 is also quite different from clusters 3 and 4. The latter two clusters are most different on EPS5. Thus the four health companies comprising cluster 2 seem to clearly average higher SALESGR5 and P/E. This cluster mean profile is a particularly appropriate display of the $K$-means output, because the objective of the analysis is to make these means as widely separated as possible and because there are usually a small number of clusters to be plotted.

## Use of F Tests

For quantification of the relative importance of each variable the univariate $F$ ratio for testing the equality of each variable's means in the $K$ clusters is given in the BMDPKM output. A large $F$ value is an indication that the corresponding variable is useful in separating the clusters. For example, the largest $F$ value is for P/E, indicating that the clusters are

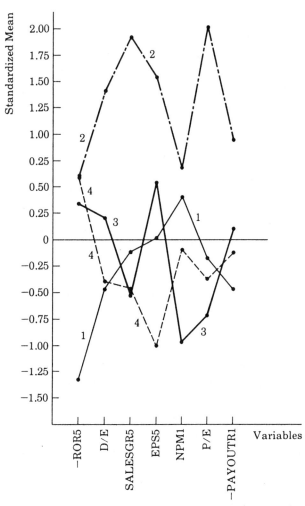

**FIGURE 16.10.** Profile of Cluster Means for Four Clusters (Financial Performance Data Set)

most different from each other in terms of this variable. Conversely, the *F* ratio for PAYOUTR1 is very small, indicating that this variable does not play an important role in defining the clusters for the particular industry groups and companies analyzed. Note that these *F* ratios are used for comparing the variables only, not for hypothesis-testing purposes.

The above example illustrates how these techniques are used in an exploratory manner. The results serve to point out some possible natural groupings as a first step in generating scientific hypotheses for the purpose of further analyses. A great deal of judgment is required for an interpretation of the results since the outputs of different runs may be different and sometimes even contradictory. Indeed, this result points out the desirability of making several runs on the same data in an attempt to reach some consensus.

## 16.7 DISCUSSION OF COMPUTER PROGRAMS

The SAS package offers five procedures for cluster analysis: (1) ACECLUS obtains estimates of the pooled within-cluster covariance matrix which can be used by the CLUSTER procedure; (2) CLUSTER performs hierarchical clustering using one of eleven methods; (3) TREE takes the results of either CLUSTER or VARCLUS and produces a dendrogram; (4) FASTCLUS uses the $K$-means method and is especially well-suited for large sample sizes; and (5) VARCLUS clusters variables, a topic not discussed in this chapter.

The BMDP package offers four cluster analysis programs: (1) 1M clusters variables; (2) 2M does hierarchical clustering using one of eleven distance measures (many of which are for discrete data, a topic not discussed here); (3) KM performs $K$-means clusters; and (4) 3M constructs block clusters for categorical data (also not discussed here).

There are two cluster analysis procedures in SPSS. CLUSTER performs hierarchical clustering, offering a choice of seven methods and six distance measures. QUICK CLUSTER performs $K$-means clustering and several variants of this method. The NOINITIAL option on the CRITERIA subcommand is used to specify the $K$-means method. This procedure is the preferable one for large data sets. Neither CLUSTER nor QUICK CLUSTER standardize variables. The data can be standardized first, however, using the DESCRIPTIVES command, and the standardized variables can be saved and then used in either cluster analysis procedure.

The cluster analysis programs offer additional options beyond those given in Table 16.5. In fact, some of the programs offer such a range of options that the choice may be difficult. We have restricted the listings in Table 16.5 to either those discussed in this chapter or those that are similar to output from other chapters.

**TABLE 16.5.** Summary of Computer Output for Cluster Analysis

| Options | BMDP | SAS | SPSS-X |
|---|---|---|---|
| **Hierarchical Cluster Analysis: Standardization** | | | |
| Not standardized | 2M | CLUSTER | CLUSTER |
| Standardized | 2M | CLUSTER | |
| Estimation of within covariances | | ACECLUS | |
| **Distance** | | | |
| Euclidian | 2M | CLUSTER | CLUSTER |
| Squared Euclidian distance | 2M | CLUSTER | CLUSTER |
| Sum of $p$th power of absolute distances | 2M | CLUSTER | |
| Distances suitable for counted data | 2M | | |
| Other options | | | CLUSTER |
| **Linkage Methods** | | | |
| Centroid | 2M | CLUSTER | CLUSTER |
| Nearest neighbor | 2M | CLUSTER | CLUSTER |
| $K$th nearest neighbor | 2M | CLUSTER | |
| Complete linkage | | CLUSTER | CLUSTER |
| Ward's method | | CLUSTER | CLUSTER |
| Other options | | CLUSTER | CLUSTER |
| **Printed Output** | | | |
| Distances between combined clusters | 2M | CLUSTER | CLUSTER |
| Dendrogram | 2M | TREE | CLUSTER |
| Initial distances between observations | 2M | CLUSTER | |
| Number of observations per cluster | | TREE | |
| **K-Means Cluster Analysis: Standardization** | | | |
| Not standardized | KM | CLUSTER | QC* |
| Standardized | KM | STANDARD | |
| Estimation of within covariances | KM | ACECLUS | |
| **Distance** | | | |
| Euclidian | KM | FASTCLUS | |
| Squared Euclidean | | | QC |
| Mahalanobis | KM | | |
| **Distance Measured From** | | | |
| Leader or seed points | KM | FASTCLUS | |
| Centroid or center | KM | FASTCLUS | QC |
| **Printed Output** | | | |
| Distances between cluster centers | KM | FASTCLUS | QC |
| Distances between points and center | KM | FASTCLUS | QC |
| Cluster means and standard deviations | KM | FASTCLUS | |
| $F$ ratios | KM | FASTCLUS | QC |
| Cluster mean profiles | KM | | |
| Plot of cluster membership | KM | FASTCLUS | |
| Save cluster membership | KM | FASTCLUS | QC |

* QC = QUICK CLUSTER

If the city block is used, then BMDP2M is appropriate for hierarchical analysis. This analysis is achieved by using $p$th-power distance option with $p = 1$. For $K$-means clustering BMDPKM offers a variety of optional printed output that may be useful in assessing the results. On the other hand, FASTCLUS is designed to be economical when used on large data sets. For very large data sets it offers options that could cut the cost even further.

It may also be appropriate to combine two analyses into one. For example, a principal components analysis may be performed on the entire data set and a small number of components selected. A cluster analysis can then be run on the values of the selected principal components. Similarly, factor analysis could be performed first, followed by a cluster analysis of the factor scores. This procedure has the advantage of reducing the original number of variables, making the interpretation possibly easier, especially if the meaningful factors have been selected.

As noted earlier, cluster analysis can be used to cluster variables instead of cases. For example, BMDP1M is designed to perform such an analysis. If the default option is used, the absolute value of the correlation between two variables measures the "distance" between them: The higher the absolute value of the correlation, the closer the two variables are. The program then proceeds to cluster variables in a hierarchical fashion. The VARCLUS procedure in SAS offers several options. It attempts to divide a set of variables into nonoverlapping clusters. A discussion of its use to reduce the number of variables employed in analysis is given in the SAS manual.

The following statements are used to run FASTCLUS on the five hypothetical observations:

```
DATA;
INPUT X1 1.0 X2 1.0;
CARDS;
11
12
36
57
85;
PROC FASTCLUS LIST MAXCLUSTERS = 2;
PROC PRINT;
PROC MEANS;
```

The first set of statements defines this simple data set of two variables and five observations. Both X1 and X2 are listed in a single column that follow on cards. The FASTCLUS procedure is used with only the LIST and MAXCLUSTERS option. Two clusters are specified by MAXCLUSTERS = 2. The LIST option results in a list of the ID of all the observations along with the number of the cluster to which the observation is assigned and the distance between the observation and the final cluster seed. A large number of statistics can be printed by using the various available printout options. The PROC PRINT statement results in the values of the original data being printed, and PROC MEANS results in the means, standard deviations, and other univariate statistics being printed for X1 and X2.

The FASTCLUS procedure given in Table 16.4 for the financial data was run by using the default options. The following control statements were supplied:

```
DATA;
INPUT @ 3 ROR5 3.1 D-E 2.1 SALESGR5 3.1
EPS5 4.1 NPN1 3.1 EPS1 3.2
DIV1 3.2 P E 2.0;
PAYOUTR1 = DIV1/EPS1;
DROP DIV1 EPS1;
IF _N_ < 15 THEN TYPE = 'CHEM.';
ELSE IF _N_ < 20 THEN TYPE = 'HEAL.';
ELSE TYPE = 'GROC';
CARDS;

(DATA)

PROC STANDARD MEAN = 0 STD = 1 OUT = STAN1;
PROC PRINT;
PROC FASTCLUS DATA = STAN1 LIST MAXCLUSTERS = 3;
```

The "@ 3" tells the computer to go to column 3 and start reading the first variable from that column. This pointer control simplifies the use of a subset of the variables. The format of each variable is given following the variables. For example, ROR5 is a three-digit variable with a single number following the decimal point. It is the first variable used in our data. The first two columns are simply ID numbers going from 1 to 25. The variable PAYOUTR1 is computed by the program as a ratio of dividends to earnings per share during the last year.

The companies are labeled by type (chem, heal, or groc), depending on their order. The first 14 companies are chemical companies, so the statement IF $\_N\_ < 15$ followed by the labeling "CHEM" is used; etc.

The original data are transformed into standardized data since these variables do not have the same units (PROC STANDARD), and a printout of the standardized data is requested (PROC PRINT). Finally, the FASTCLUS procedure is called to be run on the standardized data. Three clusters are requested by using the MAXCLUSTERS = 3 statement. Additional output can be obtained by following the instructions given in the manual. The final assignments are shown in Table 16.4, column (1), and were discussed earlier in the text.

For a run of a similar program using BMDPKM, the usual INPUT and VARIABLE paragraphs are followed by a CLUSTER paragraph, as follows:

```
/CLUSTER   STAND=WVAR.
           NUMBER=3.
```

The first sentence requests that the data be standardized, using the within-cluster variances. This statement results in the standardization changing at each stage of the clustering. If the sentence STAND = VAR. had been used, then each variable would have been divided by its own standard deviation, using all cases as a single sample. The default option is no standardization. The second sentence requests three clusters. Columns (2), (3), (4), and (6) of Table 16.4 show the outputs of various runs of BMDPKM as discussed in Section 16.6.

## 16.8 WHAT TO WATCH OUT FOR

Cluster analysis is a heuristic technique for classifying cases into groups when knowledge of the actual group membership is unknown. There are numerous methods for performing the analysis, without good guidelines for choosing among them.

*1.* Unless there is considerable separation among the inherent groups, it is not realistic to expect very clear results with cluster analysis. In particular, if the observations are distributed in a non-linear manner, it may be difficult to achieve distinct groups. For example, if we have two variables and cluster 1 is distributed such that a scatter plot has a U or crescent shape, while cluster 2 is in the empty space in the middle of the crescent, then most cluster analysis techniques would not be successful in separating the two groups.

**2.** Cluster analysis is quite sensitive to outliers. In fact, it is sometimes used to find outliers. The data should be carefully screened before running cluster programs.

**3.** Some of the computer programs supply $F$ ratios, but it should be kept in mind that the data have been separated in a particular fashion prior to making the test, so the interpretation of the significance level of the tests is suspect. Usually, the major decision as to whether the cluster analysis has been successful should depend on whether the results make intuitive sense.

**4.** Most of the methods are not scale invariant, so it makes sense to try several methods of standardization and observe the effect on the results. If the program offers the opportunity of standardizing by the pooled within-cluster covariance matrix, this option should be tried. Note that for large samples, this could increase the computer time considerably.

**5.** Some investigators split their data and run cluster analysis on both halves to see if they obtain similar results. Such agreement would give further credence to the conclusions.

## SUMMARY

In this chapter we presented a topic still in its evolutionary form. Unlike subjects discussed in previous chapters, cluster analysis has not yet gained a standard methodology. Nonetheless, a number of techniques have been developed for dividing a multivariate sample, the composition of which is not known in advance, into several groups.

In view of the state of the subject, we opted for presenting some of the techniques that have been incorporated into the standard packaged programs. These include a hierarchical and nonhierarchical technique. We also explained the use of profile plots for small data sets. Since the results of any clustering procedure are often not definitive, it is advisable to perform more than one analysis and attempt to collate the results.

## BIBLIOGRAPHY

Afifi, A. A., and Azen, S. P. 1979. *Statistical analysis: A computer oriented approach.* 2nd ed. New York: Academic Press.

Aldenderfer, M. S., and Blashfield, R. K. 1986. *Cluster analysis.* Newbury Park, Calif.: Sage.

Anderberg, M. R. 1973. *Cluster analysis for applications.* New York: Academic Press.

Andreasen, N. C., and Grove, W. M. 1982. The classification of depression: Traditional versus mathematical approaches. *American Journal of Psychiatry* 139:45–52.

Cormack, R. M. 1971. A review of classification. *Journal of the Royal Statistical Society, Series A* 134:321–367.

Everitt, B. 1980. *Cluster analysis.* London: Heinemann Educational Books.

Gordon, A. D. 1981. *Classification.* London: Chapman and Hall.

Gower, J. C. 1971. A general coefficient of similarity and some of its properties. *Biometrics* 27:857–874.

Hand, D. J. 1981. *Discrimination and classification.* New York: Wiley.

Hartigan, J. A. 1975. *Clustering algorithms.* New York: Wiley.

Johnson, R. A., and Wichern, D. W. 1988. *Applied multivariate statistical analysis.* 2nd ed. Englewood Cliffs, N. J.: Prentice-Hall.

Johnson, S. C. 1967. Hierarchical clustering schemes. *Psychometrika* 32:241–244.

MacQueen, J. B. 1967. Some methods for classification and analysis of multivariate observations. *Proceedings of the Fifth Berkeley Symposium on Mathematical Statistics and Probability* 1:281–297.

Sneath, P. H., and Sokal, R. R. 1973. *Numerical taxonomy.* San Francisco: Freeman.

Ton, J. T., and Gonzalez, R. C. 1974. *Pattern recognition principles.* Reading, Mass.: Addison-Wesley.

Ward, J. H. 1963. Hierarchical grouping to optimize an objective function. *Journal of the American Statistical Association* 58:236–244.

Wilson, E. O., Eisner, T., Briggs, W. R., Dickerson, R. E., Metzenberg, R. L., O'Brien, R. D., Susman, M., and Boggs, W. E. 1973. *Life on earth.* Sunderland, Mass.: Sinauer Associates.

## PROBLEMS

16.1  For the depression data set, use the last seven variables to perform a cluster analysis producing two groups. Compare the distribution of CESD and cases in the groups. Compare also the distribution of sex in the groups. Try two different programs with different options. Comment on your results.

16.2  For the situation described in Problem 7.7, modify the data for X1, X2, ... , X9 as follows:

- For the first 25 cases, add 10 to X1, X2, X3.
- For the next 25 cases, add 10 to X4, X5, X6.
- For the next 25 cases, add 10 to X7, X8.
- For the last 25 cases, add 10 to X9.

Now perform a cluster analysis to produce four clusters. Use two different programs with different options. Compare the derived clusters with the above groups.

16.3  Perform a cluster analysis on the chemical company data in Table 8.1, using the $K$-means method for $K = 2, 3, 4$.

16.4  For the accompanying small hypothetical data set, plot the data by using methods given in this chapter, and perform both hierarchical and $K$-means clustering with $K = 2$.

| Cases | X1 | X2 |
|-------|-----|-----|
| 1 | 11 | 10 |
| 2 | 8 | 10 |
| 3 | 9 | 11 |
| 4 | 5 | 4 |
| 5 | 3 | 4 |
| 6 | 8 | 5 |
| 7 | 11 | 11 |
| 8 | 10 | 12 |

16.5  Describe how you would expect guards, forwards, and centers in basketball to cluster on the basis of size or other variables. Which variables should be measured?

16.6  For the family lung function data in Appendix B, perform a hierarchical cluster analysis using the variables FEV1 and FVC for mothers, fathers, and oldest children. Compare the distribution of area of residence in the resulting groups.

16.7 Repeat Problem 16.6, using AREA as an additional clustering variable. Comment on your results.

16.8 Repeat Problem 16.6, using the $K$-means method for $K = 4$. Compare the results with the four clusters produced in Problem 16.6.

16.9 Create a data set from the family lung function data in Appendix B as follows: It will have three times the number of observations that the original data set has—the first third of the observations will be the mothers' measurements, the second third those of the fathers, and the final third those of the oldest children. Perform a cluster analysis, first producing three groups and then two groups, using the variables AGE, HEIGHT, WEIGHT, FEV1, and FVC. Do the mothers, fathers, and oldest children cluster on the basis of any of these variables?

16.10 Repeat Problem 16.1, using the variables age, income, and education instead of the last seven variables.

# APPENDIX A
# STATISTICAL TABLES

**TABLE A.1.** Cumulative Normal Distribution

*Example:* Area to the left of $z = -2.04$ is 0.0207.

| z | 0.00 | 0.01 | 0.02 | 0.03 | 0.04 | 0.05 | 0.06 | 0.07 | 0.08 | 0.09 |
|---|------|------|------|------|------|------|------|------|------|------|
| −3.8 | 0.0001 | 0.0001 | 0.0001 | 0.0001 | 0.0001 | 0.0001 | 0.0001 | 0.0001 | 0.0001 | 0.0001 |
| −3.7 | 0.0001 | 0.0001 | 0.0001 | 0.0001 | 0.0001 | 0.0001 | 0.0001 | 0.0001 | 0.0001 | 0.0001 |
| −3.6 | 0.0002 | 0.0002 | 0.0001 | 0.0001 | 0.0001 | 0.0001 | 0.0001 | 0.0001 | 0.0001 | 0.0001 |
| −3.5 | 0.0002 | 0.0002 | 0.0002 | 0.0002 | 0.0002 | 0.0002 | 0.0002 | 0.0002 | 0.0002 | 0.0002 |
| −3.4 | 0.0003 | 0.0003 | 0.0003 | 0.0003 | 0.0003 | 0.0003 | 0.0003 | 0.0003 | 0.0003 | 0.0002 |
| −3.3 | 0.0005 | 0.0005 | 0.0005 | 0.0004 | 0.0004 | 0.0004 | 0.0004 | 0.0004 | 0.0004 | 0.0003 |
| −3.2 | 0.0007 | 0.0007 | 0.0006 | 0.0006 | 0.0006 | 0.0006 | 0.0006 | 0.0005 | 0.0005 | 0.0005 |
| −3.1 | 0.0010 | 0.0009 | 0.0009 | 0.0009 | 0.0008 | 0.0008 | 0.0008 | 0.0008 | 0.0007 | 0.0007 |
| −3.0 | 0.0014 | 0.0013 | 0.0013 | 0.0012 | 0.0012 | 0.0011 | 0.0011 | 0.0011 | 0.0010 | 0.0010 |
| −2.9 | 0.0019 | 0.0018 | 0.0018 | 0.0017 | 0.0016 | 0.0016 | 0.0015 | 0.0015 | 0.0014 | 0.0014 |
| −2.8 | 0.0026 | 0.0025 | 0.0024 | 0.0023 | 0.0023 | 0.0022 | 0.0021 | 0.0021 | 0.0020 | 0.0019 |
| −2.7 | 0.0035 | 0.0034 | 0.0033 | 0.0032 | 0.0031 | 0.0030 | 0.0029 | 0.0028 | 0.0027 | 0.0026 |
| −2.6 | 0.0047 | 0.0045 | 0.0044 | 0.0043 | 0.0041 | 0.0040 | 0.0039 | 0.0038 | 0.0037 | 0.0036 |
| −2.5 | 0.0062 | 0.0060 | 0.0059 | 0.0059 | 0.0057 | 0.0055 | 0.0054 | 0.0052 | 0.0051 | 0.0049 | 0.0048 |
| −2.4 | 0.0082 | 0.0080 | 0.0078 | 0.0076 | 0.0073 | 0.0071 | 0.0069 | 0.0068 | 0.0068 | 0.0064 |
| −2.3 | 0.0107 | 0.0104 | 0.0102 | 0.0099 | 0.0096 | 0.0094 | 0.0091 | 0.0089 | 0.0087 | 0.0084 |
| −2.2 | 0.0139 | 0.0136 | 0.0132 | 0.0129 | 0.0125 | 0.0122 | 0.0119 | 0.0116 | 0.0113 | 0.0110 |
| −2.1 | 0.0179 | 0.0174 | 0.0170 | 0.0166 | 0.0162 | 0.0158 | 0.0154 | 0.0150 | 0.0146 | 0.0143 |
| −2.0 | 0.0228 | 0.0222 | 0.0217 | 0.0212 | 0.0207 | 0.0202 | 0.0197 | 0.0192 | 0.0188 | 0.0183 |
| −1.9 | 0.0287 | 0.0281 | 0.0274 | 0.0268 | 0.0262 | 0.0256 | 0.0250 | 0.0244 | 0.0239 | 0.0233 |
| −1.8 | 0.0359 | 0.0351 | 0.0344 | 0.0336 | 0.0329 | 0.0322 | 0.0314 | 0.0307 | 0.0301 | 0.0294 |
| −1.7 | 0.0446 | 0.0436 | 0.0427 | 0.0418 | 0.0409 | 0.0401 | 0.0392 | 0.0384 | 0.0375 | 0.0367 |
| −1.6 | 0.0548 | 0.0537 | 0.0526 | 0.0516 | 0.0505 | 0.0495 | 0.0485 | 0.0475 | 0.0465 | 0.0455 |
| −1.5 | 0.0668 | 0.0655 | 0.0643 | 0.0630 | 0.0618 | 0.0606 | 0.0594 | 0.0582 | 0.0571 | 0.0559 |

**TABLE A.1.** (*Continued*)

| z | 0.00 | 0.01 | 0.02 | 0.03 | 0.04 | 0.05 | 0.06 | 0.07 | 0.08 | 0.09 |
|---|------|------|------|------|------|------|------|------|------|------|
| −1.4 | 0.0808 | 0.0793 | 0.0778 | 0.0764 | 0.0749 | 0.0735 | 0.0721 | 0.0708 | 0.0694 | 0.0681 |
| −1.3 | 0.0968 | 0.0951 | 0.0934 | 0.0918 | 0.0901 | 0.0885 | 0.0869 | 0.0853 | 0.0838 | 0.0823 |
| −1.2 | 0.1151 | 0.1131 | 0.1112 | 0.1093 | 0.1075 | 0.1057 | 0.1038 | 0.1020 | 0.1003 | 0.0985 |
| −1.1 | 0.1357 | 0.1335 | 0.1314 | 0.1292 | 0.1271 | 0.1251 | 0.1230 | 0.1210 | 0.1190 | 0.1170 |
| −1.0 | 0.1587 | 0.1562 | 0.1539 | 0.1515 | 0.1492 | 0.1469 | 0.1446 | 0.1423 | 0.1401 | 0.1379 |
| −0.9 | 0.1841 | 0.1814 | 0.1788 | 0.1762 | 0.1736 | 0.1711 | 0.1685 | 0.1660 | 0.1635 | 0.1611 |
| −0.8 | 0.2119 | 0.2090 | 0.2061 | 0.2033 | 0.2005 | 0.1977 | 0.1949 | 0.1922 | 0.1894 | 0.1867 |
| −0.7 | 0.2420 | 0.2389 | 0.2358 | 0.2327 | 0.2297 | 0.2266 | 0.2236 | 0.2206 | 0.2177 | 0.2148 |
| −0.6 | 0.2743 | 0.2709 | 0.2676 | 0.2643 | 0.2611 | 0.2578 | 0.2546 | 0.2514 | 0.2483 | 0.2451 |
| −0.5 | 0.3085 | 0.3050 | 0.3015 | 0.2981 | 0.2946 | 0.2912 | 0.2877 | 0.2843 | 0.2810 | 0.2776 |
| −0.4 | 0.3446 | 0.3409 | 0.3372 | 0.3336 | 0.3300 | 0.3264 | 0.3228 | 0.3192 | 0.3156 | 0.3121 |
| −0.3 | 0.3821 | 0.3783 | 0.3745 | 0.3707 | 0.3669 | 0.3632 | 0.3594 | 0.3557 | 0.3520 | 0.3483 |
| −0.2 | 0.4207 | 0.4168 | 0.4129 | 0.4090 | 0.4052 | 0.4013 | 0.3974 | 0.3936 | 0.3897 | 0.3859 |
| −0.1 | 0.4602 | 0.4562 | 0.4522 | 0.4483 | 0.4443 | 0.4404 | 0.4364 | 0.4325 | 0.4286 | 0.4247 |
| −0.0 | 0.5000 | 0.4960 | 0.4920 | 0.4880 | 0.4840 | 0.4801 | 0.4761 | 0.4721 | 0.4681 | 0.4641 |
| 0.0 | 0.5000 | 0.5040 | 0.5080 | 0.5120 | 0.5160 | 0.5199 | 0.5239 | 0.5279 | 0.5319 | 0.5359 |
| 0.1 | 0.5398 | 0.5438 | 0.5478 | 0.5517 | 0.5557 | 0.5596 | 0.5636 | 0.5675 | 0.5714 | 0.5753 |
| 0.2 | 0.5793 | 0.5832 | 0.5871 | 0.5910 | 0.5948 | 0.5987 | 0.6026 | 0.6064 | 0.6103 | 0.6141 |
| 0.3 | 0.6179 | 0.6217 | 0.6255 | 0.6293 | 0.6331 | 0.6368 | 0.6406 | 0.6443 | 0.6480 | 0.6517 |
| 0.4 | 0.6554 | 0.6591 | 0.6628 | 0.6664 | 0.6700 | 0.6736 | 0.6772 | 0.6808 | 0.6844 | 0.6879 |
| 0.5 | 0.6915 | 0.6950 | 0.6985 | 0.7019 | 0.7054 | 0.7088 | 0.7123 | 0.7157 | 0.7190 | 0.7224 |
| 0.6 | 0.7257 | 0.7291 | 0.7324 | 0.7357 | 0.7389 | 0.7422 | 0.7454 | 0.7486 | 0.7517 | 0.7549 |
| 0.7 | 0.7580 | 0.7611 | 0.7642 | 0.7673 | 0.7703 | 0.7734 | 0.7764 | 0.7794 | 0.7823 | 0.7852 |
| 0.8 | 0.7881 | 0.7910 | 0.7939 | 0.7967 | 0.7995 | 0.8023 | 0.8051 | 0.8078 | 0.8106 | 0.8133 |
| 0.9 | 0.8159 | 0.8186 | 0.8212 | 0.8238 | 0.8264 | 0.8289 | 0.8315 | 0.8340 | 0.8365 | 0.8389 |
| 1.0 | 0.8413 | 0.8438 | 0.8461 | 0.8485 | 0.8508 | 0.8531 | 0.8554 | 0.8577 | 0.8599 | 0.8621 |
| 1.1 | 0.8643 | 0.8665 | 0.8686 | 0.8708 | 0.8729 | 0.8749 | 0.8770 | 0.8790 | 0.8810 | 0.8830 |
| 1.2 | 0.8849 | 0.8869 | 0.8888 | 0.8907 | 0.8925 | 0.8943 | 0.8962 | 0.8980 | 0.8997 | 0.9015 |
| 1.3 | 0.9032 | 0.9049 | 0.9066 | 0.9082 | 0.9099 | 0.9115 | 0.9131 | 0.9147 | 0.9162 | 0.9177 |
| 1.4 | 0.9192 | 0.9207 | 0.9222 | 0.9236 | 0.9251 | 0.9265 | 0.9279 | 0.9292 | 0.9306 | 0.9319 |
| 1.5 | 0.9332 | 0.9345 | 0.9357 | 0.9370 | 0.9382 | 0.9394 | 0.9406 | 0.9418 | 0.9429 | 0.9441 |
| 1.6 | 0.9452 | 0.9463 | 0.9474 | 0.9484 | 0.9495 | 0.9505 | 0.9515 | 0.9525 | 0.9535 | 0.9545 |
| 1.7 | 0.9554 | 0.9564 | 0.9573 | 0.9582 | 0.9591 | 0.9599 | 0.9608 | 0.9616 | 0.9625 | 0.9633 |
| 1.8 | 0.9641 | 0.9649 | 0.9656 | 0.9664 | 0.9671 | 0.9678 | 0.9686 | 0.9693 | 0.9699 | 0.9706 |
| 1.9 | 0.9713 | 0.9719 | 0.9726 | 0.9732 | 0.9738 | 0.9744 | 0.9750 | 0.9756 | 0.9761 | 0.9767 |
| 2.0 | 0.9772 | 0.9778 | 0.9783 | 0.9788 | 0.9793 | 0.9798 | 0.9803 | 0.9808 | 0.9812 | 0.9817 |
| 2.1 | 0.9821 | 0.9826 | 0.9830 | 0.9834 | 0.9838 | 0.9842 | 0.9846 | 0.9850 | 0.9854 | 0.9857 |
| 2.2 | 0.9861 | 0.9864 | 0.9868 | 0.9871 | 0.9875 | 0.9878 | 0.9881 | 0.9884 | 0.9887 | 0.9890 |
| 2.3 | 0.9893 | 0.9896 | 0.9898 | 0.9901 | 0.9904 | 0.9906 | 0.9909 | 0.9911 | 0.9913 | 0.9916 |
| 2.4 | 0.9918 | 0.9920 | 0.9922 | 0.9924 | 0.9927 | 0.9929 | 0.9931 | 0.9932 | 0.9934 | 0.9936 |
| 2.5 | 0.9938 | 0.9940 | 0.9941 | 0.9943 | 0.9945 | 0.9946 | 0.9948 | 0.9949 | 0.9951 | 0.9952 |
| 2.6 | 0.9953 | 0.9955 | 0.9956 | 0.9957 | 0.9959 | 0.9960 | 0.9961 | 0.9962 | 0.9963 | 0.9964 |
| 2.7 | 0.9965 | 0.9966 | 0.9967 | 0.9968 | 0.9969 | 0.9970 | 0.9971 | 0.9972 | 0.9973 | 0.9974 |

**TABLE A.1.** (*Concluded*)

| z | 0.00 | 0.01 | 0.02 | 0.03 | 0.04 | 0.05 | 0.06 | 0.07 | 0.08 | 0.09 |
|-----|--------|--------|--------|--------|--------|--------|--------|--------|--------|--------|
| 2.8 | 0.9974 | 0.9975 | 0.9976 | 0.9977 | 0.9977 | 0.9978 | 0.9979 | 0.9979 | 0.9980 | 0.9981 |
| 2.9 | 0.9981 | 0.9982 | 0.9982 | 0.9983 | 0.9984 | 0.9984 | 0.9985 | 0.9985 | 0.9986 | 0.9986 |
| 3.0 | 0.9986 | 0.9987 | 0.9987 | 0.9988 | 0.9988 | 0.9989 | 0.9989 | 0.9989 | 0.9990 | 0.9990 |
| 3.1 | 0.9990 | 0.9991 | 0.9991 | 0.9991 | 0.9992 | 0.9992 | 0.9992 | 0.9992 | 0.9993 | 0.9993 |
| 3.2 | 0.9993 | 0.9993 | 0.9994 | 0.9994 | 0.9994 | 0.9994 | 0.9994 | 0.9995 | 0.9995 | 0.9995 |
| 3.3 | 0.9995 | 0.9995 | 0.9995 | 0.9996 | 0.9996 | 0.9996 | 0.9996 | 0.9996 | 0.9996 | 0.9997 |
| 3.4 | 0.9997 | 0.9997 | 0.9997 | 0.9997 | 0.9997 | 0.9997 | 0.9997 | 0.9997 | 0.9997 | 0.9998 |
| 3.5 | 0.9998 | 0.9998 | 0.9998 | 0.9998 | 0.9998 | 0.9998 | 0.9998 | 0.9998 | 0.9998 | 0.9998 |
| 3.6 | 0.9998 | 0.9998 | 0.9999 | 0.9999 | 0.9999 | 0.9999 | 0.9999 | 0.9999 | 0.9999 | 0.9999 |
| 3.7 | 0.9999 | 0.9999 | 0.9999 | 0.9999 | 0.9999 | 0.9999 | 0.9999 | 0.9999 | 0.9999 | 0.9999 |
| 3.8 | 0.9999 | 0.9999 | 0.9999 | 0.9999 | 0.9999 | 0.9999 | 0.9999 | 0.9999 | 0.9999 | 0.9999 |
| 3.9 | 1.0000 | | | | | | | | | |

**TABLE A.2.** Percentiles of the Student's *t* Distribution

*Example*: The $q$ = 95th percentile for $\nu$ = 10 degrees of freedom is 1.812.

| $q$ \ $\nu$ | 60 | 75 | 90 | 95 | 97.5 | 99 | 99.5 | 99.95 |
|---|---|---|---|---|---|---|---|---|
| 1 | .325 | 1.000 | 3.078 | 6.314 | 12.706 | 31.821 | 63.657 | 636.619 |
| 2 | .289 | .816 | 1.886 | 2.920 | 4.303 | 6.965 | 9.925 | 31.598 |
| 3 | .277 | .765 | 1.638 | 2.353 | 3.182 | 4.541 | 5.841 | 12.941 |
| 4 | .271 | .741 | 1.533 | 2.132 | 2.776 | 3.747 | 4.604 | 8.610 |
| 5 | .267 | .727 | 1.476 | 2.015 | 2.571 | 3.365 | 4.032 | 6.859 |
| 6 | .265 | .718 | 1.440 | 1.943 | 2.447 | 3.143 | 3.707 | 5.959 |
| 7 | .263 | .711 | 1.415 | 1.895 | 2.365 | 2.998 | 3.499 | 5.405 |
| 8 | .262 | .706 | 1.397 | 1.860 | 2.306 | 2.896 | 3.355 | 5.041 |
| 9 | .261 | .703 | 1.383 | 1.833 | 2.262 | 2.821 | 3.250 | 4.781 |
| 10 | .260 | .700 | 1.372 | 1.812 | 2.228 | 2.764 | 3.169 | 4.587 |
| 11 | .260 | .697 | 1.363 | 1.796 | 2.201 | 2.718 | 3.106 | 4.437 |
| 12 | .259 | .695 | 1.356 | 1.782 | 2.179 | 2.681 | 3.055 | 4.318 |
| 13 | .259 | .694 | 1.350 | 1.771 | 2.160 | 2.650 | 3.012 | 4.221 |
| 14 | .258 | .692 | 1.345 | 1.761 | 2.145 | 2.624 | 2.977 | 4.140 |
| 15 | .258 | .691 | 1.341 | 1.753 | 2.131 | 2.602 | 2.947 | 4.073 |
| 16 | .258 | .690 | 1.337 | 1.746 | 2.120 | 2.583 | 2.921 | 4.015 |
| 17 | .257 | .689 | 1.333 | 1.740 | 2.110 | 2.567 | 2.898 | 3.965 |
| 18 | .257 | .688 | 1.330 | 1.734 | 2.101 | 2.552 | 2.878 | 3.922 |
| 19 | .257 | .688 | 1.328 | 1.729 | 2.093 | 2.539 | 2.861 | 3.883 |
| 20 | .257 | .687 | 1.325 | 1.725 | 2.086 | 2.528 | 2.845 | 3.850 |
| 21 | .257 | .686 | 1.323 | 1.721 | 2.080 | 2.518 | 2.831 | 3.819 |
| 22 | .256 | .686 | 1.321 | 1.717 | 2.074 | 2.508 | 2.819 | 3.792 |
| 23 | .256 | .685 | 1.319 | 1.714 | 2.069 | 2.500 | 2.807 | 3.767 |
| 24 | .256 | .685 | 1.318 | 1.711 | 2.064 | 2.492 | 2.797 | 3.745 |
| 25 | .256 | .684 | 1.316 | 1.708 | 2.060 | 2.485 | 2.787 | 3.725 |
| 26 | .256 | .684 | 1.315 | 1.706 | 2.056 | 2.479 | 2.779 | 3.707 |
| 27 | .256 | .684 | 1.314 | 1.703 | 2.052 | 2.473 | 2.771 | 3.690 |
| 28 | .256 | .683 | 1.313 | 1.701 | 2.048 | 2.467 | 2.763 | 3.674 |
| 29 | .256 | .683 | 1.311 | 1.699 | 2.045 | 2.462 | 2.756 | 3.659 |
| 30 | .256 | .683 | 1.310 | 1.697 | 2.042 | 2.457 | 2.750 | 3.646 |
| 40 | .255 | .681 | 1.303 | 1.684 | 2.021 | 2.423 | 2.704 | 3.551 |
| 60 | .254 | .679 | 1.296 | 1.671 | 2.000 | 2.390 | 2.660 | 3.460 |
| 120 | .254 | .677 | 1.289 | 1.658 | 1.980 | 2.358 | 2.617 | 3.373 |
| $\infty$ | .253 | .674 | 1.282 | 1.645 | 1.960 | 2.326 | 2.576 | 3.291 |

Reproduced with kind permission of the publisher and authors from Afifi and Azen, *Statistical Analysis: A Computer Oriented Approach*, 2nd ed., Academic Press, New York, 1979.

**TABLE A.3.** Percentiles of the Chi-Square Distribution

*Example*: The 90th percentile for d.f. = 15 is 22.307.

| d.f. \ % | 0.5 | 1 | 2.5 | 5 | 10 | 20 | 30 | 40 | 50 |
|---|---|---|---|---|---|---|---|---|---|
| 1 | 0.0001 | 0.0002 | 0.001 | 0.004 | 0.016 | 0.064 | 0.148 | 0.275 | 0.455 |
| 2 | 0.010 | 0.020 | 0.051 | 0.103 | 0.211 | 0.446 | 0.713 | 1.022 | 1.386 |
| 3 | 0.072 | 0.115 | 0.216 | 0.352 | 0.584 | 1.005 | 1.424 | 1.869 | 2.366 |
| 4 | 0.207 | 0.297 | 0.484 | 0.711 | 1.064 | 1.649 | 2.195 | 2.753 | 3.357 |
| 5 | 0.412 | 0.554 | 0.831 | 1.145 | 1.610 | 2.343 | 3.000 | 3.655 | 4.351 |
| 6 | 0.676 | 0.872 | 1.237 | 1.635 | 2.204 | 3.070 | 3.828 | 4.570 | 5.348 |
| 7 | 0.989 | 1.239 | 1.690 | 2.167 | 2.833 | 3.822 | 4.671 | 5.493 | 6.346 |
| 8 | 1.344 | 1.646 | 2.180 | 2.733 | 3.490 | 4.594 | 5.527 | 6.423 | 7.344 |
| 9 | 1.735 | 2.088 | 2.700 | 3.325 | 4.168 | 5.380 | 6.393 | 7.357 | 8.343 |
| 10 | 2.156 | 2.558 | 3.247 | 3.940 | 4.865 | 6.179 | 7.267 | 8.295 | 9.342 |
| 11 | 2.603 | 3.053 | 3.816 | 4.575 | 5.578 | 6.989 | 8.148 | 9.237 | 10.341 |
| 12 | 3.074 | 3.571 | 4.404 | 5.226 | 6.304 | 7.807 | 9.034 | 10.182 | 11.340 |
| 13 | 3.565 | 4.107 | 5.009 | 5.892 | 7.042 | 8.634 | 9.926 | 11.129 | 12.340 |
| 14 | 4.075 | 4.660 | 5.629 | 6.571 | 7.790 | 9.467 | 10.821 | 12.078 | 13.339 |
| 15 | 4.601 | 5.229 | 6.262 | 7.261 | 8.547 | 10.307 | 11.721 | 13.030 | 14.339 |
| 16 | 5.142 | 5.812 | 6.908 | 7.962 | 9.312 | 11.152 | 12.624 | 13.983 | 15.338 |
| 17 | 5.697 | 6.408 | 7.564 | 8.672 | 10.085 | 12.002 | 13.531 | 14.937 | 16.338 |
| 18 | 6.265 | 7.015 | 8.231 | 9.390 | 10.865 | 12.857 | 14.440 | 15.893 | 17.338 |
| 19 | 6.844 | 7.633 | 8.907 | 10.117 | 11.651 | 13.716 | 15.352 | 16.850 | 18.338 |
| 20 | 7.434 | 8.260 | 9.591 | 10.851 | 12.443 | 14.578 | 16.266 | 17.809 | 19.337 |
| 21 | 8.034 | 8.897 | 10.283 | 11.591 | 13.240 | 15.445 | 17.182 | 18.768 | 20.337 |
| 22 | 8.643 | 9.542 | 10.982 | 12.338 | 14.041 | 16.314 | 18.101 | 19.729 | 21.337 |
| 23 | 9.260 | 10.196 | 11.689 | 13.091 | 14.848 | 17.187 | 19.021 | 20.690 | 22.337 |
| 24 | 9.886 | 10.856 | 12.401 | 13.848 | 15.659 | 18.062 | 19.943 | 21.752 | 23.337 |
| 25 | 10.520 | 11.524 | 13.120 | 14.611 | 16.473 | 18.940 | 20.867 | 22.616 | 24.337 |
| 26 | 11.160 | 12.198 | 13.844 | 15.379 | 17.292 | 19.820 | 21.792 | 23.579 | 25.336 |
| 27 | 11.808 | 12.879 | 14.573 | 16.151 | 18.114 | 20.703 | 22.719 | 24.544 | 26.336 |
| 28 | 12.461 | 13.565 | 15.308 | 16.928 | 18.939 | 21.588 | 23.647 | 25.509 | 27.336 |
| 29 | 13.121 | 14.256 | 16.047 | 17.708 | 19.768 | 22.475 | 24.577 | 26.475 | 28.336 |
| 30 | 13.787 | 14.953 | 16.791 | 18.493 | 20.599 | 23.364 | 25.508 | 27.442 | 29.336 |
| 35 | 17.192 | 18.509 | 20.569 | 22.465 | 24.797 | 27.836 | 30.178 | 32.282 | 34.336 |
| 40 | 20.707 | 22.164 | 24.433 | 26.509 | 29.051 | 32.345 | 34.872 | 37.134 | 39.335 |
| 45 | 24.311 | 25.901 | 28.366 | 30.612 | 33.350 | 36.884 | 39.585 | 41.995 | 44.335 |
| 50 | 27.991 | 29.707 | 32.357 | 34.764 | 37.689 | 41.449 | 44.313 | 46.864 | 49.335 |
| 60 | 35.534 | 37.485 | 40.482 | 43.188 | 46.459 | 50.641 | 53.809 | 56.620 | 59.335 |
| 70 | 43.275 | 45.442 | 48.758 | 51.739 | 55.329 | 59.898 | 63.346 | 66.396 | 69.334 |
| 80 | 51.172 | 53.540 | 57.153 | 60.391 | 64.278 | 69.207 | 72.915 | 76.188 | 79.334 |

**TABLE A.3.** (*Continued*)

| d.f. \ % | 0.5 | 1 | 2.5 | 5 | 10 | 20 | 30 | 40 | 50 |
|---|---|---|---|---|---|---|---|---|---|
| 90 | 59.196 | 61.754 | 65.647 | 69.126 | 73.291 | 78.558 | 82.511 | 85.993 | 89.334 |
| 100 | 67.328 | 70.065 | 74.222 | 77.929 | ⁻82.358 | 87.945 | 92.129 | 95.808 | 99.334 |
| 120 | 83.852 | 86.923 | 91.573 | 95.705 | 100.624 | 106.806 | 111.419 | 115.465 | 119.334 |
| 140 | 100.655 | 104.034 | 109.137 | 113.659 | 119.029 | 125.758 | 130.766 | 135.149 | 139.334 |
| 160 | 117.679 | 121.346 | 126.870 | 131.756 | 137.546 | 144.783 | 150.158 | 154.856 | 159.334 |
| 180 | 134.884 | 138.820 | 144.741 | 149.969 | 156.153 | 163.868 | 169.588 | 174.580 | 179.334 |
| 200 | 152.241 | 156.432 | 162.728 | 168.279 | 174.835 | 183.003 | 189.049 | 194.319 | 199.344 |

| d.f. \ % | 60 | 70 | 80 | 90 | 95 | 97.5 | 99 | 99.5 | 99.95 |
|---|---|---|---|---|---|---|---|---|---|
| 1 | 0.708 | 1.074 | 1.642 | 2.706 | 3.841 | 5.024 | 6.635 | 7.879 | 12.116 |
| 2 | 1.833 | 2.408 | 3.219 | 4.605 | 5.991 | 7.378 | 9.210 | 10.597 | 15.202 |
| 3 | 2.946 | 3.665 | 4.642 | 6.251 | 7.815 | 9.348 | 11.345 | 12.838 | 17.730 |
| 4 | 4.045 | 4.878 | 5.989 | 7.779 | 9.488 | 11.143 | 13.277 | 14.860 | 19.997 |
| 5 | 5.132 | 6.064 | 7.289 | 9.236 | 11.070 | 12.833 | 15.086 | 16.750 | 22.105 |
| 6 | 6.211 | 7.231 | 8.558 | 10.645 | 12.592 | 14.449 | 16.812 | 18.548 | 24.103 |
| 7 | 7.283 | 8.383 | 9.803 | 12.017 | 14.067 | 16.013 | 18.475 | 20.278 | 26.018 |
| 8 | 8.351 | 9.524 | 11.030 | 13.362 | 15.507 | 17.535 | 20.090 | 21.955 | 27.868 |
| 9 | 9.414 | 10.656 | 12.242 | 14.684 | 16.919 | 19.023 | 21.666 | 23.589 | 29.666 |
| 10 | 10.473 | 11.781 | 13.442 | 15.987 | 18.307 | 20.483 | 23.209 | 25.188 | 31.420 |
| 11 | 11.530 | 12.899 | 14.631 | 17.275 | 19.675 | 21.920 | 24.725 | 26.757 | 33.137 |
| 12 | 12.584 | 14.011 | 15.812 | 18.549 | 21.026 | 23.337 | 26.217 | 28.300 | 34.821 |
| 13 | 13.636 | 15.119 | 16.985 | 19.812 | 22.362 | 24.736 | 27.688 | 29.819 | 36.478 |
| 14 | 14.685 | 16.222 | 18.151 | 21.064 | 23.685 | 26.119 | 29.141 | 31.319 | 38.109 |
| 15 | 15.733 | 17.322 | 19.311 | 22.307 | 24.996 | 27.488 | 30.578 | 32.801 | 39.719 |
| 16 | 16.780 | 18.418 | 20.465 | 23.542 | 26.296 | 28.845 | 32.000 | 34.267 | 41.308 |
| 17 | 17.824 | 19.511 | 21.615 | 24.769 | 27.587 | 30.191 | 33.409 | 35.718 | 42.879 |
| 18 | 18.868 | 20.601 | 22.760 | 25.989 | 28.869 | 31.526 | 34.805 | 37.156 | 44.434 |
| 19 | 19.910 | 21.689 | 23.900 | 27.204 | 30.144 | 32.852 | 36.191 | 38.582 | 45.973 |
| 20 | 20.951 | 22.775 | 25.038 | 28.412 | 31.410 | 34.170 | 37.566 | 39.997 | 47.498 |
| 21 | 21.991 | 23.858 | 26.171 | 29.615 | 32.671 | 35.479 | 38.932 | 41.401 | 49.011 |
| 22 | 23.031 | 24.939 | 27.301 | 30.813 | 33.924 | 36.781 | 40.289 | 42.796 | 50.511 |
| 23 | 24.069 | 26.018 | 28.429 | 32.007 | 35.172 | 38.076 | 41.638 | 44.181 | 52.000 |
| 24 | 25.106 | 27.096 | 29.553 | 33.196 | 36.415 | 39.364 | 42.980 | 45.559 | 53.479 |
| 25 | 26.143 | 28.172 | 30.675 | 34.382 | 37.652 | 40.646 | 44.314 | 46.928 | 54.947 |
| 26 | 27.179 | 29.246 | 31.795 | 35.563 | 38.885 | 41.923 | 45.642 | 48.290 | 56.407 |
| 27 | 28.214 | 30.319 | 32.912 | 36.741 | 40.113 | 43.195 | 46.963 | 49.645 | 57.858 |
| 28 | 29.249 | 31.391 | 34.027 | 37.916 | 41.337 | 44.461 | 48.278 | 50.993 | 59.300 |
| 29 | 30.283 | 32.461 | 35.139 | 39.087 | 42.557 | 45.722 | 49.588 | 52.336 | 60.735 |
| 30 | 31.316 | 33.530 | 36.250 | 40.256 | 43.773 | 46.979 | 50.892 | 53.672 | 62.162 |
| 35 | 36.475 | 38.859 | 41.778 | 46.059 | 49.802 | 53.203 | 57.342 | 60.275 | 69.199 |
| 40 | 41.622 | 44.165 | 47.269 | 51.805 | 56.758 | 59.342 | 63.691 | 66.766 | 76.095 |
| 45 | 46.761 | 49.452 | 52.729 | 57.505 | 61.656 | 65.410 | 69.957 | 73.166 | 82.876 |
| 50 | 51.892 | 54.723 | 58.164 | 63.167 | 67.505 | 71.420 | 76.154 | 79.490 | 89.561 |
| 60 | 62.135 | 65.227 | 68.972 | 74.397 | 79.082 | 83.298 | 88.379 | 91.952 | 102.695 |
| 70 | 72.358 | 75.689 | 79.715 | 85.527 | 90.531 | 95.023 | 100.425 | 104.215 | 115.578 |

**TABLE A.3.** (*Concluded*)

| %<br>d.f. | 60 | 70 | 80 | 90 | 95 | 97.5 | 99 | 99.5 | 99.95 |
|---|---|---|---|---|---|---|---|---|---|
| 80 | 82.566 | 86.120 | 90.405 | 96.578 | 101.879 | 106.629 | 112.329 | 116.321 | 128.261 |
| 90 | 92.761 | 96.524 | 101.054 | 107.565 | 113.145 | 118.136 | 124.116 | 128.299 | 140.782 |
| 100 | 102.946 | 106.906 | 111.667 | 118.498 | 124.342 | 129.561 | 135.807 | 140.169 | 153.167 |
| 120 | 123.289 | 127.616 | 132.806 | 140.233 | 146.567 | 152.211 | 158.950 | 163.648 | 177.603 |
| 140 | 143.604 | 148.269 | 153.854 | 161.827 | 168.613 | 174.648 | 181.840 | 186.847 | 201.683 |
| 160 | 163.898 | 168.876 | 174.828 | 183.311 | 190.516 | 196.915 | 204.530 | 209.824 | 225.481 |
| 180 | 184.173 | 189.446 | 195.743 | 204.704 | 212.304 | 219.044 | 227.056 | 232.620 | 249.048 |
| 200 | 204.434 | 209.985 | 216.609 | 226.021 | 233.994 | 241.058 | 249.445 | 255.264 | 272.423 |

**TABLE A.4.** Percentiles of the $F$ Distribution

a. Percentiles for $P = 0.25$, 0.15, and 0.10, with 1 and $v_2$ Degrees of Freedom.

*Example*: The 75th percentile ($P = 0.25$) with 1 and $v_2 = 10$ df is 1.4915.

| $v_2$ | $v_1 = 1$ | | |
|---|---|---|---|
| | $P = 0.25$ | $P = 0.15$ | $P = 0.10$ |
| 1 | 5.8285 | 17.3497 | 39.864 |
| 2 | 2.5714 | 5.2072 | 8.5263 |
| 3 | 2.0239 | 3.7030 | 5.5383 |
| 4 | 1.8074 | 3.1620 | 4.5448 |
| 5 | 1.6925 | 2.8878 | 4.0604 |
| 6 | 1.6214 | 2.7231 | 3.7760 |
| 7 | 1.5732 | 2.6134 | 3.5894 |
| 8 | 1.5384 | 2.5352 | 3.4579 |
| 9 | 1.5121 | 2.4766 | 3.3603 |
| 10 | 1.4915 | 2.4312 | 3.2850 |
| 11 | 1.4749 | 2.3949 | 3.2252 |
| 12 | 1.4613 | 2.3653 | 3.1765 |
| 13 | 1.4500 | 2.3407 | 3.1362 |
| 14 | 1.4403 | 2.3198 | 3.1022 |
| 15 | 1.4321 | 2.3020 | 3.0732 |
| 16 | 1.4249 | 2.2865 | 3.0481 |
| 17 | 1.4186 | 2.2730 | 3.0262 |
| 18 | 1.4130 | 2.2611 | 3.0070 |
| 19 | 1.4081 | 2.2506 | 2.9899 |
| 20 | 1.4037 | 2.2411 | 2.9747 |
| 21 | 1.3997 | 2.2326 | 2.9609 |
| 22 | 1.3961 | 2.2249 | 2.9486 |
| 23 | 1.3928 | 2.2179 | 2.9374 |
| 24 | 1.3898 | 2.2116 | 2.9271 |
| 25 | 1.3870 | 2.2057 | 2.9177 |
| 26 | 1.3845 | 2.2004 | 2.9091 |
| 27 | 1.3822 | 2.1954 | 2.9012 |
| 28 | 1.3800 | 2.1908 | 2.8939 |
| 29 | 1.3780 | 2.1866 | 2.8871 |
| 30 | 1.3761 | 2.1826 | 2.8807 |
| 40 | 1.3626 | 2.1542 | 2.8354 |
| 60 | 1.3493 | 2.1264 | 2.7914 |
| 120 | 1.3362 | 2.0990 | 2.7478 |
| $\infty$ | 1.3233 | 2.0722 | 2.7055 |

**TABLE A.4.** (*Continued*)

b. Percentiles for $P = 0.05$, $0.01$, and $0.005$

*Example*: The 95th percentile ($P = 0.05$) for $v_1 = 4$ and $v_2 = 15$ is 3.0556.

### P = 0.05

| $v_2$ \ $v_1$ | 1 | 2 | 3 | 4 | 5 | 6 | 7 | 8 | 9 |
|---|---|---|---|---|---|---|---|---|---|
| 1 | 161·45 | 199·50 | 215·71 | 224·58 | 230·16 | 233·99 | 236·77 | 238·88 | 240·54 |
| 2 | 18·513 | 19·000 | 19·164 | 19·247 | 19·296 | 19·330 | 19·353 | 19·371 | 19·385 |
| 3 | 10·128 | 9·5521 | 9·2766 | 9·1172 | 9·0135 | 8·9406 | 8·8868 | 8·8452 | 8·8123 |
| 4 | 7·7086 | 6·9443 | 6·5914 | 6·3883 | 6·2560 | 6·1631 | 6·0942 | 6·0410 | 5·9988 |
| 5 | 6·6079 | 5·7861 | 5·4095 | 5·1922 | 5·0503 | 4·9503 | 4·8759 | 4·8183 | 4·7725 |
| 6 | 5·9874 | 5·1433 | 4·7571 | 4·5337 | 4·3874 | 4·2839 | 4·2066 | 4·1468 | 4·0990 |
| 7 | 5·5914 | 4·7374 | 4·3468 | 4·1203 | 3·9715 | 3·8660 | 3·7870 | 3·7257 | 3·6767 |
| 8 | 5·3177 | 4·4590 | 4·0662 | 3·8378 | 3·6875 | 3·5806 | 3·5005 | 3·4381 | 3·3881 |
| 9 | 5·1174 | 4·2565 | 3·8626 | 3·6331 | 3·4817 | 3·3738 | 3·2927 | 3·2296 | 3·1789 |
| 10 | 4·9646 | 4·1028 | 3·7083 | 3·4780 | 3·3258 | 3·2172 | 3·1355 | 3·0717 | 3·0204 |
| 11 | 4·8443 | 3·9823 | 3·5874 | 3·3567 | 3·2039 | 3·0946 | 3·0123 | 2·9480 | 2·8962 |
| 12 | 4·7472 | 3·8853 | 3·4903 | 3·2592 | 3·1059 | 2·9961 | 2·9134 | 2·8486 | 2·7964 |
| 13 | 4·6672 | 3·8056 | 3·4105 | 3·1791 | 3·0254 | 2·9153 | 2·8321 | 2·7669 | 2·7144 |
| 14 | 4·6001 | 3·7389 | 3·3439 | 3·1122 | 2·9582 | 2·8477 | 2·7642 | 2·6987 | 2·6458 |
| 15 | 4·5431 | 3·6823 | 3·2874 | 3·0556 | 2·9013 | 2·7905 | 2·7066 | 2·6408 | 2·5876 |
| 16 | 4·4940 | 3·6337 | 3·2389 | 3·0069 | 2·8524 | 2·7413 | 2·6572 | 2·5911 | 2·5377 |
| 17 | 4·4513 | 3·5915 | 3·1968 | 2·9647 | 2·8100 | 2·6987 | 2·6143 | 2·5480 | 2·4943 |
| 18 | 4·4139 | 3·5546 | 3·1599 | 2·9277 | 2·7729 | 2·6613 | 2·5767 | 2·5102 | 2·4563 |
| 19 | 4·3808 | 3·5219 | 3·1274 | 2·8951 | 2·7401 | 2·6283 | 2·5435 | 2·4768 | 2·4227 |
| 20 | 4·3513 | 3·4928 | 3·0984 | 2·8661 | 2·7109 | 2·5990 | 2·5140 | 2·4471 | 2·3928 |
| 21 | 4·3248 | 3·4668 | 3·0725 | 2·8401 | 2·6848 | 2·5727 | 2·4876 | 2·4205 | 2·3661 |
| 22 | 4·3009 | 3·4434 | 3·0491 | 2·8167 | 2·6613 | 2·5491 | 2·4638 | 2·3965 | 2·3419 |
| 23 | 4·2793 | 3·4221 | 3·0280 | 2·7955 | 2·6400 | 2·5277 | 2·4422 | 2·3748 | 2·3201 |
| 24 | 4·2597 | 3·4028 | 3·0088 | 2·7763 | 2·6207 | 2·5082 | 2·4226 | 2·3551 | 2·3002 |
| 25 | 4·2417 | 3·3852 | 2·9912 | 2·7587 | 2·6030 | 2·4904 | 2·4047 | 2·3371 | 2·2821 |
| 26 | 4·2252 | 3·3690 | 2·9751 | 2·7426 | 2·5868 | 2·4741 | 2·3883 | 2·3205 | 2·2655 |
| 27 | 4·2100 | 3·3541 | 2·9604 | 2·7278 | 2·5719 | 2·4591 | 2·3732 | 2·3053 | 2·2501 |
| 28 | 4·1960 | 3·3404 | 2·9467 | 2·7141 | 2·5581 | 2·4453 | 2·3593 | 2·2913 | 2·2360 |
| 29 | 4·1830 | 3·3277 | 2·9340 | 2·7014 | 2·5454 | 2·4324 | 2·3463 | 2·2782 | 2·2229 |
| 30 | 4·1709 | 3·3158 | 2·9223 | 2·6896 | 2·5336 | 2·4205 | 2·3343 | 2·2662 | 2·2107 |
| 40 | 4·0848 | 3·2317 | 2·8387 | 2·6060 | 2·4495 | 2·3359 | 2·2490 | 2·1802 | 2·1240 |
| 60 | 4·0012 | 3·1504 | 2·7581 | 2·5252 | 2·3683 | 2·2540 | 2·1665 | 2·0970 | 2·0401 |
| 120 | 3·9201 | 3·0718 | 2·6802 | 2·4472 | 2·2900 | 2·1750 | 2·0867 | 2·0164 | 1·9588 |
| ∞ | 3·8415 | 2·9957 | 2·6049 | 2·3719 | 2·2141 | 2·0986 | 2·0096 | 1·9384 | 1·8799 |

**TABLE A.4.** (*Continued*)

P = 0.05

| $\nu_1$ / $\nu_2$ | 10 | 12 | 15 | 20 | 24 | 30 | 40 | 60 | 120 | ∞ |
|---|---|---|---|---|---|---|---|---|---|---|
| 1 | 241·88 | 243·91 | 245·95 | 248·01 | 249·05 | 250·09 | 251·14 | 252·20 | 253·25 | 254·32 |
| 2 | 19·396 | 19·413 | 19·429 | 19·446 | 19·454 | 19·462 | 19·471 | 19·479 | 19·487 | 19·496 |
| 3 | 8·7855 | 8·7446 | 8·7029 | 8·6602 | 8·6385 | 8·6166 | 8·5944 | 8·5720 | 8·5494 | 8·5265 |
| 4 | 5·9644 | 5·9117 | 5·8578 | 5·8025 | 5·7744 | 5·7459 | 5·7170 | 5·6878 | 5·6581 | 5·6281 |
| 5 | 4·7351 | 4·6777 | 4·6188 | 4·5581 | 4·5272 | 4·4957 | 4·4638 | 4·4314 | 4·3984 | 4·3650 |
| 6 | 4·0600 | 3·9999 | 3·9381 | 3·8742 | 3·8415 | 3·8082 | 3·7743 | 3·7398 | 3·7047 | 3·6688 |
| 7 | 3·6365 | 3·5747 | 3·5108 | 3·4445 | 3·4105 | 3·3758 | 3·3404 | 3·3043 | 3·2674 | 3·2298 |
| 8 | 3·3472 | 3·2840 | 3·2184 | 3·1503 | 3·1152 | 3·0794 | 3·0428 | 3·0053 | 2·9669 | 2·9276 |
| 9 | 3·1373 | 3·0729 | 3·0061 | 2·9365 | 2·9005 | 2·8637 | 2·8259 | 2·7872 | 2·7475 | 2·7067 |
| 10 | 2·9782 | 2·9130 | 2·8450 | 2·7740 | 2·7372 | 2·6996 | 2·6609 | 2·6211 | 2·5801 | 2·5379 |
| 11 | 2·8536 | 2·7876 | 2·7186 | 2·6464 | 2·6090 | 2·5705 | 2·5309 | 2·4901 | 2·4480 | 2·4045 |
| 12 | 2·7534 | 2·6866 | 2·6169 | 2·5436 | 2·5055 | 2·4663 | 2·4259 | 2·3842 | 2·3410 | 2·2962 |
| 13 | 2·6710 | 2·6037 | 2·5331 | 2·4589 | 2·4202 | 2·3803 | 2·3392 | 2·2966 | 2·2524 | 2·2064 |
| 14 | 2·6021 | 2·5342 | 2·4630 | 2·3879 | 2·3487 | 2·3082 | 2·2664 | 2·2230 | 2·1778 | 2·1307 |
| 15 | 2·5437 | 2·4753 | 2·4035 | 2·3275 | 2·2878 | 2·2468 | 2·2043 | 2·1601 | 2·1141 | 2·0658 |
| 16 | 2·4935 | 2·4247 | 2·3522 | 2·2756 | 2·2354 | 2·1938 | 2·1507 | 2·1058 | 2·0589 | 2·0096 |
| 17 | 2·4499 | 2·3807 | 2·3077 | 2·2304 | 2·1898 | 2·1477 | 2·1040 | 2·0584 | 2·0107 | 1·9604 |
| 18 | 2·4117 | 2·3421 | 2·2686 | 2·1906 | 2·1497 | 2·1071 | 2·0629 | 2·0166 | 1·9681 | 1·9168 |
| 19 | 2·3779 | 2·3080 | 2·2341 | 2·1555 | 2·1141 | 2·0712 | 2·0264 | 1·9796 | 1·9302 | 1·8780 |
| 20 | 2·3479 | 2·2776 | 2·2033 | 2·1242 | 2·0825 | 2·0391 | 1·9938 | 1·9464 | 1·8963 | 1·8432 |
| 21 | 2·3210 | 2·2504 | 2·1757 | 2·0960 | 2·0540 | 2·0102 | 1·9645 | 1·9165 | 1·8657 | 1·8117 |
| 22 | 2·2967 | 2·2258 | 2·1508 | 2·0707 | 2·0283 | 1·9842 | 1·9380 | 1·8895 | 1·8380 | 1·7831 |
| 23 | 2·2747 | 2·2036 | 2·1282 | 2·0476 | 2·0050 | 1·9605 | 1·9139 | 1·8649 | 1·8128 | 1·7570 |
| 24 | 2·2547 | 2·1834 | 2·1077 | 2·0267 | 1·9838 | 1·9390 | 1·8920 | 1·8424 | 1·7897 | 1·7331 |
| 25 | 2·2365 | 2·1649 | 2·0889 | 2·0075 | 1·9643 | 1·9192 | 1·8718 | 1·8217 | 1·7684 | 1·7110 |
| 26 | 2·2197 | 2·1479 | 2·0716 | 1·9898 | 1·9464 | 1·9010 | 1·8533 | 1·8027 | 1·7488 | 1·6906 |
| 27 | 2·2043 | 2·1323 | 2·0558 | 1·9736 | 1·929C | 1·8842 | 1·8361 | 1·7851 | 1·7307 | 1·6717 |
| 28 | 2·1900 | 2·1179 | 2·0411 | 1·9586 | 1·9147 | 1·8687 | 1·8203 | 1·7689 | 1·7138 | 1·6541 |
| 29 | 2·1768 | 2·1045 | 2·0275 | 1·9446 | 1·9005 | 1·8543 | 1·8055 | 1·7537 | 1·6981 | 1·6377 |
| 30 | 2·1646 | 2·0921 | 2·0148 | 1·9317 | 1·8874 | 1·8409 | 1·7918 | 1·7396 | 1·6835 | 1·6223 |
| 40 | 2·0772 | 2·0035 | 1·9245 | 1·8389 | 1·7929 | 1·7444 | 1·6928 | 1·6373 | 1·5766 | 1·5089 |
| 60 | 1·9926 | 1·9174 | 1·8364 | 1·7480 | 1·7001 | 1·6491 | 1·5943 | 1·5343 | 1·4673 | 1·3893 |
| 120 | 1·9105 | 1·8337 | 1·7505 | 1·6587 | 1·6084 | 1·5543 | 1·4952 | 1·4290 | 1·3519 | 1·2539 |
| ∞ | 1·8307 | 1·7522 | 1·6664 | 1·5705 | 1·5173 | 1·4591 | 1·3940 | 1·3180 | 1·2214 | 1·0000 |

**TABLE A.4.** (*Continued*)

$$P = 0.01$$

| $v_2$ \ $v_1$ | 1 | 2 | 3 | 4 | 5 | 6 | 7 | 8 | 9 |
|---|---|---|---|---|---|---|---|---|---|
| 1 | 4052·2 | 4999·5 | 5403·3 | 5624·6 | 5763·7 | 5859·0 | 5928·3 | 5981·6 | 6022·5 |
| 2 | 98·503 | 99·000 | 99·166 | 99·249 | 99·299 | 99·332 | 99·356 | 99·374 | 99·388 |
| 3 | 34·116 | 30·817 | 29·457 | 28·710 | 28·237 | 27·911 | 27·672 | 27·489 | 27·345 |
| 4 | 21·198 | 18·000 | 16·694 | 15·977 | 15·522 | 15·207 | 14·976 | 14·799 | 14·659 |
| 5 | 16·258 | 13·274 | 12·060 | 11·392 | 10·967 | 10·672 | 10·456 | 10·289 | 10·158 |
| 6 | 13·745 | 10·925 | 9·7795 | 9·1483 | 8·7459 | 8·4661 | 8·2600 | 8·1016 | 7·9761 |
| 7 | 12·246 | 9·5466 | 8·4513 | 7·8467 | 7·4604 | 7·1914 | 6·9928 | 6·8401 | 6·7188 |
| 8 | 11·259 | 8·6491 | 7·5910 | 7·0060 | 6·6318 | 6·3707 | 6·1776 | 6·0289 | 5·9106 |
| 9 | 10·561 | 8·0215 | 6·9919 | 6·4221 | 6·0569 | 5·8018 | 5·6129 | 5·4671 | 5·3511 |
| 10 | 10·044 | 7·5594 | 6·5523 | 5·9943 | 5·6363 | 5·3858 | 5·2001 | 5·0567 | 4·9424 |
| 11 | 9·6460 | 7·2057 | 6·2167 | 5·6683 | 5·3160 | 5·0692 | 4·8861 | 4·7445 | 4·6315 |
| 12 | 9·3302 | 6·9266 | 5·9526 | 5·4119 | 5·0643 | 4·8206 | 4·6395 | 4·4994 | 4·3875 |
| 13 | 9·0738 | 6·7010 | 5·7394 | 5·2053 | 4·8616 | 4·6204 | 4·4410 | 4·3021 | 4·1911 |
| 14 | 8·8616 | 6·5149 | 5·5639 | 5·0354 | 4·6950 | 4·4558 | 4·2779 | 4·1399 | 4·0297 |
| 15 | 8·6831 | 6·3589 | 5·4170 | 4·8932 | 4·5556 | 4·3183 | 4·1415 | 4·0045 | 3·8948 |
| 16 | 8·5310 | 6·2262 | 5·2922 | 4·7726 | 4·4374 | 4·2016 | 4·0259 | 3·8896 | 3·7804 |
| 17 | 8·3997 | 6·1121 | 5·1850 | 4·6690 | 4·3359 | 4·1015 | 3·9267 | 3·7910 | 3·6822 |
| 18 | 8·2854 | 6·0129 | 5·0919 | 4·5790 | 4·2479 | 4·0146 | 3·8406 | 3·7054 | 3·5971 |
| 19 | 8·1850 | 5·9259 | 5·0103 | 4·5003 | 4·1708 | 3·9386 | 3·7653 | 3·6305 | 3·5225 |
| 20 | 8·0960 | 5·8489 | 4·9382 | 4·4307 | 4·1027 | 3·8714 | 3·6987 | 3·5644 | 3·4567 |
| 21 | 8·0166 | 5·7804 | 4·8740 | 4·3688 | 4·0421 | 3·8117 | 3·6396 | 3·5056 | 3·3981 |
| 22 | 7·9454 | 5·7190 | 4·8166 | 4·3134 | 3·9880 | 3·7583 | 3·5867 | 3·4530 | 3·3458 |
| 23 | 7·8811 | 5·6637 | 4·7649 | 4·2635 | 3·9392 | 3·7102 | 3·5390 | 3·4057 | 3·2986 |
| 24 | 7·8229 | 5·6136 | 4·7181 | 4·2184 | 3·8951 | 3·6667 | 3·4959 | 3·3629 | 3·2560 |
| 25 | 7·7698 | 5·5680 | 4·6755 | 4·1774 | 3·8550 | 3·6272 | 3·4568 | 3;3239 | 3·2172 |
| 26 | 7·7213 | 5·5263 | 4·6366 | 4·1400 | 3·8183 | 3·5911 | 3·4210 | 3·2884 | 3·1818 |
| 27 | 7·6767 | 5·4881˙ | 4·6009 | 4·1056 | 3·7848 | 3·5580 | 3·3882 | 3·2558 | 3·1494 |
| 28 | 7·6356 | 5·4529 | 4·5681 | 4·0740 | 3·7539 | 3·5276 | 3·3581 | 3·2259 | 3·1195 |
| 29 | 7·5976 | 5·4205 | 4·5378 | 4·0449 | 3·7254 | 3·4995 | 3·3302 | 3·1982 | 3·0920 |
| 30 | 7·5625 | 5·3904 | 4·5097 | 4·0179 | 3·6990 | 3·4735 | 3·3045 | 3·1726 | 3·0665 |
| 40 | 7·3141 | 5·1785 | 4·3126 | 3·8283 | 3·5138 | 3·2910 | 3·1238 | 2·9930 | 2·8876 |
| 60 | 7·0771 | 4·9774 | 4·1259 | 3·6491 | 3·3389 | 3·1187 | 2·9530 | 2·8233 | 2·7185 |
| 120 | 6·8510 | 4·7865 | 3·9493 | 3·4796 | 3·1735 | 2·9559 | 2·7918 | 2·6629 | 2·5586 |
| ∞ | 6·6349 | 4·6052 | 3·7816 | 3·3192 | 3·0173 | 2·8020 | 2·6393 | 2·5113 | 2·4073 |

**TABLE A.4.** (*Continued*)

$$P = 0.01$$

| $\nu_1$ / $\nu_2$ | 10 | 12 | 15 | 20 | 24 | 30 | 40 | 60 | 120 | ∞ |
|---|---|---|---|---|---|---|---|---|---|---|
| 1 | 6055·8 | 6106·3 | 6157·3 | 6208·7 | 6234·6 | 6260·7 | 6286·8 | 6313·0 | 6339·4 | 6366·0 |
| 2 | 99·399 | 99·416 | 99·432 | 99·449 | 99·458 | 99·466 | 99·474 | 99·483 | 99·491 | 99·501 |
| 3 | 27·229 | 27·052 | 26·872 | 26·690 | 26·598 | 26·505 | 26·411 | 26·316 | 26·221 | 26·125 |
| 4 | 14·546 | 14·374 | 14·198 | 14·020 | 13·929 | 13·838 | 13·745 | 13·652 | 13·558 | 13·463 |
| 5 | 10·051 | 9·8883 | 9·7222 | 9·5527 | 9·4665 | 9·3793 | 9·2912 | 9·2020 | 9·1118 | 9·0204 |
| 6 | 7·8741 | 7·7183 | 7·5590 | 7·3958 | 7·3127 | 7·2285 | 7·1432 | 7·0568 | 6·9690 | 6·8801 |
| 7 | 6·6201 | 6·4691 | 6·3143 | 6·1554 | 6·0743 | 5·9921 | 5·9084 | 5·8236 | 5·7372 | 5·6495 |
| 8 | 5·8143 | 5·6668 | 5·5151 | 5·3591 | 5·2793 | 5·1981 | 5·1156 | 5·0316 | 4·9460 | 4·8588 |
| 9 | 5·2565 | 5·1114 | 4·9621 | 4·8080 | 4·7290 | 4·6486 | 4·5667 | 4·4831 | 4·3978 | 4·3105 |
| 10 | 4·8492 | 4·7059 | 4·5582 | 4·4054 | 4·3269 | 4·2469 | 4·1653 | 4·0819 | 3·9965 | 3·9090 |
| 11 | 4·5393 | 4·3974 | 4·2509 | 4·0990 | 4·0209 | 3·9411 | 3·8596 | 3·7761 | 3·6904 | 3·6025 |
| 12 | 4·2961 | 4·1553 | 4·0096 | 3·8584 | 3·7805 | 3·7008 | 3·6192 | 3·5355 | 3·4494 | 3·3608 |
| 13 | 4·1003 | 3·9603 | 3·8154 | 3·6646 | 3·5868 | 3·5070 | 3·4253 | 3·3413 | 3·2548 | 3·1654 |
| 14 | 3·9394 | 3·8001 | 3·6557 | 3·5052 | 3·4274 | 3·3476 | 3·2656 | 3·1813 | 3·0942 | 3·0040 |
| 15 | 3·8049 | 3·6662 | 3·5222 | 3·3719 | 3·2940 | 3·2141 | 3·1319 | 3·0471 | 2·9595 | 2·8684 |
| 16 | 3·6909 | 3·5527 | 3·4089 | 3·2588 | 3·1808 | 3·1007 | 3·0182 | 2·9330 | 2·8447 | 2·7528 |
| 17 | 3·5931 | 3·4552 | 3·3117 | 3·1615 | 3·0835 | 3·0032 | 2·9205 | 2·8348 | 2·7459 | 2·6530 |
| 18 | 3·5082 | 3·3706 | 3·2273 | 3·0771 | 2·9990 | 2·9185 | 2·8354 | 2·7493 | 2·6597 | 2·5660 |
| 19 | 3·4338 | 3·2965 | 3·1533 | 3·0031 | 2·9249 | 2·8442 | 2·7608 | 2·6742 | 2·5839 | 2·4893 |
| 20 | 3·3682 | 3·2311 | 3·0880 | 2·9377 | 2·8594 | 2·7785 | 2·6947 | 2·6077 | 2·5168 | 2·4212 |
| 21 | 3·3098 | 3·1729 | 3·0299 | 2·8796 | 2·8011 | 2·7200 | 2·6359 | 2·5484 | 2·4568 | 2·3603 |
| 22 | 3·2576 | 3·1209 | 2·9780 | 2·8274 | 2·7488 | 2·6675 | 2·5831 | 2·4951 | 2·4029 | 2·3055 |
| 23 | 3·2106 | 3·0740 | 2·9311 | 2·7805 | 2·7017 | 2·6202 | 2·5355 | 2·4471 | 2·3542 | 2·2559 |
| 24 | 3·1681 | 3·0316 | 2·8887 | 2·7380 | 2·6591 | 2·5773 | 2·4923 | 2·4035 | 2·3099 | 2·2107 |
| 25 | 3·1294 | 2·9931 | 2·8502 | 2·6993 | 2·6203 | 2·5383 | 2·4530 | 2·3637 | 2·2695 | 2·1694 |
| 26 | 3·0941 | 2·9579 | 2·8150 | 2·6640 | 2·5848 | 2·5026 | 2·4170 | 2·3273 | 2·2325 | 2·1315 |
| 27 | 3·0618 | 2·9256 | 2·7827 | 2·6316 | 2·5522 | 2·4699 | 2·3840 | 2·2938 | 2·1984 | 2·0965 |
| 28 | 3·0320 | 2·8959 | 2·7530 | 2·6017 | 2·5223 | 2·4397 | 2·3535 | 2·2629 | 2·1670 | 2·0642 |
| 29 | 3·0045 | 2·8685 | 2·7256 | 2·5742 | 2·4946 | 2·4118 | 2·3253 | 2·2344 | 2·1378 | 2·0342 |
| 30 | 2·9791 | 2·8431 | 2·7002 | 2·5487 | 2·4689 | 2·3860 | 2·2992 | 2·2079 | 2·1107 | 2·0062 |
| 40 | 2·8005 | 2·6648 | 2·5216 | 2·3689 | 2·2880 | 2·2034 | 2·1142 | 2·0194 | 1·9172 | 1·8047 |
| 60 | 2·6318 | 2·4961 | 2·3523 | 2·1978 | 2·1154 | 2·0285 | 1·9360 | 1·8363 | 1·7263 | 1·6006 |
| 120 | 2·4721 | 2·3363 | 2·1915 | 2·0346 | 1·9500 | 1·8600 | 1·7628 | 1·6557 | 1·5330 | 1·3805 |
| ∞ | 2·3209 | 2·1848 | 2·0385 | 1·8783 | 1·7908 | 1·6964 | 1·5923 | 1·4730 | 1·3246 | 1·0000 |

**TABLE A.4.** (*Continued*)

P = 0.005

| $\nu_1$ $\nu_2$ | 1 | 2 | 3 | 4 | 5 | 6 | 7 | 8 | 9 |
|---|---|---|---|---|---|---|---|---|---|
| 1 | 16211 | 20000 | 21615 | 22500 | 23056 | 23437 | 23715 | 23925 | 24091 |
| 2 | 198·50 | 199·00 | 199·17 | 199·25 | 199·30 | 199·33 | 199·36 | 199·37 | 199·39 |
| 3 | 55·552 | 49·799 | 47·467 | 46·195 | 45·392 | 44·838 | 44·434 | 44·126 | 43·882 |
| 4 | 31·333 | 26·284 | 24·259 | 23·155 | 22·456 | 21·975 | 21·622 | 21·352 | 21·139 |
| 5 | 22·785 | 18·314 | 16·530 | 15·556 | 14·940 | 14·513 | 14·200 | 13·961 | 13·772 |
| 6 | 18·635 | 14·544 | 12·917 | 12·028 | 11·464 | 11·073 | 10·786 | 10·566 | 10·391 |
| 7 | 16·236 | 12·404 | 10·882 | 10·050 | 9·5221 | 9·1554 | 8·8854 | 8·6781 | 8·5138 |
| 8 | 14·688 | 11·042 | 9·5965 | 8·8051 | 8·3018 | 7·9520 | 7·6942 | 7·4960 | 7·3386 |
| 9 | 13·614 | 10·107 | 8·7171 | 7·9559 | 7·4711 | 7·1338 | 6·8849 | 6·6933 | 6·5411 |
| 10 | 12·826 | 9·4270 | 8·0807 | 7·3428 | 6·8723 | 6·5446 | 6·3025 | 6·1159 | 5·9676 |
| 11 | 12·226 | 8·9122 | 7·6004 | 6·8809 | 6·4217 | 6·1015 | 5·8648 | 5·6821 | 5·5368 |
| 12 | 11·754 | 8·5096 | 7·2258 | 6·5211 | 6·0711 | 5·7570 | 5·5245 | 5·3451 | 5·2021 |
| 13 | 11·374 | 8·1865 | 6·9257 | 6·2335 | 5·7910 | 5·4819 | 5·2529 | 5·0761 | 4·9351 |
| 14 | 11·060 | 7·9217 | 6·6803 | 5·9984 | 5·5623 | 5·2574 | 5·0313 | 4·8566 | 4·7173 |
| 15 | 10·798 | 7·7008 | 6·4760 | 5·8029 | 5·3721 | 5·0708 | 4·8473 | 4·6743 | 4·5364 |
| 16 | 10·575 | 7·5138 | 6·3034 | 5·6378 | 5·2117 | 4·9134 | 4·6920 | 4·5207 | 4·3838 |
| 17 | 10·384 | 7·3536 | 6·1556 | 5·4967 | 5·0746 | 4·7789 | 4·5594 | 4·3893 | 4·2535 |
| 18 | 10·218 | 7·2148 | 6·0277 | 5·3746 | 4·9560 | 4·6627 | 4·4448 | 4·2759 | 4·1410 |
| 19 | 10·073 | 7·0935 | 5·9161 | 5·2681 | 4·8526 | 4·5614 | 4·3448 | 4·1770 | 4·0428 |
| 20 | 9·9439 | 6·9865 | 5·8177 | 5·1743 | 4·7616 | 4·4721 | 4·2569 | 4·0900 | 3·9564 |
| 21 | 9·8295 | 6·8914 | 5·7304 | 5·0911 | 4·6808 | 4·3931 | 4·1789 | 4·0128 | 3·8799 |
| 22 | 9·7271 | 6·8064 | 5·6524 | 5·0168 | 4·6088 | 4·3225 | 4·1094 | 3·9440 | 3·8116 |
| 23 | 9·6348 | 6·7300 | 5·5823 | 4·9500 | 4·5441 | 4·2591 | 4·0469 | 3·8822 | 3·7502 |
| 24 | 9·5513 | 6·6610 | 5·5190 | 4·8898 | 4·4857 | 4·2019 | 3·9905 | 3·8264 | 3·6949 |
| 25 | 9·4753 | 6·5982 | 5·4615 | 4·8351 | 4·4327 | 4·1500 | 3·9394 | 3·7758 | 3·6447 |
| 26 | 9·4059 | 6·5409 | 5·4091 | 4·7852 | 4·3844 | 4·1027 | 3·8928 | 3·7297 | 3·5989 |
| 27 | 9·3423 | 6·4885 | 5·3611 | 4·7396 | 4·3402 | 4·0594 | 3·8501 | 3·6875 | 3·5571 |
| 28 | 9·2838 | 6·4403 | 5·3170 | 4·6977 | 4·2996 | 4·0197 | 3·8110 | 3·6487 | 3·5186 |
| 29 | 9·2297 | 6·3958 | 5·2764 | 4·6591 | 4·2622 | 3·9830 | 3·7749 | 3·6130 | 3·4832 |
| 30 | 9·1797 | 6·3547 | 5·2388 | 4·6233 | 4·2276 | 3·9492 | 3·7416 | 3·5801 | 3·4505 |
| 40 | 8·8278 | 6·0664 | 4·9759 | 4·3738 | 3·9860 | 3·7129 | 3·5088 | 3·3498 | 3·2220 |
| 60 | 8·4946 | 5·7950 | 4·7290 | 4·1399 | 3·7600 | 3·4918 | 3·2911 | 3·1344 | 3·0083 |
| 120 | 8·1790 | 5·5393 | 4·4973 | 3·9207 | 3·5482 | 3·2849 | 3·0874 | 2·9330 | 2·8083 |
| ∞ | 7·8794 | 5·2983 | 4·2794 | 3·7151 | 3·3499 | 3·0913 | 2·8968 | 2·7444 | 2·6210 |

**TABLE A.4.** (*Concluded*)

## P = 0.005

| $\nu_1$ / $\nu_2$ | 10 | 12 | 15 | 20 | 24 | 30 | 40 | 60 | 120 | ∞ |
|---|---|---|---|---|---|---|---|---|---|---|
| 1 | 24224 | 24426 | 24630 | 24836 | 24940 | 25044 | 25148 | 25253 | 25359 | 25465 |
| 2 | 199·40 | 199·42 | 199·43 | 199·45 | 199·46 | 199·47 | 199·47 | 199·48 | 199·49 | 199·51 |
| 3 | 43·686 | 43·387 | 43·085 | 42·778 | 42·622 | 42·466 | 42·308 | 42·149 | 41·989 | 41·829 |
| 4 | 20·967 | 20·705 | 20·438 | 20·167 | 20·030 | 19·892 | 19·752 | 19·611 | 19·468 | 19·325 |
| 5 | 13·618 | 13·384 | 13·146 | 12·903 | 12·780 | 12·656 | 12·530 | 12·402 | 12·274 | 12·144 |
| 6 | 10·250 | 10·034 | 9·8140 | 9·5888 | 9·4741 | 9·3583 | 9·2408 | 9·1219 | 9·0015 | 8·8793 |
| 7 | 8·3803 | 8·1764 | 7·9678 | 7·7540 | 7·6450 | 7·5345 | 7·4225 | 7·3088 | 7·1933 | 7·0760 |
| 8 | 7·2107 | 7·0149 | 6·8143 | 6·6082 | 6·5029 | 6·3961 | 6·2875 | 6·1772 | 6·0649 | 5·9505 |
| 9 | 6·4171 | 6·2274 | 6·0325 | 5·8318 | 5·7292 | 5·6248 | 5·5186 | 5·4104 | 5·3001 | 5·1875 |
| 10 | 5·8467 | 5·6613 | 5·4707 | 5·2740 | 5·1732 | 5·0705 | 4·9659 | 4·8592 | 4·7501 | 4·6385 |
| 11 | 5·4182 | 5·2363 | 5·0489 | 4·8552 | 4·7557 | 4·6543 | 4·5508 | 4·4450 | 4·3367 | 4·2256 |
| 12 | 5·0855 | 4·9063 | 4·7214 | 4·5299 | 4·4315 | 4·3309 | 4·2282 | 4·1229 | 4·0149 | 3·9039 |
| 13 | 4·8199 | 4·6429 | 4·4600 | 4·2703 | 4·1726 | 4·0727 | 3·9704 | 3·8655 | 3·7577 | 3·6465 |
| 14 | 4·6034 | 4·4281 | 4·2468 | 4·0585 | 3·9614 | 3·8619 | 3·7600 | 3·6553 | 3·5473 | 3·4359 |
| 15 | 4·4236 | 4·2498 | 4·0698 | 3·8826 | 3·7859 | 3·6867 | 3·5850 | 3·4803 | 3·3722 | 3·2602 |
| 16 | 4·2719 | 4·0994 | 3·9205 | 3·7342 | 3·6378 | 3·5388 | 3·4372 | 3·3324 | 3·2240 | 3·1115 |
| 17 | 4·1423 | 3·9709 | 3·7929 | 3·6073 | 3·5112 | 3·4124 | 3·3107 | 3·2058 | 3·0971 | 2·9839 |
| 18 | 4·0305 | 3·8599 | 3·6827 | 3·4977 | 3·4017 | 3·3030 | 3·2014 | 3·0962 | 2·9871 | 2·8732 |
| 19 | 3·9329 | 3·7631 | 3·5866 | 3·4020 | 3·3062 | 3·2075 | 3·1058 | 3·0004 | 2·8908 | 2·7762 |
| 20 | 3·8470 | 3·6779 | 3·5020 | 3·3178 | 3·2220 | 3·1234 | 3·0215 | 2·9159 | 2·8058 | 2·6904 |
| 21 | 3·7709 | 3·6024 | 3·4270 | 3·2431 | 3·1474 | 3·0488 | 2·9467 | 2·8408 | 2·7302 | 2·6140 |
| 22 | 3·7030 | 3·5350 | 3·3600 | 3·1764 | 3·0807 | 2·9821 | 2·8799 | 2·7736 | 2·6625 | 2·5455 |
| 23 | 3·6420 | 3·4745 | 3·2999 | 3·1165 | 3·0208 | 2·9221 | 2·8198 | 2·7132 | 2·6016 | 2·4837 |
| 24 | 3·5870 | 3·4199 | 3·2456 | 3·0624 | 2·9667 | 2·8679 | 2·7654 | 2·6585 | 2·5463 | 2·4276 |
| 25 | 3·5370 | 3·3704 | 3·1963 | 3·0133 | 2·9176 | 2·8187 | 2·7160 | 2·6088 | 2·4960 | 2·3765 |
| 26 | 3·4916 | 3·3252 | 3·1515 | 2·9685 | 2·8728 | 2·7738 | 2·6709 | 2·5633 | 2·4501 | 2·3297 |
| 27 | 3·4499 | 3·2839 | 3·1104 | 2·9275 | 2·8318 | 2·7327 | 2·6296 | 2·5217 | 2·4078 | 2·2867 |
| 28 | 3·4117 | 3·2460 | 3·0727 | 2·8899 | 2·7941 | 2·6949 | 2·5916 | 2·4834 | 2·3689 | 2·2469 |
| 29 | 3·3765 | 3·2111 | 3·0379 | 2·8551 | 2·7594 | 2·6601 | 2·5565 | 2·4479 | 2·3330 | 2·2102 |
| 30 | 3·3440 | 3·1787 | 3·0057 | 2·8230 | 2·7272 | 2·6278 | 2·5241 | 2·4151 | 2·2997 | 2·1760 |
| 40 | 3·1167 | 2·9531 | 2·7811 | 2·5984 | 2·5020 | 2·4015 | 2·2958 | 2·1838 | 2·0635 | 1·9318 |
| 60 | 2·9042 | 2·7419 | 2·5705 | 2·3872 | 2·2898 | 2·1874 | 2·0789 | 1·9622 | 1·8341 | 1·6885 |
| 120 | 2·7052 | 2·5439 | 2·3727 | 2·1881 | 2·0890 | 1·9839 | 1·8709 | 1·7469 | 1·6055 | 1·4311 |
| ∞ | 2·5188 | 2·3583 | 2·1868 | 1·9998 | 1·8983 | 1·7891 | 1·6691 | 1·5325 | 1·3637 | 1·0000 |

Reproduced with kind permission of the editors of *Biometrika* from M. Merrington and C. M. Thompson, *Biometrika* 33 : 73–88, 1943.

# APPENDIX B
# LUNG FUNCTION
# DATA

The lung function data set includes information on nonsmoking families from the UCLA study of chronic obstructive respiratory disease (CORD). In the CORD study persons seven years old and older from four areas (Burbank, Lancaster, Long Beach, and Glendora) were sampled, and information was obtained from them at two time periods. The data set presented here is a subset including families with both a mother and father, and one, two, or three children between the ages of 7 and 17 who answered the questionnaire and took the lung function tests at the first time period. The purpose of the CORD study was to determine the effects of different types of air pollutants on respiratory function, but numerous other types of studies have been performed on this data set. Further information concerning the CORD study can be found in the Bibliography at the end of this appendix.

The code book in Table B.1 summarizes the information in this subset of data. After the first two variables, note that the same set of data is given for fathers (F), mothers (M), oldest child (OC), middle child (MC), and youngest child (YC).

The format is F3.0, F1.0, 5(F1.0, 2F2.0, F3.0, 2F3.2). Note that in the listing of the data in Table B.2 columns of blanks have been added to isolate the data for different members of the family. The format statement and the code book assume no blanks. Some families have only one or two children between the ages of 7 and 17. If there is only one child, it is listed as the oldest child. Thus there are numerous missing values in the data for the middle and youngest child.

Forced vital capacity (FVC) is the volume of air, in liters, that can be expelled by the participant after having breathed in as deeply as possible, that is, full expiration regardless of how long it takes. The measurement of FVC is affected by the ability of the participant to understand the instructions and by the amount of effort made to breathe in deeply and to expel as much air as possible. For this reason seven years is the practical lower age limit at which valid and reliable data can be obtained. Variable FEV1 is a measure of the volume of the air expelled in the first second after the start of expiration. Variable FEV1 or the ratio FEV1/FVC has been used as an outcome variable in many studies of the effects of smoking on lung function.

The authors wish to thank Dean Roger Detels, principal investigator of the CORD project, for the use of this data set and Miss Cathleen Reems for assembling it.

**TABLE B.1.** Code Book for Lung Function Data Set

| Variable Number | Variable Location | Variable Name | Description |
|---|---|---|---|
| 1 | 1–3 | ID | 1–150 |
| 2 | 4 | AREA | 1 = Burbank |
| | | | 2 = Lancaster |
| | | | 3 = Long Beach |
| | | | 4 = Glendora |
| 3 | 5 | FSEX | 1 = male |
| 4 | 6–7 | FAGE | Age, father |
| 5 | 8–9 | FHEIGHT | Height (in), father |
| 6 | 10–12 | FWEIGHT | Weight (lb), father |
| 7 | 13–15 | FFVC | FVC father |
| 8 | 16–18 | FFEV1 | FEV1 father |
| 9 | 19 | MSEX | 2 = female |
| 10 | 20–21 | MAGE | Age, mother |
| 11 | 22–23 | MHEIGHT | Height (in), mother |
| 12 | 24–26 | MWEIGHT | Weight (lb), mother |
| 13 | 27–29 | MFVC | FVC mother |
| 14 | 30–32 | MFEV1 | FEV1 mother |
| 15 | 33 | OCSEX | Sex, oldest child |
| | | | 1 = male |
| | | | 2 = female |
| 16 | 34–35 | OCAGE | Age, oldest child |
| 17 | 36–37 | OCHEIGHT | Height, oldest child |
| 18 | 38–40 | OCWEIGHT | Weight, oldest child |
| 19 | 41–43 | OCFVC | FVC oldest child |
| 20 | 44–46 | OCFEV1 | FEV1 oldest child |
| 21 | 47 | MCSEX | Sex, middle child |
| 22 | 48–49 | MCAGE | Age, middle child |
| 23 | 50–51 | MCHEIGHT | Height, middle child |
| 24 | 52–54 | MCWEIGHT | Weight, middle child |
| 25 | 55–57 | MCFVC | FVC middle child |
| 26 | 58–60 | MCFEV1 | FEV1 middle child |
| 27 | 61 | YCSEX | Sex, youngest child |
| 28 | 62–63 | YCAGE | Age, youngest child |
| 29 | 64–65 | YCHEIGHT | Height, youngest child |
| 30 | 66–68 | YCWEIGHT | Weight, youngest child |
| 31 | 69–71 | YCFVC | FVC youngest child |
| 32 | 72–74 | YCFEV1 | FEV1 youngest child |

**TABLE B.2.** Lung Function Data Set

| columns 123 | 4 | 1 — 5678901234 5678 (father) | 2 — 9012345678 9012 (mother) | 3 — 3456789012 3456 (child #1) | 4 5 — 7890123456 7890 (child #2) | 6 7 — 1234567890 1234 (child #3) |
|---|---|---|---|---|---|---|
| 1 | 1 | 15361161391323 | 24362136370331 | 21259115296279 | | |
| 2 | 1 | 14072198441395 | 23866160411347 | 11056 66323239 | | |
| 3 | 1 | 12669210445347 | 22759114309265 | 1 850 59114111 | | |
| 4 | 1 | 13468187433374 | 23658132265206 | 21157106256185 | 1 949 56159130 | |
| 5 | 1 | 14661121354290 | 23962128245233 | 11661 88260247 | 21260 85268234 | 21050 53154143 |
| 6 | 1 | 14472153610491 | 23666125349306 | 11567100388355 | 11357 87276237 | 21055 72195169 |
| 7 | 1 | 13564145345339 | 22768206492425 | 21154 70218163 | | |
| 8 | 1 | 14591664844419 | 24563115342271 | 11567153460388 | 2 954 81193187 | |
| 9 | 1 | 14568180489429 | 24168144357313 | 21465144289272 | 11262108257235 | |
| 10 | 1 | 13066166550449 | 22667156364345 | 2 949 52192142 | | |
| 11 | 1 | 14670188473390 | 24461136348243 | 11768145501381 | | |
| 12 | 1 | 15068179391334 | 24864179232186 | 21150 54152144 | 1 750 61236153 | |
| 13 | 1 | 13174210526398 | 23066143399365 | 2 852 61148132 | | |
| 14 | 1 | 13767195468401 | 23267164508357 | 21055 61224205 | 1 854 87241165 | |
| 15 | 1 | 14070190522456 | 23559112303257 | 21156 92244206 | | |
| 16 | 1 | 13271170501411 | 23268138448354 | 2 955 61169149 | 2 749 50118108 | |
| 17 | 1 | 13774198595414 | 23562145360292 | 11361103296250 | 21058 85249218 | |
| 18 | 1 | 15472223361285 | 24765122363275 | 11159 84216213 | 2 954 58202184 | |
| 19 | 1 | 13466176467383 | 23057110260204 | 2 849 52171142 | | |
| 20 | 1 | 14069178618479 | 23763131350287 | 21664117371353 | 11257 85254214 | |
| 21 | 1 | 13369176651472 | 23165142428319 | 21162101297258 | 2 850 52134131 | |
| 22 | 1 | 13172163512463 | 23066147417355 | 1 749 55164157 | | |
| 23 | 1 | 15173215499416 | 24963122299294 | 11770180496436 | 11567148391287 | 11264105326299 |
| 24 | 1 | 13071163518470 | 22965125206202 | 1 850 50158144 | | |
| 25 | 2 | 13967151458377 | 23464176418351 | 11771164567449 | 21361145266242 | |

**TABLE B.2.** (*Continued*)

| columns 123 | 4 | 1 56789012345678 | 2 90123456789012 | 3 90123456789012 | 4 34567890123456 | 5 78901234567890 | 6 78901234567890 | 7 12345678901234 |
|---|---|---|---|---|---|---|---|---|
| 26 | 2 | 13668151488425 | 23363124319286 | 11360104313299 | 2 954 61195169 | | | |
| 27 | 2 | 13366178421375 | 22966170397325 | 11055 73233190 | 21053 55114 90 | | | |
| 28 | 2 | 13473209544470 | 23263108344285 | 11359 83235215 | | | | |
| 29 | 2 | 15268182365318 | 24464128356298 | 21463115303291 | 21361102226198 | | | 21156 68192168 |
| 30 | 2 | 13968209417362 | 23863133339291 | 11463103330253 | 11359105261219 | | | |
| 31 | 2 | 14167115381250 | 23864132365334 | 21561102374235 | 21666141356305 | | | 11466130473405 |
| 32 | 2 | 14370196470413 | 24467145411328 | 21766139366282 | 2 850 55174136 | | | |
| 33 | 2 | 13270194592501 | 23160168215175 | 21054104238199 | 21054 61185169 | | | |
| 34 | 2 | 13371178586466 | 23264122310297 | 21259 93226218 | | | | 1 746 44194112 |
| 35 | 2 | 13674191598496 | 23664133301257 | 1 956 82240203 | 2 750 51149126 | | | 2 750 51131127 |
| 36 | 2 | 14766150465385 | 24665131394303 | 21359 85336233 | | | | |
| 37 | 2 | 12666158473377 | 22667151344302 | 2 951 57184166 | | | | |
| 38 | 2 | 13569184400350 | 24265179362305 | 11768154482441 | 21260109266227 | | | |
| 39 | 2 | 15273187611472 | 25064140380304 | 11770170503439 | 11257 85314173 | | | 11053 65179162 |
| 40 | 2 | 13566169302251 | 23461 90233194 | 11361 90250191 | | | | |
| 41 | 2 | 15070146510429 | 24867159317308 | 21669138468424 | 21264203305262 | | | |
| 42 | 2 | 14172234508409 | 23965241298254 | 21664147381303 | 1 850 60194139 | | | 11267191355273 |
| 43 | 2 | 13465159477401 | 23061150362297 | 21154 85217183 | | | | |
| 44 | 2 | 15469176519424 | 25261125303258 | 11460 96361309 | | | | |
| 45 | 2 | 14667186449405 | 24661231288248 | 21763174321287 | 11054 72207193 | | | |
| 46 | 2 | 14866172437320 | 24765200252248 | 11463133403312 | | | | |
| 47 | 2 | 14268152391329 | 24262116281256 | 21611382294288 | 21562113334293 | | | |
| 48 | 2 | 13870214462408 | 23865129407294 | 11672141527431 | 11368124413383 | | | |
| 49 | 2 | 13873196606497 | 23666130422366 | 21566128353314 | 21365114278264 | | | |
| 50 | 2 | 14575186666585 | 24366151280227 | 11675171689545 | | | | |

| columns 123 | 4 | 1<br>5678901234 5678<br>(father) | 2<br>9012345678 9012<br>(mother) | 4<br>3456789012 3456<br>(child #1) | 5<br>7890123456 7890<br>(child #2) | 7<br>1234567890 1234<br>(child #3) |
|---|---|---|---|---|---|---|
| 51 | 2 | 151711232448350 | 24862150345289 | 11769185680499 | 21567141371288 | |
| 52 | 2 | 14066187452397 | 24264140423369 | 11769161581544 | 11466131459419 | |
| 53 | 2 | 14367150438356 | 23866138275262 | 2 954 58128120 | | |
| 54 | 2 | 14971172398325 | 23961150309268 | 21663111262245 | 11157 76217199 | |
| 55 | 2 | 13567158447404 | 23164123351311 | 11256 75236204 | 21055 68189182 | |
| 56 | 2 | 15973180512410 | 25667138418321 | 21769131463429 | 11674136439381 | |
| 57 | 2 | 13873172504449 | 23565146335309 | 11054 60153147 | 1 849 49136122 | |
| 58 | 2 | 14671194600489 | 24567128343283 | 21661105338310 | 11363 94318293 | |
| 59 | 2 | 14071176563476 | 24065128337300 | 11671148488453 | 21565106264212 | 11161 91337269 |
| 60 | 2 | 13571187566500 | 23665142356307 | 11056 69201180 | | |
| 61 | 2 | 14065144475409 | 23863108302256 | 11256 90226189 | 21153 68175155 | |
| 62 | 2 | 15067170383306 | 24168145421305 | 11159122280226 | | |
| 63 | 2 | 14368200496385 | 23863137349277 | 21767122379349 | | |
| 64 | 2 | 15068205373309 | 25362160275225 | 11567186410334 | | |
| 65 | 2 | 14270160513450 | 23962143412362 | 11665125325311 | 21357 86296253 | |
| 66 | 2 | 13966188488390 | 23262124257206 | 11049 50107 98 | | |
| 67 | 2 | 13270235537430 | 23262181362316 | 11367140434356 | 1125910301256 | 1 956 82255213 |
| 68 | 2 | 14069213462395 | 23962134386285 | 11768151490456 | 21763111317304 | 11467126438372 |
| 69 | 2 | 15469179510384 | 25262121336318 | 11666118392380 | 11154 64202179 | |
| 70 | 2 | 13666173421352 | 23363113292278 | 21361104293269 | 21160 96264255 | |
| 71 | 2 | 14473178602450 | 24265161316295 | 21160 78149133 | | |
| 72 | 2 | 14568173413339 | 24066125374307 | 21467116368336 | 21159 77208176 | |
| 73 | 2 | 14269220411372 | 24061164283256 | 11664125325272 | 21360109292258 | 11056 87280238 |
| 74 | 3 | 15173158657499 | 24766116404333 | 11772163604521 | 21162 98322269 | |
| 75 | 3 | 154661153340305 | 25264152296245 | 21663 95242238 | | |

TABLE B.2. (Continued)

| columns 123 | 4 | 1<br>56789012345678 | 2<br>90123456789012 | 3 | 4<br>34567890123456 | 5<br>78901234567890 | 6<br>78901234567890 | 7<br>12345678901234 |
|---|---|---|---|---|---|---|---|---|
| 76 | 3 | 14669192407343 | 24867205325283 | | 21667142326301 | 21466136294275 | | 11360134257209 |
| 77 | 3 | 12872172638545 | 22862124330328 | | 1 952 64150140 | | | |
| 78 | 3 | 15370194553443 | 25662133282224 | | 11465117376261 | | | |
| 79 | 3 | 12869194495407 | 22762128302259 | | 2 848 53128108 | | | |
| 80 | 3 | 13967226319293 | 23462126245222 | | 21462113289262 | | | |
| 81 | 3 | 14374216600490 | 24164141316266 | | 21060 94335216 | | | |
| 82 | 3 | 13569164526402 | 23462144349305 | | 11259100269240 | | 1 849 55172144 | |
| 83 | 3 | 13870171569451 | 23564145432356 | | 11264111406331 | | | |
| 84 | 3 | 13768138624537 | 23667128327307 | | 11470156563474 | | 1 957 84230206 | |
| 85 | 3 | 14271169407329 | 24169145447331 | | 11259 80201197 | | | |
| 86 | 3 | 13367183511445 | 23363124287218 | | 2 749 55164138 | | 2 856 99272238 | |
| 87 | 3 | 14564153559471 | 24660150309252 | | 11058 80240225 | | | |
| 88 | 3 | 13367175540399 | 23257108279221 | | 21355 72152110 | | | |
| 89 | 3 | 14173213434362 | 24067170370305 | | 21364108314289 | | | |
| 90 | 3 | 14169188466415 | 24268159243215 | | 11160 98288228 | | 2 850 50189172 | |
| 91 | 3 | 19474200585486 | 23168154567426 | | 2 854 64193183 | | | |
| 92 | 3 | 18866143441384 | 24067130355308 | | 2 846 45163127 | 11061 91296233 | | 1 955 74205183 |
| 93 | 4 | 14270197568458 | 23768147398296 | | 11260103321274 | | | |
| 94 | 4 | 14367171410327 | 24166260349289 | | 2 952 66185160 | 11362 95309251 | | 2 954 70197144 |
| 95 | 4 | 14070164606389 | 24065163335270 | | 11665112391315 | | | |
| 96 | 4 | 13671205587398 | 23464195388321 | | 2 952 58192155 | | | |
| 97 | 4 | 14473145413345 | 23763133284437 | | 21459 86274247 | 21057 82236212 | | 2 750 65175142 |
| 98 | 4 | 14270177509422 | 24266155359931 | | 21364117334311 | | | |
| 99 | 4 | 13071156620525 | 22962143362299 | | 1 748 55155138 | | | |
| 100 | 4 | 14669207375280 | 24461140352294 | | 21660111279277 | | | |

TABLE B.2. (Continued)

| columns<br>123 | 4 | 1<br>56789012345678<br>(father) | 2 3<br>90123456789012<br>(mother) | 4<br>34567890123456<br>(child #1) | 5 6<br>78901234567890<br>(child #2) | 7<br>12345678901234<br>(child #3) |
|---|---|---|---|---|---|---|
| 101 | 4 | 14474202479385 | 23161589420380 | 2 851 62119116 |  |  |
| 102 | 4 | 14371198652501 | 24065122402324 | 21566162392362 | 21364145399351 |  |
| 103 | 4 | 14266162409309 | 23863120403343 | 1 953 60194155 | 2 748 55153136 |  |
| 104 | 4 | 13370180532441 | 23061186357331 | 1 848 55167142 |  |  |
| 105 | 4 | 14076245524455 | 23666185496460 | 11572172521462 | 11364115389310 | 21058 81194156 |
| 106 | 4 | 13371168477321 | 23266138407325 | 21466115361297 | 11259 91276218 |  |
| 107 | 4 | 14369200447399 | 23869234447349 | 11670207498409 | 11572176454402 |  |
| 108 | 4 | 14469151513438 | 23467140373325 | 21259 73299268 | 1 854 58156154 |  |
| 109 | 4 | 13170183509441 | 23065132399375 | 2 853 66202182 |  |  |
| 110 | 4 | 14070228650560 | 23465140301300 | 2 849 47147136 |  |  |
| 111 | 4 | 14469205509428 | 24261684204377 | 21766172443402 |  |  |
| 112 | 4 | 13969191468366 | 23761171371313 | 11568150474367 | 21361139274250 | 11260 95257216 |
| 113 | 4 | 13867174591497 | 23161108357318 | 1 954 83331203 |  |  |
| 114 | 4 | 13768206496409 | 22862138405345 | 2 850 70148119 |  |  |
| 115 | 4 | 13669184536423 | 23567220438380 | 11261 76319280 |  |  |
| 116 | 4 | 13269190580518 | 23162118242241 | 21157 83264220 |  |  |
| 117 | 4 | 13070210586480 | 22967211368295 | 1 850 59150139 |  |  |
| 118 | 4 | 13472193642498 | 23262 98340327 | 2 851 64172155 |  |  |
| 119 | 4 | 13871191466369 | 23666139423332 | 21363110310267 | 11365138835264 |  |
| 120 | 4 | 13572209519426 | 23269184402307 | 21365104229209 |  |  |
| 121 | 4 | 14771178521445 | 24563146319256 | 21766104376324 | 21665113385369 |  |
| 122 | 4 | 15268193512439 | 24963135290219 | 11666128464383 |  | 21565124460393 |
| 123 | 4 | 14669169536462 | 24064146348264 | 11670206477389 |  |  |
| 124 | 4 | 14268167427333 | 23565186333279 | 11461 98284233 | 21464123280251 |  |
| 125 | 4 | 14468209452380 | 23562124304242 | 2 953 64186172 |  | 11258 76222191 |

**TABLE B.2.** (*Concluded*)

| columns 123 | 4 | 1 567890123456 78 | 2 901234567890 12 | 3 901234567890 12 | 4 345678901234 56 | 5 789012345678 90 | 6 123456789012 34 | 7 1234567890 1234 |
|---|---|---|---|---|---|---|---|---|
| 126 | 4 | 13472198563440 | 23266152306263 |  | 2 952 58218137 |  |  |  |
| 127 | 4 | 14068131487430 | 23762175357301 |  | 21463117321320 |  | 21262 85316254 | 1 954 66224191 |
| 128 | 4 | 13971169562468 | 23762175315262 |  | 11366122435368 |  | 2 954 60168156 | 1 954 70220197 |
| 129 | 4 | 13770225497412 | 23068194442375 |  | 11157 75234220 |  | 11055 69224205 |  |
| 130 | 4 | 12975222640558 | 22969130334296 |  | 1 750 50157155 |  |  |  |
| 131 | 4 | 13765167410329 | 23862110331290 |  | 11769144498453 |  |  |  |
| 132 | 4 | 13964162409364 | 24062141289267 |  | 21560104288277 |  | 21261101257244 | 21156 73249210 |
| 133 | 4 | 13270167592485 | 23365135425389 |  | 21154 69191175 |  | 2 748 51144142 |  |
| 134 | 4 | 13771180516451 | 23666131408328 |  | 11771170566446 |  |  |  |
| 135 | 4 | 13672198452400 | 23565132328295 |  | 11263117267263 |  | 2 856 72201190 |  |
| 136 | 4 | 13068166525481 | 23064124371329 |  | 2 851 53157146 |  |  |  |
| 137 | 4 | 14569164461398 | 24262160305273 |  | 21766115329296 |  |  |  |
| 138 | 4 | 13368132426372 | 23065176386338 |  | 21058 72198177 |  |  |  |
| 139 | 4 | 15268195463399 | 24965152392306 |  | 11769122440335 |  |  |  |
| 140 | 4 | 14574196555434 | 24464 93333299 |  | 21665100308292 |  | 11156 72186164 |  |
| 141 | 4 | 12868198534451 | 22764220406348 |  | 11055 91259202 |  |  |  |
| 142 | 4 | 14467149401347 | 23964143306253 |  | 2 951 60118114 |  |  |  |
| 143 | 4 | 13670191440359 | 23367142378332 |  | 2 852 67160139 |  |  |  |
| 144 | 4 | 14471220462376 | 23664267301296 |  | 11769136452407 |  |  |  |
| 145 | 4 | 14073214537448 | 23561130307263 |  | 11463117346279 |  | 2 852 68182145 |  |
| 146 | 4 | 15369162441340 | 25065147279245 |  | 21766127360347 |  |  |  |
| 147 | 4 | 13772195473418 | 23764145404346 |  | 11155 66223169 |  |  |  |
| 148 | 4 | 13967181649450 | 23363132360313 |  | 11156 96232211 |  |  |  |
| 149 | 4 | 13666129495374 | 22960110380325 |  | 1 746 49161136 |  | 1 849 55181165 |  |
| 150 | 4 | 14864170351292 | 24463150355289 |  | 21561115353285 |  |  |  |

# BIBLIOGRAPHY

Detels, R., Rokaw, S. N., Coulson, A. H., Tashkin, D. P., Sayre, J. W., Massey, F. J. The UCLA population studies of chronic obstructive respiratory disease. I. Methodology and comparison of lung function in areas of high and low pollution. *American Journal of Epidemiology* 109:33–58, 1979.

Detels, R., Sayre, J. W., Coulson, A. H., Rokaw, S. N., Massey, F. J., Tashkin, D. P., Wu, M. The UCLA population studies of chronic obstructive respiratory disease. IV. Respiratory effects of long-term exposure to photochemical oxidants, nitrogen dioxide, and sulfates on current and never smokers. *American Review of Respiratory Disease* 124:673–680, 1981.

Rokaw, S. N., Detels, R., Coulson, A. H., Sayre, J. W., Tashkin, D. P., Allwright, S. S., Massey, F. J. The UCLA population studies of chronic respiratory disease. III. Comparison of pulmonary function in three communities exposed to photochemical oxidants, multiple primary pollutants, or minimal pollutants. *Chest* 78:252–262, 1980.

Tashkin, D. P., Detels, R., Coulson, A. H., Rokaw, S. N., Sayre, J. W. The UCLA population studies of chronic obstructive respiratory disease. II. Determination of reliability and estimation of sensitivity and specificity. *Environmental Research* 20:403–424, 1979.

# APPENDIX C
# LUNG CANCER
# SURVIVAL DATA

Chapter 13 includes a description of this data set. Table 13.1, page 348, presents a code book for the data.

The authors wish to thank Dr. E. Carmack Holmes, Professor of Surgical Oncology at the UCLA School of Medicine, for his kind permission to use the data.

**TABLE C.1.** Lung Cancer Data

| i d | s t a g e t | s t a g e n | h i s t | t r e a t | p e r f b l | p o n i f | s m o k f u | s m o k b l | d a y s | d e a t h |
|---|---|---|---|---|---|---|---|---|---|---|
| 1 | 1 | 0 | 1 | 0 | 0 | 0 | 2 | 1 | 2926 | 0 |
| 2 | 1 | 0 | 1 | 1 | 0 | 0 | 1 | 1 | 590 | 1 |
| 3 | 0 | 0 | 1 | 0 | 0 | 0 | 1 | 2 | 2803 | 0 |
| 4 | 1 | 0 | 1 | 1 | 0 | 0 | 2 | 1 | 2762 | 0 |
| 5 | 1 | 0 | 1 | 0 | 0 | 0 | 2 | 1 | 2616 | 0 |
| 6 | 0 | 0 | 1 | 1 | 0 | 0 | MIS | 1 | 2716 | 0 |
| 7 | 0 | 0 | 1 | 1 | 0 | 0 | 1 | 1 | 1485 | 1 |
| 8 | 1 | 0 | 1 | 0 | 0 | 0 | MIS | 1 | 2456 | 0 |
| 9 | 1 | 0 | 1 | 0 | 0 | 0 | MIS | 2 | 500 | 0 |
| 10 | 1 | 0 | 1 | 1 | 0 | 0 | 2 | 2 | 2307 | 0 |
| 11 | 0 | 1 | 2 | 0 | 0 | 0 | 3 | 3 | 1200 | 1 |
| 12 | 0 | 0 | 2 | 1 | 0 | 0 | 1 | 1 | 1073 | 0 |
| 13 | 0 | 0 | 2 | 1 | 0 | 0 | 1 | 1 | 2209 | 0 |
| 14 | 1 | 0 | 2 | 0 | 0 | 0 | 1 | 1 | 2394 | 0 |
| 15 | 1 | 0 | 2 | 1 | 0 | 0 | 2 | 1 | 2383 | 0 |
| 16 | 1 | 0 | 1 | 0 | 0 | 0 | 1 | 1 | 3425 | 0 |
| 17 | 1 | 0 | 1 | 1 | 0 | 0 | MIS | 1 | 694 | 1 |
| 18 | 1 | 0 | 1 | 0 | 0 | 0 | 2 | 2 | 2738 | 0 |
| 19 | 0 | 0 | 1 | 1 | 0 | 0 | 3 | 3 | 3180 | 0 |
| 20 | 1 | 0 | 1 | 0 | 0 | 0 | MIS | 2 | 204 | 1 |
| 21 | 1 | 0 | 1 | 1 | 0 | 0 | MIS | 2 | 3180 | 0 |
| 22 | 0 | 0 | 1 | 0 | 0 | 0 | 1 | MIS | 2847 | 0 |
| 23 | 1 | 0 | 1 | 0 | 0 | 0 | 2 | 1 | 849 | 1 |
| 24 | 0 | 0 | 1 | 1 | 0 | 0 | MIS | 1 | 2568 | 0 |
| 25 | 1 | 0 | 2 | 0 | 1 | 0 | MIS | 1 | 524 | 1 |
| 26 | 0 | 0 | 2 | 1 | 0 | 0 | 2 | 2 | 2504 | 0 |
| 27 | 1 | 0 | 2 | 1 | 0 | 0 | 2 | 2 | 3183 | 0 |
| 28 | 1 | 0 | 2 | 0 | 0 | 0 | 2 | 2 | 3116 | 1 |
| 29 | 0 | 0 | 2 | 1 | 0 | 0 | MIS | 1 | 2940 | 0 |
| 30 | 1 | 0 | 2 | 0 | 0 | 0 | 3 | 3 | 2638 | 0 |
| 31 | 0 | 0 | 2 | 1 | 0 | 0 | 1 | 1 | 504 | 1 |
| 32 | 1 | 0 | 2 | 0 | 0 | 0 | 1 | 1 | 726 | 1 |
| 33 | 1 | 0 | 2 | 1 | 0 | 0 | MIS | 1 | 287 | 1 |
| 34 | 1 | 0 | 2 | 0 | 0 | 0 | 2 | 1 | 744 | 1 |
| 35 | 1 | 0 | 2 | 1 | 0 | 0 | 2 | 2 | 1376 | 1 |
| 36 | 1 | 0 | 1 | 0 | 0 | 0 | MIS | 2 | 314 | 1 |
| 37 | 0 | 0 | 1 | 1 | 0 | 0 | 2 | 1 | 3532 | 0 |
| 38 | 1 | 0 | 1 | 0 | 0 | 0 | MIS | 1 | 187 | 1 |
| 39 | 1 | 0 | 1 | 0 | 0 | 0 | 2 | 2 | 2924 | 0 |
| 40 | 1 | 0 | 1 | 1 | 0 | 0 | 2 | 2 | 1259 | 1 |

**TABLE C.1.** (*Continued*)

| i d | s t a g e t | s t a g e n | h i s t | t r e a t | p e r f b l | p o i n f | s m o k f u | s m o k b l | d a y s | d e a t h |
|---|---|---|---|---|---|---|---|---|---|---|
| 41 | 1 | 0 | 1 | 1 | 0 | 0 | 2 | 2 | 2464 | 0 |
| 42 | 0 | 0 | 1 | 0 | 0 | 0 | 2 | 2 | 2349 | 0 |
| 43 | 0 | 0 | 2 | 1 | 0 | 0 | 2 | 1 | 3172 | 0 |
| 44 | 0 | 0 | 2 | 0 | 0 | 0 | 2 | 1 | 1836 | 1 |
| 45 | 0 | 0 | 2 | 1 | 0 | 0 | 2 | 1 | 3376 | 0 |
| 46 | 0 | 0 | 2 | 0 | 0 | MIS | 3 | 3 | 3398 | 0 |
| 47 | 1 | 0 | 2 | 0 | 0 | 0 | MIS | 1 | 274 | 1 |
| 48 | 0 | 0 | 2 | 1 | 1 | 0 | 2 | 1 | 3265 | 0 |
| 49 | 1 | 0 | 2 | 1 | 0 | 0 | 2 | 1 | 1598 | 1 |
| 50 | 0 | 0 | 2 | 0 | 1 | 0 | 1 | 1 | 2985 | 0 |
| 51 | 0 | 0 | 2 | 0 | 0 | 0 | 2 | 2 | 2784 | 0 |
| 52 | 0 | 1 | 2 | 1 | 1 | 0 | MIS | 1 | 287 | 1 |
| 53 | 1 | 0 | 2 | 1 | 0 | 0 | 2 | 1 | 2480 | 0 |
| 54 | 0 | 0 | 2 | 0 | 0 | 0 | 3 | 3 | 2310 | 0 |
| 55 | 0 | 0 | 1 | 0 | 1 | 0 | 3 | 3 | 2259 | 1 |
| 56 | 1 | 0 | 1 | 1 | 0 | 0 | 1 | 1 | 847 | 1 |
| 57 | 0 | 0 | 1 | 0 | 0 | 0 | 2 | 2 | 3443 | 0 |
| 58 | 0 | 0 | 1 | 0 | 0 | 0 | 1 | 1 | 3210 | 0 |
| 59 | 1 | 0 | 1 | 1 | 0 | 0 | 2 | 1 | 3115 | 0 |
| 60 | 0 | 0 | 1 | 1 | 1 | 0 | 2 | 2 | 1811 | 1 |
| 61 | 1 | 0 | 1 | 1 | 0 | 0 | 3 | 3 | 454 | 1 |
| 62 | 1 | 0 | 1 | 0 | 0 | 0 | 1 | 1 | 2017 | 1 |
| 63 | 0 | 0 | 1 | 0 | 0 | 0 | 1 | 1 | 595 | 1 |
| 64 | 0 | 0 | 2 | 0 | 0 | 0 | 1 | 2 | 3445 | 0 |
| 65 | 1 | 0 | 2 | 1 | 0 | 0 | 2 | 2 | 2253 | 1 |
| 66 | 0 | 0 | 2 | 1 | 0 | 0 | 3 | 3 | 3353 | 0 |
| 67 | 1 | 0 | 2 | 0 | 0 | 0 | 2 | 1 | 1216 | 1 |
| 68 | 1 | 0 | 2 | 0 | 0 | 0 | MIS | 1 | 3181 | 0 |
| 69 | 1 | 0 | 2 | 1 | 1 | 0 | 2 | 2 | 782 | 1 |
| 70 | 1 | 0 | 2 | 0 | 1 | 0 | MIS | 2 | 106 | 1 |
| 71 | 0 | 0 | 2 | 1 | 0 | 0 | 2 | 1 | 2821 | 0 |
| 72 | 0 | 0 | 2 | 0 | 0 | 0 | 2 | 2 | 2732 | 0 |
| 73 | 0 | 0 | 2 | 0 | 0 | 0 | 1 | 1 | 2198 | 1 |
| 74 | 0 | 0 | 2 | 0 | 0 | 0 | 2 | 2 | 270 | 1 |
| 75 | 0 | 1 | 2 | 1 | 0 | 0 | 3 | 3 | 2600 | 0 |
| 76 | 0 | 0 | 2 | 0 | 0 | 0 | 2 | 2 | 765 | 1 |
| 77 | 0 | 0 | 1 | 0 | 0 | 0 | 2 | 2 | 585 | 1 |
| 78 | 1 | 0 | 1 | 1 | 1 | 0 | 1 | MIS | 2050 | 1 |
| 79 | 1 | 0 | 1 | 1 | 1 | 0 | 1 | 1 | 1463 | 1 |
| 80 | 0 | 0 | 1 | 0 | 0 | 0 | 1 | MIS | 1904 | 0 |

**TABLE C.1.** (*Continued*)

| i d | s t a g e t | s t a g e n | h i s t | t r e a t | p e r f b l | p o i n f | s m o k f u | s m o k b l | d a y s | d e a t h |
|---|---|---|---|---|---|---|---|---|---|---|
| 81 | 1 | 0 | 1 | 1 | 0 | 0 | 2 | 2 | 3346 | 0 |
| 82 | 0 | 0 | 1 | 1 | 0 | 0 | 1 | 1 | 3297 | 0 |
| 83 | 1 | 0 | 1 | 1 | 0 | 0 | 1 | 1 | 2779 | 0 |
| 84 | 0 | 0 | 2 | 0 | 1 | 1 | 1 | 1 | 2773 | 0 |
| 85 | 0 | 1 | 1 | 1 | 0 | 0 | 2 | 1 | 2662 | 0 |
| 86 | 1 | 0 | 2 | 1 | 0 | 0 | 2 | 1 | 757 | 1 |
| 87 | 1 | 0 | 2 | 1 | 0 | 0 | 2 | 2 | 1863 | 1 |
| 88 | 0 | 0 | 2 | 0 | 0 | 0 | 2 | 2 | 3381 | 0 |
| 89 | 1 | 0 | 2 | 1 | 0 | 0 | 1 | 1 | 3211 | 0 |
| 90 | 1 | 0 | 2 | 0 | 0 | 0 | 2 | 1 | 2904 | 0 |
| 91 | 1 | 0 | 2 | 0 | 0 | 0 | 2 | 1 | 3211 | 0 |
| 92 | 1 | 0 | 2 | 1 | 0 | 0 | 2 | 2 | 2835 | 0 |
| 93 | 0 | 0 | 2 | 0 | 0 | 0 | 1 | 1 | 2607 | 0 |
| 94 | 0 | 0 | 2 | 1 | 0 | 0 | 1 | 1 | 2541 | 1 |
| 95 | 1 | 0 | 2 | 1 | 1 | 1 | 1 | 1 | 743 | 1 |
| 96 | 0 | 0 | 2 | 0 | 0 | 0 | 1 | 1 | 530 | 1 |
| 97 | 0 | 0 | 2 | 1 | 1 | 1 | 2 | 2 | 1620 | 1 |
| 98 | 1 | 0 | 2 | 1 | 0 | 0 | 1 | 1 | 2559 | 0 |
| 99 | 0 | 0 | 2 | 0 | 1 | 0 | 2 | 1 | 2231 | 0 |
| 100 | 1 | 0 | 2 | 0 | 0 | 0 | 3 | 3 | 2372 | 0 |
| 101 | 1 | 0 | 1 | 0 | 0 | 0 | 1 | MIS | 567 | 1 |
| 102 | 1 | 0 | 1 | 1 | 1 | 0 | MIS | MIS | 208 | 1 |
| 103 | 0 | 1 | 1 | 0 | 0 | 0 | 2 | 1 | 2915 | 0 |
| 104 | 1 | 0 | 1 | 1 | 0 | 0 | 2 | MIS | 888 | 1 |
| 105 | 0 | 1 | 1 | 1 | 0 | 0 | 2 | 2 | 1403 | 1 |
| 106 | 1 | 0 | 1 | 0 | 0 | 0 | 2 | 1 | 2610 | 0 |
| 107 | 0 | 1 | 1 | 1 | 0 | 0 | 2 | 2 | 2516 | 1 |
| 108 | 0 | 0 | 1 | 1 | 0 | 0 | 1 | 1 | 3407 | 0 |
| 109 | 1 | 0 | 1 | 1 | 0 | 0 | 2 | 2 | 3303 | 0 |
| 110 | 0 | 0 | 1 | 0 | 0 | 0 | 2 | 2 | 3246 | 0 |
| 111 | 0 | 0 | 1 | 0 | 0 | 0 | 2 | 1 | 3348 | 0 |
| 112 | 1 | 0 | 1 | 1 | 0 | 0 | 2 | 2 | 1948 | 1 |
| 113 | 1 | 0 | 1 | 0 | 0 | 0 | 2 | 2 | 1360 | 1 |
| 114 | 1 | 0 | 1 | 0 | 0 | 0 | 2 | 2 | 1778 | 1 |
| 115 | 1 | 0 | 1 | 1 | 0 | 0 | 2 | 1 | 1834 | 1 |
| 116 | 0 | 0 | 1 | 0 | 0 | 0 | 1 | 2 | 3072 | 0 |
| 117 | 0 | 0 | 1 | 0 | 0 | 0 | 2 | 2 | 3052 | 0 |
| 118 | 0 | 0 | 2 | 1 | 0 | 0 | 1 | 1 | 2998 | 0 |
| 119 | 1 | 0 | 1 | 0 | 0 | 0 | 2 | 1 | 1219 | 1 |
| 120 | 1 | 0 | 1 | 0 | 1 | 1 | MIS | 2 | 29 | 1 |

**TABLE C.1.** (*Continued*)

| i d | s t a g e t | s t a g e n | h i s t | t r e a t | p e r f b l | p o i n f | s m o k f u | s m o k b l | d a y s | d e a t h |
|-----|---|---|---|---|---|---|------|------|------|---|
| 121 | 0 | 0 | 1 | 1 | 0 | 0 | 2   | 1   | 2879 | 0 |
| 122 | 0 | 0 | 1 | 1 | 0 | 0 | 2   | 1   | 2814 | 0 |
| 123 | 0 | 0 | 1 | 0 | 0 | 1 | 2   | 1   | 1621 | 1 |
| 124 | 0 | 0 | 1 | 1 | 0 | 0 | 1   | 1   | 2804 | 0 |
| 125 | 1 | 0 | 1 | 0 | 0 | 0 | 1   | 2   | 1237 | 1 |
| 126 | 0 | 0 | 1 | 1 | 1 | 0 | 1   | 1   | 2530 | 1 |
| 127 | 0 | 0 | 1 | 0 | 0 | 0 | 2   | 2   | 2549 | 0 |
| 128 | 0 | 0 | 1 | 0 | 1 | 0 | 2   | 2   | 228  | 1 |
| 129 | 0 | 0 | 2 | 1 | 1 | 0 | 1   | 1   | 541  | 1 |
| 130 | 1 | 0 | 1 | 0 | 0 | 0 | 1   | 1   | 2386 | 0 |
| 131 | 1 | 0 | 1 | 1 | 1 | 0 | 1   | 1   | 2487 | 1 |
| 132 | 0 | 0 | 2 | 1 | 0 | 1 | MIS | MIS | 641  | 1 |
| 133 | 1 | 0 | 2 | 0 | 0 | 0 | 1   | 1   | 1577 | 1 |
| 134 | 0 | 0 | 2 | 1 | 0 | 0 | 2   | 1   | 1600 | 1 |
| 135 | 0 | 0 | 2 | 0 | 1 | 0 | 1   | 1   | 3013 | 0 |
| 136 | 1 | 0 | 2 | 0 | 0 | 0 | 1   | MIS | 2089 | 1 |
| 137 | 1 | 0 | 2 | 1 | 1 | 0 | 1   | MIS | 2492 | 1 |
| 138 | 0 | 0 | 2 | 1 | 0 | 0 | MIS | MIS | 426  | 0 |
| 139 | 0 | 0 | 2 | 0 | 1 | 0 | MIS | 1   | 8    | 0 |
| 140 | 0 | 0 | 2 | 1 | 0 | 1 | 2   | 1   | 457  | 1 |
| 141 | 1 | 0 | 2 | 1 | 1 | 0 | 1   | 1   | 2809 | 0 |
| 142 | 0 | 1 | 2 | 0 | 0 | 0 | MIS | 1   | 262  | 1 |
| 143 | 0 | 0 | 2 | 1 | 1 | 0 | 3   | 3   | 480  | 1 |
| 144 | 0 | 0 | 2 | 1 | 0 | 0 | 2   | 1   | 2526 | 0 |
| 145 | 0 | 0 | 2 | 0 | 0 | 0 | 2   | 1   | 2454 | 0 |
| 146 | 0 | 0 | 2 | 1 | 0 | 0 | 1   | 1   | 2676 | 0 |
| 147 | 0 | 0 | 2 | 0 | 0 | 0 | 2   | 2   | 2594 | 0 |
| 148 | 1 | 0 | 2 | 1 | 1 | 0 | 2   | 2   | 443  | 1 |
| 149 | 0 | 0 | 2 | 1 | 0 | 1 | MIS | 1   | 31   | 1 |
| 150 | 0 | 0 | 2 | 1 | 0 | 0 | 2   | 1   | 1041 | 1 |
| 151 | 1 | 0 | 1 | 1 | 0 | 0 | 1   | 1   | 3198 | 0 |
| 152 | 1 | 0 | 1 | 0 | 0 | 0 | 3   | 3   | 1689 | 1 |
| 153 | 0 | 1 | 1 | 1 | 0 | 0 | 1   | 1   | 2870 | 0 |
| 154 | 0 | 1 | 1 | 0 | 0 | 0 | 2   | 2   | 419  | 1 |
| 155 | 1 | 0 | 1 | 1 | 0 | 0 | 2   | 2   | 2706 | 0 |
| 156 | 0 | 1 | 1 | 0 | 0 | 0 | 1   | 1   | 2518 | 0 |
| 157 | 0 | 1 | 1 | 1 | 1 | 0 | 1   | 2   | 2528 | 0 |
| 158 | 0 | 0 | 1 | 1 | 0 | 0 | 2   | 1   | 1860 | 1 |
| 159 | 1 | 0 | 1 | 0 | 0 | 0 | 1   | 1   | 3284 | 0 |
| 160 | 0 | 0 | 1 | 1 | 0 | 0 | 2   | 1   | 509  | 1 |

**TABLE C.1.** (*Continued*)

| i d | s t a g e t | s t a g e n | h i s t | t r e a t | p e r f b l | p o i n f | s m o k f u | s m o k b l | d a y s | d e a t h |
|-----|---|---|---|---|---|---|---|---|---|---|
| 161 | 1 | 0 | 1 | 0 | 0 | 0 | 2 | 2 | 2385 | 1 |
| 162 | 0 | 0 | 1 | 1 | 0 | 0 | 2 | 2 | 3290 | 0 |
| 163 | 1 | 0 | 1 | 0 | 0 | 0 | MIS | MIS | 267 | 1 |
| 164 | 0 | 0 | 1 | 1 | 0 | 1 | 2 | 2 | 2846 | 0 |
| 165 | 0 | 1 | 1 | 1 | 0 | 0 | 2 | 2 | 2566 | 0 |
| 166 | 0 | 0 | 1 | 0 | 0 | 0 | 2 | 1 | 2562 | 0 |
| 167 | 1 | 0 | 1 | 0 | 0 | 0 | 2 | 1 | 1915 | 1 |
| 168 | 1 | 0 | 1 | 1 | 0 | 0 | MIS | 1 | 158 | 1 |
| 169 | 0 | 0 | 1 | 0 | 0 | 0 | 2 | 2 | 2313 | 0 |
| 170 | 0 | 0 | 1 | 1 | 0 | 0 | 1 | 1 | 2621 | 0 |
| 171 | 0 | 0 | 1 | 0 | 0 | 0 | 1 | 1 | 2264 | 0 |
| 172 | 0 | 0 | 1 | 0 | 0 | 0 | 2 | 2 | 706 | 1 |
| 173 | 0 | 0 | 2 | 0 | 0 | 0 | 2 | 1 | 2655 | 0 |
| 174 | 1 | 0 | 2 | 0 | 0 | 0 | 3 | 3 | 79 | 1 |
| 175 | 0 | 0 | 2 | 1 | 0 | 0 | 2 | 2 | 3441 | 0 |
| 176 | 0 | 1 | 2 | 0 | 0 | 0 | 1 | 1 | 1713 | 1 |
| 177 | 0 | 0 | 2 | 0 | 0 | 0 | 2 | 1 | 448 | 1 |
| 178 | 1 | 0 | 2 | 1 | 0 | 0 | 2 | 2 | 242 | 1 |
| 179 | 1 | 0 | 2 | 0 | 0 | 0 | 1 | 1 | 188 | 1 |
| 180 | 1 | 0 | 2 | 1 | 0 | 0 | 1 | 2 | 1017 | 1 |
| 181 | 1 | 0 | 2 | 1 | 0 | 0 | 1 | 1 | 3321 | 0 |
| 182 | 1 | 0 | 2 | 1 | 0 | 0 | 1 | MIS | 3106 | 0 |
| 183 | 0 | 0 | 2 | 1 | 0 | 0 | 2 | 2 | 2225 | 1 |
| 184 | 1 | 0 | 2 | 0 | 0 | 0 | 1 | 1 | 3259 | 0 |
| 185 | 1 | 0 | 2 | 1 | 0 | 0 | 2 | 2 | 2929 | 0 |
| 186 | 1 | 0 | 2 | 0 | 0 | 0 | MIS | 1 | 29 | 1 |
| 187 | 0 | 0 | 2 | 0 | 0 | 0 | 1 | 1 | 2084 | 1 |
| 188 | 0 | 1 | 2 | 1 | 0 | 0 | 2 | 1 | 3097 | 0 |
| 189 | 0 | 0 | 2 | 0 | 0 | 0 | 3 | 3 | 2986 | 0 |
| 190 | 0 | 0 | 2 | 1 | 0 | 0 | 2 | 2 | 3102 | 0 |
| 191 | 1 | 0 | 2 | 1 | 0 | 0 | 2 | 1 | 2884 | 0 |
| 192 | 1 | 0 | 2 | 0 | 0 | 0 | 1 | 1 | 2702 | 0 |
| 193 | 0 | 0 | 2 | 1 | 0 | 0 | 2 | 1 | 2598 | 0 |
| 194 | 0 | 0 | 2 | 0 | 0 | 0 | 1 | 1 | 1531 | 1 |
| 195 | 1 | 0 | 2 | 0 | 0 | 0 | 1 | 1 | 2938 | 0 |
| 196 | 1 | 0 | 2 | 1 | 0 | 0 | 2 | 2 | 1668 | 1 |
| 197 | 0 | 0 | 2 | 1 | 0 | 0 | 3 | 3 | 2658 | 0 |
| 198 | 1 | 0 | 2 | 0 | 0 | 0 | 1 | 1 | 2429 | 0 |
| 199 | 1 | 0 | 2 | 1 | 0 | 0 | MIS | 2 | 69 | 1 |
| 200 | 0 | 0 | 2 | 0 | 0 | 0 | 2 | 1 | 2636 | 0 |

**TABLE C.1.** (*Continued*)

| i d | s t a g e t | s t a g e n | h i s t | t r e a t | p e r f b l | p o n i f | s m o k f u | s m o k b l | d a y s | d e a t h |
|---|---|---|---|---|---|---|---|---|---|---|
| 201 | 0 | 0 | 2 | 0 | 0 | 0 | 2 | 1 | 2663 | 0 |
| 202 | 1 | 0 | 2 | 1 | 0 | 0 | 2 | 1 | 366 | 1 |
| 203 | 1 | 0 | 2 | 0 | 0 | 0 | 2 | 2 | 2515 | 0 |
| 204 | 0 | 0 | 2 | 1 | 0 | 0 | 3 | 3 | 2592 | 0 |
| 205 | 1 | 0 | 2 | 1 | 0 | 0 | MIS | 2 | 442 | 1 |
| 206 | 0 | 0 | 2 | 0 | 0 | 0 | 2 | 1 | 1438 | 1 |
| 207 | 0 | 0 | 2 | 0 | 0 | 0 | 3 | 3 | 2673 | 0 |
| 208 | 0 | 1 | 2 | 1 | 0 | 0 | 1 | 1 | 818 | 1 |
| 209 | 0 | 0 | 2 | 1 | 0 | 0 | MIS | 1 | 772 | 0 |
| 210 | 0 | 0 | 2 | 1 | 0 | 0 | 1 | 1 | 2415 | 0 |
| 211 | 1 | 0 | 2 | 0 | 0 | 0 | 1 | 1 | 2052 | 0 |
| 212 | 0 | 0 | 2 | 1 | 0 | 0 | 2 | 1 | 2174 | 0 |
| 213 | 1 | 0 | 1 | 1 | 0 | 0 | 2 | 1 | 2092 | 1 |
| 214 | 1 | 0 | 1 | 0 | 0 | 0 | 2 | 2 | 2208 | 0 |
| 215 | 1 | 0 | 1 | 1 | 0 | 0 | 2 | 2 | 2905 | 0 |
| 216 | 0 | 1 | 1 | 0 | 0 | 0 | 2 | 1 | 3066 | 0 |
| 217 | 1 | 0 | 1 | 1 | 0 | 0 | 2 | 1 | 1314 | 1 |
| 218 | 0 | 1 | 1 | 0 | 0 | 0 | 2 | 1 | 1093 | 1 |
| 219 | 0 | 1 | 1 | 1 | 0 | 0 | 1 | 1 | 221 | 1 |
| 220 | 1 | 0 | 1 | 1 | 0 | 0 | 2 | 2 | 2535 | 0 |
| 221 | 0 | 0 | 1 | 1 | 0 | 0 | 2 | 2 | 3032 | 1 |
| 222 | 0 | 1 | 1 | 0 | 0 | 0 | 2 | 1 | 2580 | 1 |
| 223 | 1 | 0 | 1 | 0 | 1 | 1 | 3 | 3 | 3349 | 0 |
| 224 | 0 | 1 | 1 | 1 | 1 | 0 | 1 | 1 | 1696 | 0 |
| 225 | 0 | 0 | 1 | 0 | 1 | 1 | 2 | MIS | 339 | 1 |
| 226 | 0 | 1 | 1 | 0 | 0 | 0 | MIS | 1 | 512 | 1 |
| 227 | 0 | 0 | 1 | 1 | 1 | 0 | 1 | 1 | 3009 | 0 |
| 228 | 1 | 0 | 1 | 0 | 0 | 1 | MIS | 1 | 7 | 1 |
| 229 | 0 | 1 | 1 | 1 | 0 | 0 | 2 | 2 | 3305 | 0 |
| 230 | 0 | 0 | 1 | 0 | 0 | 0 | 1 | 1 | 2986 | 0 |
| 231 | 1 | 0 | 1 | 1 | 0 | 0 | MIS | 1 | 149 | 1 |
| 232 | 1 | 0 | 1 | 0 | 0 | 0 | 1 | 1 | 1970 | 1 |
| 233 | 1 | 0 | 1 | 1 | 0 | 0 | 2 | 2 | 2008 | 0 |
| 234 | 0 | 1 | 1 | 0 | 0 | 0 | 2 | 2 | 3276 | 0 |
| 235 | 0 | 0 | 1 | 1 | 0 | 0 | 2 | 2 | 1359 | 1 |
| 236 | 0 | 0 | 1 | 1 | 0 | 0 | 2 | 2 | 870 | 1 |
| 237 | 1 | 0 | 1 | 1 | 0 | 0 | MIS | 2 | 161 | 1 |
| 238 | 0 | 0 | 1 | 0 | 0 | 0 | 1 | 1 | 936 | 1 |
| 239 | 1 | 0 | 1 | 0 | 0 | 0 | 2 | 2 | 2998 | 0 |
| 240 | 0 | 0 | 1 | 1 | 0 | 0 | 1 | 2 | 3047 | 0 |

**TABLE C.1.** (*Continued*)

| i d | s t a g e t | s t a g e n | h i s t | t r e a t | p e r f b l | p o i n f | s m o k f u | s m o k b l | d a y s | d e a t h |
|---|---|---|---|---|---|---|---|---|---|---|
| 241 | 1 | 0 | 1 | 1 | 0 | 0 | 3 | 3 | 2864 | 0 |
| 242 | 1 | 0 | 1 | 0 | 0 | 0 | 2 | 1 | 2892 | 0 |
| 243 | 1 | 0 | 1 | 0 | 0 | 0 | 1 | 1 | 2972 | 0 |
| 244 | 1 | 0 | 1 | 1 | 0 | 0 | 1 | 1 | 160 | 1 |
| 245 | 0 | 0 | 1 | 0 | 0 | 0 | 1 | 1 | 2646 | 0 |
| 246 | 0 | 0 | 1 | 1 | 0 | 0 | 1 | 1 | 2084 | 1 |
| 247 | 0 | 0 | 1 | 1 | 0 | 0 | 2 | 1 | 1657 | 0 |
| 248 | 0 | 0 | 1 | 0 | 0 | 0 | 2 | 2 | 2434 | 0 |
| 249 | 0 | 0 | 1 | 0 | 0 | 0 | 2 | 2 | 2347 | 1 |
| 250 | 1 | 0 | 1 | 1 | 0 | 0 | 2 | 1 | 2069 | 1 |
| 251 | 1 | 0 | 1 | 1 | 0 | 0 | 1 | 1 | 723 | 0 |
| 252 | 0 | 0 | 1 | 0 | 0 | 0 | 2 | 2 | 2580 | 0 |
| 253 | 1 | 0 | 1 | 0 | 0 | 0 | MIS | 1 | 140 | 1 |
| 254 | 0 | 0 | 1 | 1 | 0 | 0 | 2 | 2 | 2424 | 0 |
| 255 | 1 | 0 | 1 | 0 | 0 | 0 | 2 | 1 | 2641 | 0 |
| 256 | 1 | 0 | 1 | 1 | 1 | 0 | 2 | 1 | 1253 | 0 |
| 257 | 1 | 0 | 1 | 1 | 0 | 0 | 2 | 2 | 600 | 1 |
| 258 | 0 | 0 | 1 | 0 | 0 | 0 | 1 | 1 | 1475 | 1 |
| 259 | 1 | 0 | 1 | 1 | 0 | 0 | 1 | 1 | 1659 | 1 |
| 260 | 0 | 0 | 1 | 0 | 0 | 0 | 2 | 1 | 2622 | 0 |
| 261 | 1 | 0 | 1 | 0 | 1 | 1 | 2 | 1 | 329 | 1 |
| 262 | 1 | 0 | 1 | 1 | 0 | 0 | 2 | 1 | 954 | 1 |
| 263 | 1 | 0 | 1 | 0 | 0 | 0 | 2 | 2 | 550 | 1 |
| 264 | 0 | 0 | 1 | 1 | 0 | 0 | 2 | 2 | 2359 | 0 |
| 265 | 1 | 0 | 1 | 0 | 0 | 1 | 2 | 2 | 2531 | 0 |
| 266 | 0 | 0 | 1 | 1 | MIS | 0 | 2 | 1 | 2531 | 0 |
| 267 | 1 | 0 | 1 | 0 | 0 | 0 | 2 | 1 | 468 | 1 |
| 268 | 1 | 0 | 1 | 1 | 0 | 0 | 2 | 1 | 162 | 0 |
| 269 | 1 | 0 | 1 | 1 | 0 | 0 | 2 | 2 | 1002 | 1 |
| 270 | 1 | 0 | 1 | 0 | 1 | 0 | 2 | 2 | 717 | 1 |
| 271 | 0 | 0 | 2 | 0 | 0 | 1 | 2 | 1 | 1088 | 1 |
| 272 | 1 | 0 | 2 | 0 | 0 | 0 | 2 | 2 | 3539 | 0 |
| 273 | 0 | 0 | 2 | 1 | 1 | 0 | 1 | 1 | 1851 | 0 |
| 274 | 1 | 0 | 2 | 1 | 0 | 0 | 1 | 2 | 285 | 1 |
| 275 | 1 | 0 | 2 | 0 | 1 | 0 | MIS | 2 | 1497 | 1 |
| 276 | 0 | 1 | 2 | 0 | 0 | 0 | 1 | 2 | 3113 | 0 |
| 277 | 0 | 0 | 2 | 1 | 0 | 0 | MIS | 2 | 2942 | 0 |
| 278 | 0 | 1 | 2 | 1 | 0 | 1 | 2 | MIS | 405 | 1 |
| 279 | 0 | 0 | 2 | 0 | 0 | 0 | 2 | 2 | 935 | 1 |
| 280 | 0 | 0 | 2 | 0 | 1 | 0 | 1 | 2 | 3122 | 0 |

**TABLE C.1.** (*Continued*)

| i d | s t a g e t | s t a g e n | h i s t | t r e a t | p e r f b l | p o i n f | s m o k f u | s m o k b l | d a y s | d e a t h |
|---|---|---|---|---|---|---|---|---|---|---|
| 281 | 0 | 0 | 2 | 1 | 0 | 1 | MIS | 1 | 4 | 1 |
| 282 | 0 | 0 | 2 | 0 | 0 | 0 | 2 | 2 | 197 | 1 |
| 283 | 1 | 0 | 2 | 1 | 0 | 0 | 2 | 1 | 2860 | 0 |
| 284 | 1 | 0 | 2 | 1 | 0 | 0 | 2 | 2 | 2938 | 0 |
| 285 | 0 | 0 | 2 | 0 | 0 | 0 | 1 | 2 | 2864 | 0 |
| 286 | 1 | 0 | 2 | 0 | 0 | 0 | 2 | 1 | 451 | 1 |
| 287 | 0 | 0 | 2 | 1 | 0 | 0 | 2 | 2 | 240 | 1 |
| 288 | 1 | 0 | 2 | 0 | 0 | 0 | 2 | 2 | 2605 | 0 |
| 289 | 0 | 0 | 2 | 1 | 0 | 0 | 2 | 1 | 1277 | 1 |
| 290 | 1 | 0 | 2 | 1 | 0 | 0 | 2 | 1 | 959 | 1 |
| 291 | 1 | 0 | 2 | 0 | 0 | 0 | 2 | 2 | 1294 | 1 |
| 292 | 1 | 0 | 2 | 0 | 0 | 0 | MIS | 1 | 130 | 1 |
| 293 | 0 | 0 | 2 | 1 | 0 | 0 | 1 | 1 | 743 | 1 |
| 294 | 1 | 0 | 2 | 0 | 0 | 0 | 2 | 2 | 2236 | 1 |
| 295 | 0 | 0 | 2 | 1 | 0 | 0 | 1 | 1 | 2662 | 0 |
| 296 | 1 | 0 | 2 | 1 | 0 | 0 | 2 | 1 | 2521 | 0 |
| 297 | 1 | 0 | 2 | 0 | 0 | 0 | 2 | 1 | 1207 | 1 |
| 298 | 0 | 0 | 2 | 0 | 0 | 0 | 1 | 3 | 1352 | 1 |
| 299 | 0 | 0 | 2 | 1 | 0 | 0 | 2 | 2 | 695 | 1 |
| 300 | 1 | 0 | 2 | 0 | 0 | 0 | MIS | 1 | 65 | 1 |
| 301 | 1 | 0 | 2 | 1 | 0 | 0 | 2 | 2 | 631 | 1 |
| 302 | 0 | 0 | 2 | 0 | 0 | 0 | 1 | 1 | 1571 | 1 |
| 303 | 0 | 1 | 2 | 1 | 0 | 0 | 2 | 1 | 482 | 1 |
| 304 | 0 | 0 | 2 | 1 | 0 | 0 | 1 | 1 | 2250 | 0 |
| 305 | 0 | 0 | 2 | 0 | 0 | 0 | 1 | 1 | 1134 | 0 |
| 306 | 0 | 0 | 1 | 0 | 0 | 0 | 2 | 1 | 861 | 0 |
| 307 | 0 | 0 | 1 | 1 | 1 | 0 | 2 | 1 | 619 | 1 |
| 308 | 1 | 0 | 1 | 0 | 0 | 0 | 2 | 2 | 290 | 1 |
| 309 | 1 | 0 | 1 | 1 | 1 | 0 | 1 | 1 | 1088 | 1 |
| 310 | 0 | 0 | 1 | 0 | 0 | 0 | 2 | 1 | 2972 | 0 |
| 311 | 1 | 0 | 1 | 1 | 0 | 0 | 1 | 1 | 2957 | 0 |
| 312 | 0 | 0 | 1 | 0 | 0 | 1 | 1 | 1 | 3241 | 0 |
| 313 | 1 | 0 | 1 | 0 | 0 | 0 | MIS | 1 | 1049 | 1 |
| 314 | 0 | 0 | 2 | 0 | 0 | 0 | 2 | 2 | 655 | 1 |
| 315 | 0 | 0 | 2 | 1 | 1 | 0 | 2 | 2 | 977 | 1 |
| 316 | 1 | 0 | 2 | 0 | 0 | 0 | 2 | MIS | 3454 | 0 |
| 317 | 0 | 0 | 2 | 1 | 0 | 0 | MIS | 1 | 248 | 0 |
| 318 | 1 | 0 | 2 | 1 | 0 | 0 | 1 | 2 | 2418 | 1 |
| 319 | 0 | 0 | 2 | 0 | 0 | 0 | 1 | 1 | 2219 | 1 |
| 320 | 1 | 0 | 2 | 0 | 0 | 0 | MIS | 1 | 1287 | 1 |

**TABLE C.1.** (*Continued*)

| i d | s t a g e t | s t a g e n | h i s t | t r e a t | p e r f b l | p o i n f | s m o k f u | s m o k b l | d a y s | d e a t h |
|---|---|---|---|---|---|---|---|---|---|---|
| 321 | 0 | 0 | 2 | 1 | 0 | 0 | 1 | 1 | 2121 | 1 |
| 322 | 0 | 0 | 2 | 1 | 0 | 0 | 2 | 1 | 2864 | 0 |
| 323 | 0 | 0 | 2 | 1 | 0 | 0 | MIS | 1 | 686 | 0 |
| 324 | 1 | 0 | 2 | 0 | 0 | 0 | 2 | 2 | 2463 | 0 |
| 325 | 1 | 0 | 2 | 1 | 0 | 0 | 1 | 1 | 2645 | 0 |
| 326 | 0 | 0 | 2 | 0 | 0 | 0 | 1 | 1 | 637 | 1 |
| 327 | 1 | 0 | 2 | 0 | 0 | 0 | 1 | 1 | 714 | 1 |
| 328 | 0 | 0 | 2 | 1 | 0 | 0 | 1 | 1 | 461 | 1 |
| 329 | 0 | 0 | 2 | 1 | 1 | 0 | 1 | 1 | 2449 | 1 |
| 330 | 0 | 0 | 2 | 1 | 0 | 0 | 1 | 1 | 1695 | 1 |
| 331 | 0 | 0 | 2 | 0 | 0 | 0 | 2 | 2 | 822 | 1 |
| 332 | 1 | 0 | 2 | 1 | 0 | 0 | 1 | 1 | 2517 | 0 |
| 333 | 0 | 0 | 1 | 0 | 0 | 0 | MIS | 1 | 142 | 1 |
| 334 | 0 | 1 | 1 | 1 | 0 | 0 | 1 | 2 | 1540 | 1 |
| 335 | 0 | 0 | 1 | 0 | 0 | 0 | 1 | 1 | 3238 | 0 |
| 336 | 1 | 0 | 1 | 0 | 0 | 0 | 2 | 2 | 2804 | 1 |
| 337 | 1 | 0 | 1 | 1 | 0 | 0 | 1 | 2 | 2300 | 1 |
| 338 | 1 | 0 | 1 | 0 | 0 | 0 | 1 | 1 | 1096 | 1 |
| 339 | 1 | 0 | 1 | 1 | 1 | 1 | MIS | MIS | 530 | 1 |
| 340 | 0 | 0 | 1 | 0 | 0 | 0 | 1 | 1 | 2143 | 1 |
| 341 | 1 | 0 | 1 | 1 | 0 | 0 | MIS | 2 | 231 | 1 |
| 342 | 1 | 0 | 2 | 1 | 0 | 0 | MIS | 2 | 129 | 1 |
| 343 | 0 | 0 | 2 | 1 | 0 | 0 | 1 | 1 | 1274 | 1 |
| 344 | 0 | 0 | 2 | 1 | 1 | 0 | MIS | 2 | 40 | 1 |
| 345 | 0 | 0 | 2 | 0 | 0 | 0 | MIS | 2 | 265 | 0 |
| 346 | 0 | 0 | 2 | 1 | 0 | 0 | MIS | 1 | 317 | 1 |
| 347 | 1 | 0 | 2 | 0 | 0 | 0 | 2 | 2 | 2218 | 1 |
| 348 | 0 | 0 | 2 | 0 | 0 | 0 | 2 | 2 | 1305 | 1 |
| 349 | 1 | 0 | 2 | 1 | 0 | 0 | MIS | 2 | 631 | 1 |
| 350 | 1 | 0 | 2 | 0 | 0 | 0 | MIS | 2 | 364 | 0 |
| 351 | 0 | 0 | 2 | 1 | 0 | 0 | 2 | 2 | 1161 | 1 |
| 352 | 0 | 0 | 2 | 0 | 0 | 0 | 2 | 2 | 2355 | 0 |
| 353 | 1 | 0 | 2 | 1 | 0 | 0 | 1 | 1 | 1239 | 1 |
| 354 | 1 | 0 | 2 | 0 | 0 | 0 | MIS | 2 | 867 | 1 |
| 355 | 0 | 0 | 2 | 1 | 0 | 0 | 2 | 2 | 2245 | 0 |
| 356 | 1 | 0 | 2 | 1 | 0 | 0 | MIS | 1 | 348 | 1 |
| 357 | 0 | 0 | 2 | 1 | 0 | 0 | 2 | 1 | 458 | 1 |
| 358 | 0 | 0 | 2 | 1 | 0 | 0 | 1 | 1 | 2218 | 0 |
| 359 | 1 | 0 | 2 | 0 | 0 | 0 | 1 | 1 | 1548 | 1 |
| 360 | 0 | 0 | 1 | 1 | 1 | 0 | 2 | 1 | 2043 | 0 |

**TABLE C.1.** (*Concluded*)

| i d | s t a g e t | s t a g e n | h i s t | t r e a t | p e r f b l | p o i n f | s m o k f u | s m o k b l | d a y s | d e a t h |
|---|---|---|---|---|---|---|---|---|---|---|
| 361 | 0 | 0 | 1 | 0 | 1 | 0 | 3 | 3 | 1872 | 0 |
| 362 | 0 | 0 | 1 | 0 | 0 | 0 | 2 | 2 | 1662 | 0 |
| 363 | 1 | 0 | 2 | 1 | 0 | 0 | 1 | 2 | 681 | 1 |
| 364 | 0 | 0 | 1 | 1 | 1 | 0 | 1 | 1 | 777 | 1 |
| 365 | 1 | 0 | 1 | 0 | 1 | 0 | 2 | 2 | 1309 | 0 |
| 366 | 1 | 0 | 1 | 1 | 0 | 0 | 2 | 2 | 1083 | 1 |
| 367 | 0 | 0 | 1 | 0 | 0 | 0 | 1 | 2 | 1042 | 0 |
| 368 | 0 | 1 | 1 | 1 | 0 | 0 | 2 | 2 | 1848 | 0 |
| 369 | 0 | 0 | 1 | 0 | 1 | 0 | 1 | 1 | 1677 | 0 |
| 370 | 0 | 0 | 1 | 1 | 1 | 0 | 1 | 1 | 1658 | 0 |
| 371 | 1 | 0 | 1 | 0 | 0 | 0 | 1 | 1 | 1482 | 0 |
| 372 | 0 | 0 | 1 | 1 | 1 | 0 | 1 | 1 | 1421 | 1 |
| 373 | 1 | 0 | 1 | 0 | 0 | 0 | 2 | 2 | 1291 | 0 |
| 374 | 1 | 0 | 1 | 1 | 0 | 0 | 2 | 2 | 1319 | 0 |
| 375 | 0 | 0 | 1 | 0 | 0 | 0 | 1 | 1 | 1323 | 0 |
| 376 | 1 | 0 | 2 | 0 | 1 | 0 | 2 | 2 | 980 | 1 |
| 377 | 1 | 0 | 2 | 1 | 0 | 0 | 2 | 2 | 1390 | 1 |
| 378 | 1 | 0 | 2 | 1 | 0 | 0 | 1 | 2 | 471 | 1 |
| 379 | 1 | 0 | 2 | 1 | 1 | 0 | 1 | 1 | 347 | 1 |
| 380 | 0 | 1 | 2 | 0 | 1 | 0 | 2 | 2 | 1517 | 0 |
| 381 | 0 | 0 | 2 | 1 | 1 | 0 | 2 | 2 | 909 | 1 |
| 382 | 0 | 0 | 2 | 0 | 1 | 0 | 1 | 2 | 1106 | 1 |
| 383 | 1 | 0 | 2 | 1 | 0 | 0 | 2 | 2 | 1138 | 0 |
| 384 | 1 | 0 | 2 | 1 | 0 | 0 | 2 | 2 | 493 | 1 |
| 385 | 1 | 0 | 2 | 0 | 0 | 0 | 1 | 1 | 1151 | 0 |
| 386 | 0 | 0 | 1 | 0 | 0 | 0 | 1 | 1 | 1337 | 0 |
| 387 | 0 | 1 | 1 | 1 | 1 | 0 | 1 | 2 | 1132 | 0 |
| 388 | 1 | 0 | 1 | 0 | 1 | 0 | MIS | 1 | 260 | 1 |
| 389 | 0 | 0 | 1 | 1 | 0 | 0 | 2 | 2 | 1830 | 0 |
| 390 | 1 | 0 | 1 | 0 | 0 | 0 | 1 | 1 | 731 | 1 |
| 391 | 0 | 0 | 1 | 1 | MIS | 0 | 2 | 2 | 1045 | 1 |
| 392 | 1 | 0 | 1 | 0 | 0 | 0 | MIS | 1 | 107 | 1 |
| 393 | 0 | 0 | 1 | 1 | 1 | 0 | MIS | 2 | 78 | 1 |
| 394 | 0 | 0 | 1 | 0 | 1 | 0 | MIS | 1 | 273 | 1 |
| 395 | 0 | 1 | 1 | 0 | 1 | 0 | 2 | 2 | 1473 | 0 |
| 396 | 0 | 0 | 1 | 1 | 0 | 0 | 2 | 1 | 1124 | 0 |
| 397 | 0 | 0 | 1 | 0 | 1 | 0 | MIS | 2 | 246 | 1 |
| 398 | 0 | 0 | 2 | 1 | 1 | 0 | MIS | 2 | 270 | 1 |
| 399 | 1 | 0 | 2 | 0 | 0 | 0 | 2 | 2 | 985 | 1 |
| 400 | 1 | 0 | 2 | 1 | 0 | 1 | 1 | 2 | 929 | 1 |
| 401 | 1 | 0 | 2 | 0 | 0 | 0 | 1 | 1 | 1291 | 0 |

# Index